Seventh Edition

Laboratory Manual of
GENERAL
ECOLOGY

Seventh Edition

Laboratory Manual of
GENERAL
ECOLOGY

George W. Cox
San Diego State University

WCB **Wm. C. Brown Publishers**

Dubuque, IA Bogota Boston Buenos Aires Caracas Chicago
Guilford, CT London Madrid Mexico City Sydney Toronto

Book Team

Editor *Margaret J. Kemp*
Developmental Editor *Kathleen R. Loewenberg*
Production Editor *Kay Driscoll*
Art Editor *Rachel Imsland*
Publishing Services Coordinator *Barbara Hodgson*

 **Wm. C. Brown Publishers**

President and Chief Executive Officer *Beverly Kolz*
Vice President, Publisher *Kevin Kane*
Vice President, Director of Sales and Marketing *Virginia S. Moffat*
Vice President, Director of Production *Colleen A. Yonda*
National Sales Manager *Douglas J. DiNardo*
Marketing Manager *Thomas C. Lyon*
Advertising Manager *Janelle Keeffer*
Production Editorial Manager *Renée Menne*
Publishing Services Manager *Karen J. Slaght*
Royalty/Permissions Manager *Connie Allendorf*

A Times Mirror Company

Cover Design by Pagecrafters, Inc.
Cover Photo by The Stock Market/Zefa Germany

Copyedited by Katherine Stevenson

Library of Congress Catalog Card Number: 94-72901

ISBN 0-697-24365-6

Printed in the United States of America by Times Mirror Higher Education Group, Inc.,
2460 Kerper Boulevard, Dubuque, IA 52001

10   9   8   7   6   5   4   3   2   1

Contents

Preface

Ecology is the science dealing with the structure and dynamics of systems that consist of organisms and their living and nonliving environments. A system, by definition, is a set of components united by regular interactions. Defined in this way, ecological systems exist at several levels of organization or complexity. An *individual organism,* interacting with its immediate physical, chemical, and biotic environment, is an ecological system. A *population*—the group of individuals of a particular species that occur together in a particular habitat—is also an ecological system, as a result of regular interactions with other populations and with its nonliving environment. A *community,* or set of species that coexist in a given location, also interacts with its environment, forming an ecological system that we term an *ecosystem.*

The objective of ecological study is to understand the mechanisms governing the structure and function of these ecological systems. When we examine an ecological system at a particular point in time, we want to account for the kinds of organisms present, their abundances, and their relationships with the many physical and chemical factors of their nonliving surroundings. Understanding these static conditions—the *structure* of the system—requires the study of *function*—the processes that occur through time in the system. We must observe the action of factors that impinge upon the system from outside. We must determine the rates of interchange of energy and matter among the components of the system, as well as their rates of entry to or loss from the system. Last, and most difficult, we must infer the nature of the controls on these rates.

Historically, ecology began as a descriptive field, as did many sciences. The recognition and naming of natural groupings of species and the classification of environmental factors controlling the distribution and abundance of organisms dominated this descriptive approach. In its most extreme form, this emphasis resulted in the characterization of ecology as ''super-description.'' Gradually, however, ecology has evolved a unique body of theory dealing with interrelations between organisms and their environments at all levels of organization, from that of the individual and its immediate microenvironment to that of the ecosystem. In recent years this body of theory has enlarged rapidly, largely as a consequence of increased emphasis on functional studies of ecological systems. This functional approach is concerned with understanding how ecological systems operate, and how this operation determines system structure at a given time, together with the pattern of change in structure through time. For the most part, modern ecology employs the concepts and techniques of descriptive ecology as ''tools'' for studying the dynamics of ecological systems.

The working tools of the ecologist are varied. From descriptive ecology have come chemical and physical methods for measuring conditions of the nonliving environment, along with qualitative and quantitative techniques for describing individuals, populations, and communities. With increasing emphasis on functional relationships, however, quantitative methods and techniques of mathematical and statistical analysis have assumed greater importance. As ecologists attempt to probe the dynamics of more and more complex assemblages, the valuable techniques of computer analysis and simulation have been drawn from systems science. Ecological study has become more rigorous in approach and more strongly oriented toward the testing of hypotheses. The formulation and testing of ecological hypotheses require quantitative methodology and the use of statistical techniques for the acceptance or rejection of hypotheses. In essence, the methods of computer simulation and analysis are simply the procedures that must be used to apply this hypothesis-testing approach to the most complex levels of ecological organization.

BASIC FEATURES OF THE MANUAL

This laboratory manual considers the field of ecology as a whole; it is a manual of general ecology. This designation rests not on the fact that exercises are included on every major group of organisms and every basic environmental type, but rather on the emphasis on key aspects of ecological description and theory. The commonly distinguished areas of ecology, such as animal ecology, plant ecology, limnology, and marine ecology, possess a common body of theory and technique that may be applied, with modification in detail only, to all taxonomic groups and all major environments. The existence of this common body of knowledge, in fact, now demands the use of a unified approach in introducing the science of ecology to the beginning student.

For example, concepts that originated in plant ecology are increasingly being adopted in studies of animal ecology, and vice versa. The concept of ecotypic variation and the methodology of ordination, which originated primarily in studies of plants and plant communities, are now in active use by animal ecologists. Likewise, temperature acclimation and life-table analyses, which originated as concepts and techniques of animal ecology, have now been extended to plant physiological and population ecology.

In addition to examining aspects of ecological theory and practice which apply to all branches of the field, I have emphasized quantitative and experimental approaches to ecological problems. Early in the manual, hypothesis formulation and testing are introduced, the designs of experiments and sampling protocols are discussed, and the basic statistical techniques used in deciding between acceptance and rejection of null hypotheses are outlined. Other skills important to practicing ecologists—library use and scientific writing—are also covered. Throughout the manual, computerized techniques of data analysis, community comparison, and systems simulation, now essential for advanced study and research in ecology, are introduced.

The bulk of the manual deals with specific techniques and ecological problems. In these exercises I have attempted to incorporate a considerable degree of plasticity, so that they may be used in different geographical regions, in various community types, or with diverse species. For exercises requiring organisms with special characteristics, I have suggested satisfactory, widely distributed forms. Although I have avoided many exercises requiring complicated or expensive equipment, I have included some exercises in physiological ecology that do require specific equipment that is becoming standard for well-equipped ecology laboratories. Research in physiological ecology and ecosystem science increasingly requires the use of high-tech equipment.

This edition contains a fold-out map of a desert plant community in Borrego Valley, San Diego County, California. This map provides a means for carrying out several exercises involving field studies of plant populations and communities when it is not possible to go into the field. The mapped area consists of a gently sloping alluvial fan (or bajada), one part of which is traversed by a braided desert wash. Locations and sizes of all perennial plants are shown on this map. This map can be used in connection with exercises on sampling design (4), distance sampling (12), vegetation analysis (14), intrapopulation dispersion (21), plant competition (27), interspecific association (31), species diversity (32), community similarity and ordination (33), and species-area curves (35).

The number and diverse nature of the exercises in this manual should allow the instructor considerable freedom in the selection of activities for use during a particular term. This diversity also allows the laboratory portion of the course to be varied from term to term. An additional element of flexibility derives from the fact that exercises can be used either in formal class activity or as the basis of individual student activities or projects.

No attempt has been made to supply detailed outlines for field trips to specific habitats or regional community types such as the deciduous forest, desert, freshwater lake, or marine intertidal zone. The responsibility is left to the instructor, who is the best authority on local ecological relationships and who has the best understanding of opportunities and constraints for field trips. Specific exercises have been described so that they may be incorporated into whatever specific format is dictated by the local situation.

At the end of each exercise, a set of pertinent references has been included. In part, these represent sources for the material of the exercise itself. However, they also include references to studies that go beyond the topic as covered in the exercise. I hope that these prove up-to-date and useful to the student in the preparation of reports on laboratory work, in the design of individual term projects and research studies, and in the formulation of problems for more advanced study. To aid students and instructors in relating exercises in this manual to textbook discussions of related material, I have included for each exercise cross-references to material in current ecology textbooks. In addition, a glossary is provided, covering terms used in the manual, and a conversion table for international metric units is included.

NEW TO THIS EDITION

A number of major changes have been made in the manual. Several exercises from the previous edition have been replaced or combined, and seven new exercises have been added. These new exercises deal with sampling design in ecological studies (4); soil arthropod sampling (9); mapping home ranges and territories (13); field measurement of photosynthesis (17); population growth, limitation, and interaction (28); metapopulation dynamics (29); and geographic information systems (39). This constitutes a greater change in content than in any earlier revision.

As many exercises as possible have been modified to encourage the use of computer programs or software packages. Six exercises are now structured around software packages that allow sophisticated analysis of questions at the population, community, and ecosystem levels: mapping home ranges and territories (13); population growth, limitation, and interaction (28); metapopulation dynamics (29); community similarity and ordination (33); ecosystem simulation (38); and geographic information systems (39). In addition, software programs or packages useful in data analysis or as adjunct activities have been identified for most other exercises. Information on the sources of these programs, as well as on the sources of hardware central to other exercises, is provided in the new appendix B.

The structure of all the exercises has been modified to include a section entitled "Suggested Activities." In this section specific kinds of field or classroom studies are listed; at least one is an activity that can be carried out indoors (e.g., in a library, laboratory, or greenhouse). A section entitled "Questions for Discussion" has also been created. The bibliographies of all exercises have been updated.

ACKNOWLEDGMENTS

I thank the many colleagues and students who have contributed to exercises in the earlier six editions of this manual, and whose feedback about many of these exercises has led to their improvement. In this edition, I am particularly indebted to Ellen Bauder, Dennis Claussen, Margaret E. Cochran, Boyd Collier, Thomas Ebert, Jim Eckblad, James Ehleringer, Richard Furnas, Steve Hastings, James E. Hines, David Hulse, Stuart Hurlbert, Craig James, John Kie, Paul M. Kotila, Charles J. Krebs, Jeff Laake, James D. Lawrey, Lee McClenaghan, Eric R. Pianka, Marion Preest, Howard F. Towner, and Anne Turhollow for comments and suggestions on exercises that are new or extensively revised in this edition. I wish to thank the reviewers of this edition, Robert M. Knutson, Luther College; Ken Hoover, Ph.D., Jacksonville University; Allen M. Moore, Western Carolina University; Bette H. Nybakken, Hartnell College. In large measure, this manual reflects the stimulation provided by my colleagues in the Ecology Program Area, as well as those in other areas of Biology, at San Diego State University.

San Diego, California
George W. Cox

Notes to Instructors and Students

Each exercise in this manual is cross-referenced to major ecology textbooks. On the first page of the exercise, these textbooks are cited by author and year of publication, and the chapters and pages with material appropriate to the topic of the exercise are listed. Instructors and students are encouraged to refer to the appropriate pages in their class text to review terminology, concepts, and issues relating to the exercise topic.

ECOLOGY TEXTBOOKS

Begon, M., J. L. Harper, and C. R. Townsend. 1990. Ecology: individuals, populations, and communities. 2nd ed. Sinauer Associates, Sunderland, Massachusetts, USA.

Brewer, R. 1994. The science of ecology. 2nd ed. Saunders College Publishing, Ft. Worth, Texas, USA.

Colinvaux, P. A. 1993. Ecology 2. John Wiley & Sons, New York, New York, USA.

Ehrlich, P. R., and J. Roughgarden. 1987. The science of ecology. Macmillan Publishing Company, New York, New York, USA.

Krebs, C. J. 1994. Ecology. The experimental analysis of distribution and abundance. 4th ed. HarperCollins College Publishers, New York, New York, USA.

Odum, E. P. 1993. Ecology and our endangered life support systems. 2nd ed. Sinauer Associates, Sunderland, Massachusetts, USA.

Pianka, E. R. 1994. Evolutionary ecology. 5th ed. HarperCollins, New York, New York, USA.

Ricklefs, R. E. 1990. Ecology. 3rd ed. W. H. Freeman and Company, New York, New York, USA.

——— . 1993. The economy of nature. 3rd ed. W. H. Freeman and Company, New York, New York, USA.

Smith, R. L. 1990. Ecology and field biology. 4th ed. Harper & Row, New York, New York, USA.

——— . 1992. Elements of ecology. 3rd ed. Harper & Row, New York, New York, USA.

Stiling, P. 1992. Introductory ecology. Prentice Hall, Englewood Cliffs, New Jersey, USA.

EXERCISE 1

Designing an Ecological Study

INTRODUCTION

Ecology can be defined as the study of ecological systems. A *system,* by dictionary definition, is any set of components that are tied together by regular interactions. Ecological systems are made up of one or more organisms, together with the nonliving environment with which they interact. Such systems exist at several different levels of organization. An ecological system can consist of a single organism and its surroundings, a population or set of interacting populations in a certain habitat, or the entire community together with the abiotic environment with which these species interact, a unit termed an *ecosystem.*

Ecologists are interested in both the structure and the function of ecological systems. *Structure* refers to measurable conditions of the system at one point in time. These include biotic attributes such as the body mass of an organism, the density of individuals in a population, the ratio of predator numbers to those of a prey species, or the biomass of all species that share some basic similarity, such as photosynthesis. Structure also refers to the physical and chemical conditions that prevail in the space occupied by organisms.

Function refers to processes that create the structure at a given instant, and how these processes are affected as the structure changes. Function includes, for example, the processes that determine the distribution and abundance of species, and thus the makeup of communities and ecosystems. It also includes the relationships that determine the growth rates of organisms, the rates of survival and reproduction of individuals in populations, the rates of predation by one species on another, and the rates of cycling of nutrients and flow of energy among the components of an ecosystem.

A major goal of modern ecology is to understand how ecological systems function, so that their behavior can be predicted, and so that they can be managed for long-term human benefit. To this end, ecologists seek to be rigorous in their techniques of investigation.

THE HYPOTHESIS-TESTING METHOD OF ECOLOGICAL STUDY

What is the hypothesis-testing method?

Most ecologists strive to use the hypothesis-testing or hypothetico-deductive approach in their studies. This approach, based largely on the ideas of Karl Popper (1968), is the statement of explicit hypotheses about ecological relationships, followed by the collection of data that lead to their acceptance or rejection. It views activities that cannot lead to the rejection of certain hypotheses and the acceptance of others as wasted efforts that do not advance ecological knowledge. The need for a rigorous hypothesis-testing approach has been stressed by many ecologists (Strong 1980, Romesburg 1981, Willson 1981, Quinn and Dunham 1983, Connor and Simberloff 1986). The effort to make ecological study more rigorous in this way is one of the key features of the current phase of ecology (Peters 1991, but also see Schrader-Frechette and McCoy 1993).

Brewer (1994): 1:9–11   Ricklefs (1993): 1:14–17   Smith (1992): 1:6–7
Krebs (1994): 1:8–15   Smith (1990): 2:16–23   Stiling (1992): 1:8–14

Where do ecologists get their hypotheses?

The stimulus for almost all ecological research, whether in the field or in the laboratory, comes initially from the observation of a distinctive pattern in nature. Usually, an initial observation is of some difference between two or more ecological situations. Usually, as well, this observation is tentative, involving a few organisms, a single population, or conditions at one time. Sometimes ecologists make their own initial observations, but often they base studies on patterns that have been observed and documented by others.

How does an initial observation lead to a comprehensive study?

Beginning with an initial observation, an idealized study consists of two general stages: (1) a descriptive stage, concerned with whether or not a distinctive structural pattern truly exists, and (2) a functional stage, in which the cause or effect of this pattern is explored (figure 1.1).

What does the descriptive phase of study involve?

The initial observation, in effect, is a *hypothesis* that a difference in structure exists. This hypothesis therefore must be tested by collecting and analyzing data to determine if the difference exists with a probability greater than that expected by chance. Such a test usually requires the collection of unbiased, quantitative data that can be analyzed statistically. These steps form the descriptive stage of an ecological study, and they determine, with a specified degree of probability, whether or not a structural pattern exists.

What does the functional phase of study involve?

If a distinctive structural pattern does exist, the stage is set for studies of function. These might relate either to the cause of the structural difference or to its consequences for the ecological system. To examine causes or consequences, one must first identify the possible cause-effect relationships that might be operating. Controlled experiments or controlled observations can then be designed to distinguish among these possibilities. These activities also must be designed to furnish quantitative data that can be tested statistically.

How can one make a fruitful initial observation?

In selecting a topic, one of the most difficult steps is finding a problem that can be carried through to the functional stage. Although intuition and experience are valuable, a systematic approach to observations can be helpful. Deliberate comparison of the different components of a single ecological system (such as the species that coexist in the same habitat) or of one component in different systems (such as the same species in different habitats) is one way to do this. Such comparisons will almost always reveal differences, some of which may provide the basis for a productive study. When such comparisons are being made, one should think about the kind of sampling and analysis that will be needed to determine whether an apparent difference is real, and about the sorts of testable functional hypotheses that might be made once a structural difference is shown to exist.

What kinds of ecological systems should be compared?

In making comparisons the student should remember that good problems need not deal only with single species of plants or animals, but can also concern interacting populations of different species, or even relationships at the ecosystem level. Although patterns at these more complex levels of organization are often difficult to recognize, they involve some of the most challenging and important questions in ecology.

How specific should comparisons be?

To make a study practical and manageable, ecological questions should be as specific as possible. In general, a good initial observation is one that clearly implies the type of measurement required to test it. Ideally, the situations compared should differ in only a few physical or biotic features, thus making it easier to postulate the causes of any differences that are noted.

DESCRIPTIVE STAGE

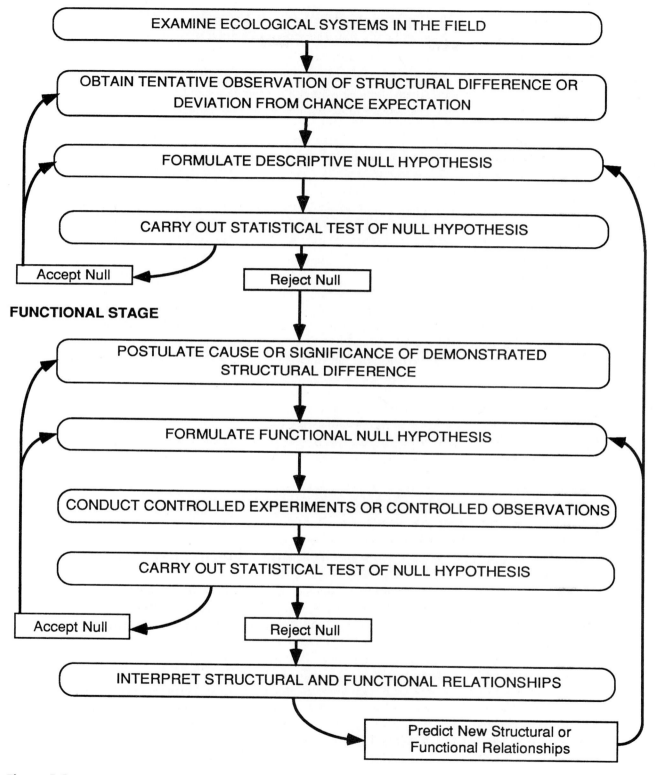

Figure 1.1.
Idealized sequence of steps leading from a tentative initial observation in the field through descriptive and functional stages of ecological study

MAKING INITIAL OBSERVATIONS

Some of the comparisons that can be made are outlined below. Some suggestions about how structural differences might be related to function also are given. These comparisons cover all levels of organization, from the organism to the ecosystem.

Compare one species in different situations.

Compare individuals or populations of one species in different geographical areas, in different habitats, or at different times. Couple these observations with notes on conditions of the physical and biotic environment.

Specific features to compare:

Abundance. Presence versus absence of a species or a difference in population density might be due to habitat-selection behavior, to requirements for specific resources, to limits of tolerance for conditions of the physical environment, or to beneficial or detrimental interactions with other species.

Morphology. Individuals may differ in size, shape, structure, color, or number and arrangement of body parts, due either to genetic differences or to the direct action of the environment during their development. Whether genetic or not, such features may have adaptive value, by increasing growth, survival, or reproduction under the respective physical or biotic challenges (e.g., climate, herbivores, predators, competitors).

Behavior. Daily or seasonal activity times, foraging behaviors, nest site selection, frequency of aggressive displays, and many other behaviors may differ in various situations. These may reflect differences in physical or biotic conditions of the habitat, the population density of the species itself, or the abundance of competitors or predators.

Population dispersion and structure. How individuals are dispersed (randomly, clumped, or uniform) may reflect the heterogeneity of habitat conditions, or the influence of positive or negative interactions among individuals. Conditions that influence reproductive success, rate of predation, or mortality from other causes may influence age structure and sex ratio.

Compare characteristics of different species in the same situation.

Coexisting species may show features that prevent detrimental interactions among them or that make their interaction favorable to one or more species.

Specific features to compare:

Morphology or behavior. Differences may serve to partition the use of resources such as space, food, or nest sites, thus reducing interspecific competition. They may also provide protection against predators (herbivores, in the case of plants).

Microspatial distribution. Differences in the exact sites occupied may reduce interspecific competition for resources, or may reflect different patterns of tolerance of conditions of the physical environment.

Patterns of co-occurrence. Individuals of different species may tend either to associate closely or to avoid each other, reflecting symbiotic, commensal, competitive, or feeding relationships. Such patterns also may result from actions of one species that modify the physical environment favorably or unfavorably for others.

Compare characteristics of ecologically similar species in different situations.

Similar species may be restricted to different habitats, or to different daily or seasonal activity times in the same habitat, by competition with each other. Such restriction also may reflect adaptation to different physical or biotic environments.

Specific features to compare:

Morphology and behavior. Species similar in resource use may have adaptations that make each species competitively superior in its own situation. Other differences may be adaptations to use resources or to tolerate environmental conditions that differ between the habitats or times.

Population dispersion and structure. Differences in the heterogeneity of habitat conditions may affect dispersion patterns. Differences in physical or biotic factors that affect survival and reproduction may be reflected in population density, age structure, or sex ratio.

Compare community or ecosystem characteristics in different situations.

The abundance of various ecological groups of organisms (e.g., annual or perennial plants, herbivores, carnivores) may vary with climate, nutrient and energy supply, or disturbing factors such as fire or flooding. Abiotic conditions such as microclimate, soil structure, and the profile of physical and chemical conditions in lakes may also differ due to such influences. Biotic or abiotic features of communities and ecosystems may differ because of biotic successional age, or because of differences in the ability of organisms to colonize the sites.

Specific features to compare:

Characteristics of species. The patterns of morphology, behavior, or population structure of species may differ (e.g., annuals versus perennials, grazers versus browsers, generalists versus specialists). These might reflect different physical conditions, different mechanisms of resource use, or different stages of biotic succession.

Number of species. Habitat heterogeneity, stability of physical conditions, constancy and level of primary production, degree of geographic isolation, abundance of predators, and stage in succession may influence the diversity of species.

Distribution of individuals among species. Differences in the frequency of rare and common species might relate to the mechanism of division of resources among community members or to the influence of disturbance.

Numbers and biomass of different trophic groups. The abundance of producers, herbivores, carnivores, and decomposers might be related to the rate of energy or nutrient input to the system, export or import of organic matter, body size-metabolism relationships of community members, patterns of nutrient cycling, stage in biotic succession, and many other factors.

FORMULATING AND TESTING NULL HYPOTHESES

Hypotheses about structure or function must be tested to determine if they have a high probability of being correct. Subjective impressions about such relationships often turn out to be wrong, due either to the inadequacy of the initial observations or to bias of the observer. A formal test of a hypothesis is carried out by stating a null hypothesis (H_O), collecting unbiased observations or experimental data, and performing a statistical test of the null hypothesis. The null hypothesis has the following form:

H_O: The difference between two or more sets of observational or experimental data is not greater than expected by chance.

     or

H_O: The values obtained in an experiment or set of observations do not differ from a theoretical expectation.

In the latter case the comparison is between a real data set and a theoretical data set in which the values are those predicted by a *model.* If the model assumes the operation of random processes (e.g., spacing, co-occurrence), it is termed a *null model.* For example, if a hypothesis deals with the spacing of individual plants in a population, the theoretical expectation could be one of random spacing. If the model contains mechanistic interactions that can be expressed as equations (e.g., an exponential or logistic equation), it is termed a *mathematical model.* For example, if a growing population of a species is measured over time, it might be compared to predictions of the logistic equation of population growth. A model of relationships within a complex system, using a series of equations or other modeling techniques, is termed a *simulation model.* Simulation models can yield predictions of many measurable features, to which data from a real system can be compared.

A statistical test of the null hypothesis leads to either its acceptance or its rejection. Accepting the null hypothesis means that either no difference exists or that the data are inadequate to demonstrate it. Rejection means that a difference exists with a probability corresponding to that of the significance level used in the statistical test. Rejection of the null hypothesis results in acceptance of a second, or alternate, hypothesis (H_a), which has the following form:

H_a: The difference between two (or more) observational or experimental situations is so great that it is very unlikely to have occurred by chance alone.

or

H_a: The set of observed values differs so greatly from the theoretical expectation that it is very unlikely to have occurred by chance alone.

A good null hypothesis is simple, suggests a characteristic that can be measured accurately, and concerns a relationship for which adequate sampling or experimental data can be obtained.

Example

The following example illustrates the sequence leading from an initial observation through an analysis of one aspect of functional significance.

Initial observation.

A plant species was observed in two locations, one on an ocean bluff, the other on a ridge 1 km inland. The ocean-bluff plants appeared to have thicker leaves.

Descriptive null hypothesis.

H_O: There is no difference in mean thickness of leaves taken from plants in the two areas.

Descriptive sampling and testing program.

In each area, 30 plants are sampled. One leaf per plant, between 3.5 and 4.0 cm in length, is taken. Leaf thickness is measured to the nearest 0.1 mm with calipers. Mean values for plants from the two areas are compared by a *t* test, and the null hypothesis is rejected.

Functional null hypothesis.

The difference might be due, proximately, to genetic differences in the plants rather than to direct effects of habitat factors during individual growth (many other functional hypotheses can be made). The following null hypothesis is then stated:

H_O: Plants grown side by side from seed taken from the two populations will show no difference in mean thickness of leaves.

Functional experimentation and testing program.

Seeds are collected in the two areas and grown until the plants reach some minimum size. Leaves are sampled and measured, and the data summarized and tested as above. If this test rejects the functional null hypothesis, the conclusion follows that the populations, with a certain probability, differ genetically in factors related to leaf thickness. If the functional null hypothesis is not rejected, the conclusion is that the data are not consistent with a genetic difference in leaf thickness, not that genetic differences are absent.

🌐 SUGGESTED ACTIVITIES

1.   Visit an area where environmental conditions are diverse enough that many of the above comparisons can be made. Examples of such situations include (1) a ridge with north-facing and south-facing slopes, (2) a sandy coastline where plant communities vary in development with distance from the shore, (3) an intertidal marine shore with sandy and rocky sections, or (4) a stream with pool and riffle zones.

Individually, or perhaps in pairs, students should make observations and prepare a hypothetical outline for a study, writing out the following items:

Descriptive Stage
- A tentative observation of a difference between two or more situations, stated in descriptive terms
- A testable null hypothesis (H_O) for this difference
- A suggested sampling procedure and statistical test

Functional Stage
- A possible cause or effect of the structural difference (assuming rejection of the above H_O)
- A testable null hypothesis of this functional relationship
- A suggested experimental or observational procedure and a statistical test for evaluating this relationship

Exchange and criticize problem outlines prepared by other students in your class. How could they be made more specific? What considerations of time and available facilities must be taken into account, in your situation, for pursuit of such studies?

2. Examine a published study such as Gotelli (1993), Madden and Young (1992), Alpine and Cloern (1992), or one suggested by your instructor. Identify the specific descriptive and functional hypotheses considered by the authors. Do they constitute a logical and comprehensive examination of the problem? Can you suggest other hypotheses that might, or should, have been tested?

3. As a class effort, analyze the articles in an issue of the journal *Ecology*. How many examine descriptive relationships, functional relationships, and both descriptive and functional relationships? How many test explicit descriptive and/or functional null hypotheses? How many make explicit tests of difference between data sets from different situations? How many test observed data against a null model, mathematical model, or simulation model?

✪ QUESTIONS FOR DISCUSSION

1. Does the concept of descriptive and functional stages parallel the overall history of the field of ecology? Was all early work in ecology descriptive? Do most current ecological studies include an explicit descriptive stage?

2. Do most current ecological studies explicitly test hypotheses? Do most modern studies carry through to an examination of functional relationships?

3. What sorts of difficulties exist in using the hypothesis-testing approach to studies in evolutionary ecology? Do you think that individuals interested in theoretical ecology would necessarily agree that ecological investigations should use the hypothesis-testing approach?

4. What role do long-term monitoring studies play in modern ecology, considering that these studies are not designed explicitly to test hypotheses?

✪ SELECTED BIBLIOGRAPHY

Alpine, A. E., and J. E. Cloern. 1992. Trophic interactions and direct physical effects control phytoplankton biomass and production in an estuary. Limnology and Oceanography 37:946-955.

Caswell, H. 1988. Theory and models in ecology: a different perspective. Bulletin of the Ecological Society of America 69:102-109.

Connor, E. F., and D. Simberloff. 1986. Competition, scientific method, and null models in ecology. American Scientist 74:155-162.

Fagerstrom, T. 1987. On theory, data and mathematics in ecology. Oikos 50:258–261.

Gotelli, N. J. 1993. Ant lion zones: causes of high-density predator aggregations. Ecology 74:226–237.

Howe, H. F., and L. C. Westley. 1988. Testing hypotheses in evolutionary ecology. Pages 17–26 *in* Ecological relationships of plants and animals. Oxford University Press, New York, New York, USA.

Krebs, C. J. 1988. The experimental approach to rodent population dynamics. Oikos 52:143–149.

Madden, D., and T. P. Young. 1992. Symbiotic ants as an alternative defense against giraffe herbivory in spinescent *Acacia drepanolobium*. Oecologia 91:235–238.

Matter, W. J., and R. W. Mannan. 1989. More on gaining reliable knowledge: a comment. Journal of Wildlife Management 53:1172–1176.

May, R. M., and J. Seger. 1986. Ideas in ecology. American Scientist 74:256–267.

Peters, R. H. 1991. A critique for ecology. Cambridge University Press, Cambridge, England.

Popper, K. R. 1968. The logic of scientific discovery. Rev. ed. Harper & Row, New York, New York, USA.

Quinn, J. F., and A. E. Dunham. 1983. On hypothesis testing in ecology and evolution. American Naturalist 122:602–617.

Romesburg, H. C. 1981. Wildlife science: gaining reliable knowledge. Journal of Wildlife Management 45:293–313.

———. 1989. More on gaining reliable knowledge: a reply. Journal of Wildlife Management 53:1177–1180.

Scheiner, S. M. 1993. Introduction: theories, hypotheses, and statistics. Pages 1–13 *in* S. M. Scheiner and J. Gurevitch, editors. Design and analysis of ecological experiments. Chapman & Hall, New York, New York, USA.

Schrader-Frechette, K. S., and E. D. McCoy. 1993. Method in ecology. Strategies for conservation. Cambridge University Press, Cambridge, England.

Wilson, E. B., Jr. 1990. An introduction to scientific research. Dover Publications, New York, New York, USA.

Literature Research in Ecology

INTRODUCTION

A practicing scientist must relate his or her research to that of other workers in the field. In planning research, one must survey the information that others have obtained on the topic, and design work that will go beyond current knowledge. Likewise, in interpreting research results, one should compare and contrast one's own findings with those of others. Many students entering science, however, do not realize how much information is available on a topic that arouses their interest, and are unfamiliar with techniques for locating this information in the enormous scientific literature. The ability to use library and literature resources, particularly those accessed and managed by computer systems, is an essential skill for anybody entering a career in science.

This exercise is a brief introduction to the use of resources for searching the ecological literature. Its primary objective is to aid students in locating books and articles pertinent to course research activities, special study or seminar topics, and thesis research projects. The techniques of literature research and library use are in very rapid flux as these procedures become computerized. Your instructor will doubtless want to supplement this exercise with information on special literature research aids on your own campus.

RECORDING REFERENCES

References on a particular topic should be recorded in a fashion that enables them to be filed logically and retrieved easily. Most ecologists have traditionally stored reference citations in card files, but computer software for storing and managing such information is now state-of-the-art.

The traditional approach is to record references on 3 × 5 inch cards. Examples of references for three basic types of materials are given in figure 2.1. In addition to the citation, content information from the book, chapter, or article can be recorded on the bottom or back of the card. Be sure to record *all* information required for a bibliographic reference (see exercise 5), and to note the *source* of the reference itself (in case you have to check it again).

A number of software packages are now available for the Macintosh, DOS, and Windows microcomputer environments. These packages facilitate storing, sorting, searching, and using bibliographic references. Specifically, they

- Store information on the author(s), title, publication date and location, and notes about content,
- Allow bibliographies to be sorted alphabetically by authors's name or chronologically by date of publication,
- Allow bibliographies to be searched by keywords to retrieve references of desired content,
- Enable references to be inserted in proper style into text created in major word-processing software, and
- Allow the *Literature Cited* sections of articles to be created automatically in the desired format within the word-processed document.

Book

> Roughgarden, Jonathan
>
> 1979
>
> Theory of population genetics and evolutionary ecology: an introduction.
>
> Macmillan, New York. x + 634 pp.
>
> Source: SDSU Card Catalog

Journal
Article

> Andersen, Douglas C.
>
> 1982
>
> Belowground herbivory: the adaptive geometry of geomyid burrows.
>
> American Naturalist 119:18-28.
>
> Source: Biol. Abstr. 74, No. 1, entry 1668

Article in
Symposium
or Edited
Monograph

> Pianka, Eric R.
>
> 1976
>
> Competition and niche theory.
>
> *In* Robert M. May (Ed.), Theoretical ecology. Blackwell, Oxford, England. Pp. 114-141.
>
> Source: SDSU Card Catalog

Figure 2.1.
Formats of reference cards for books, journal articles, and chapters in edited volumes.

Before investing in such a program, you should determine whether the software is compatible with your microcomputer hardware, memory capacity, and word-processing software, and if it has the capacity to store reference information on the range of bibliographic materials you will probably use (e.g., books, journal articles, symposium proceedings, government reports, abstracts). Popular commercially available software packages include *Pro-Cite, EndNote, RefBase,* and *Papyrus.*

Your instructor may arrange a demonstration of one of these bibliographic packages. Examine it and familiarize yourself with how references can be stored and retrieved.

Library In-house Resources

Visit your university or science library. Locate the various aids to searching the literature, and familiarize yourself with their use. Your instructor may request you to prepare an outline summary of the specific resources available on your campus, where they are located, and special details about their use.

Library Catalogs and Circulation Listings

Most major university libraries now have computerized cataloging systems that allow you to search for items by (1) title, (2) author, (3) subject, (4) keywords, and (5) call number. Separate catalogs may be available for state, U.S. government, and United Nations publications. For items located by a search, information may be provided on location and availability. For scientific journals, information is provided on location of both unbound and bound issues.

The library card catalogs provide a hard-copy backup of all holdings of the library in two listings: (1) by author and title, and (2) by subject. The *Library of Congress Subject Headings* volumes tell what subject headings are listed in the subject card catalog.

Current Periodicals Stacks, Book Stacks, and Microforms

Books, monographs, and most scientific journals are still maintained in unbound or bound form. Unbound current issues of periodicals are usually held on shelves near a reading area until a series suitable for binding has accumulated. For many periodicals, especially those that are bulky or subject to rapid deterioration, such as newspapers and news magazines, back issues are retained only on microfiche or microfilm. Locate the microforms area of your library, and familiarize yourself with the indexing system to these materials.

Abstracting and Indexing Services

There are a number of very useful indexes and guides to the periodical literature; some of these also cover material published in book form. These guides provide references and, in some cases, abstracts of the current literature. To survey the publications on a given topic fully, a researcher must utilize these guides. They also provide one of the most efficient ways to obtain a list of useful current references relating to a topic. The most serviceable of these guides for ecologists are listed in table 2.1. Locate the section of your library where these guides are kept and determine which of them are available (check those available, and note the yearly holdings for each).

The first seven of these indexes are of particular importance, and are described below.

1. *Biological Abstracts (BA).*

    BA is the most important source of references and abstracts for current research in ecology. It covers research papers disseminated in over 9,000 journals published in over 100 countries, and annually contains more than 367,000 abstracts. It is issued twice monthly, as well as in a biannual and five-year compilation. The five-year accumulations are on microfilm or microfiche and cover the period 1959–1989. These abstracts, prepared by authors of the articles, are indexed in several ways to facilitate retrieval of papers according to desired topic or author. The organization and use of the indexing systems are outlined below:

Table 2.1. Important guides to the periodical literature in ecology. Record holdings for those available in your library

Title	Holdings
1. Biological Abstracts (OL)*	_____
2. Biological Abstracts/RRM (OL)	_____
3. Zoological Record (OL)	_____
4. Science Citation Index	_____
5. Ecological Abstracts	_____
6. Aquatic Sciences and Fishery Abstracts (OL)	_____
ASFA 1. Biological Sciences and Living Resources	_____
ASFA 2. Ocean Technology, Policy and Non-living Resources	_____
ASFA 3. Aquatic Pollution and Environmental Quality	_____
7. Oceanic Abstracts	_____
8. Oceanographic Literature Review	_____
9. Deep-Sea Research	_____
Part B. Oceanographic Literature Review	_____
10. Environmental Abstracts	_____
11. Current Advances in Ecological and Environmental Sciences (OL)	_____
12. Forestry Abstracts	_____
13. Wildlife Review	_____
14. Pollution Abstracts (OL)	_____
15. Animal Behavior Abstracts	_____
16. Bibliography of Agriculture	_____
17. Biological and Agricultural Index	_____
18. Index to Scientific and Technical Proceedings	_____
19. General Science Index	_____
20. _____	_____

*OL indicates on-line availability

A. *Organization.* Abstracts are numbered and arranged according to an alphabetically organized subject-matter outline. The major organizational headings are listed at the beginning of each volume; subheadings of these areas exist within the abstracts listing. The following typical excerpt is from the April 30, 1994, issue:

A specific abstract from the *ECOLOGY, Animal* section is given below, with the information contained in the abstract noted.

Reference number — Authors — Authors' address

103972. Smith, Geoffrey R. and Royce E. Ballinger. (Sch. Biological Sciences, Univ. Nebraska, Lincoln, NE 68588, USA.) *American Midland Naturalist 131(1): 181–189. 1994.* **Temperature relationships in the high–altitude viviparous lizard,** *Sceloporus jarrovi.*—The thermal ecology of *Sceloporus jarrovi* was studied in the Chiricahua Mountains of southeastern Arizona at different elevations and times of year. Mean body temperature (T_b) was 31.8 C (range 6.6–38.8 C). Mean air temperature (T_a) was 20.6 C (range 5.0–36.2 C). The slope of the regression of T_b on T_a was 0.37. Winter T_bs were lower than summer T_bs. Low elevation lizards had higher T_bs than high elevation lizards. Males and females had similar T_bs. Pregnant females had lower T_bs than nonpregnant females. Pregnant females also had a lower slope for the T_b on T_a regression than nonpregnant females. Body temperatures of pregnant females at early stages of pregnancy (November) were not significantly different from pregnant females at late stages of pregnancy (May). *Sceloporus jarrovi* maintained lower T_bs than the sympatric lizard *Sceloporus scalaris.* These two species have different T_b on T_a regressions.

Title — Abstract —

Journal, volume, issue, pages, year

B. *Author Index.* This index lists authors alphabetically by last name, and gives abstract numbers pertaining to them. Papers with more than one author are indexed under the names of all authors:

BALLINGER R E 103972

SMITH G R 103972

C. *Biosystematic Index.* Papers dealing with a general group of plants or animals may be located by this index, which gives abstract numbers for papers listed first by taxonomic group, and second by key subject-matter terms:

● **Sauria**
Amphibia and Reptilia, General, Systematic
102633 102634 102636 102637 102638
Behavior, Animal
100136 100137
Blood Cell Studies
100792
Digestive Physiology, Biochemistry
103454
Ecology, Animal
103666 103667 103669 103678 103680
103717 103718 103731 103732 103735
103745 103753 103758 103762 103972
103983 103985
Gonads and Placenta
104870 104873
Neurology, Physiology, Biochemistry
110871 110972

D. *Generic Index.* Papers dealing in depth with particular species are listed in this index by genus and species. The major concept dealt with in the paper is also noted, and a coding with the abstract number (keyed at the page bottom) indicates the nature of any taxonomic revision included in the paper.

Genus-species	Major concept	Ref. No.
SCAPHYTOPIUS-MAGDALENSIS · · · ·	PL DIS BAC · · · · · ·	112944
SCARDAFELLA-SQUAMMATA-SQUAMMATA · · · · · · · · · · · · ·	BEHAV ANIMAL · · ·	100147
SCARUS-TAENIOPTERUS · · · · · · · ·	ORAL PHYSL · · · · ·	103020
SCELOPORUS-AENEUS-AENEUS · · · ·	REPT SYST · · · · · ·	102790
SCELOPORUS-AENEUS-SUBNIGER · ·	REPT SYST · · · · · ·	102790
SCELOPORUS-BICANTHALIS · · · · · · ·	REPT SYST · · · · · ·	102790
SCELOPORUS-JARROVI · · · · · · · · · · ·	ECOL ANIMAL · · · ·	103972
SCELOPORUS-SCALARIS · · · · · · · · · ·	ECOL ANIMAL · · · ·	103972
SCELOPORUS-SMITHI · · · · · · · · · · ·	REPT SYST · · · · · ·	102793
SCELOPORUS-TEAPENSIS · · · · · · · · ·	REPT SYST · · · · · ·	102793
SCELOPORUS-VARIABILIS · · · · · · · · ·	REPT SYST · · · · · ·	102793
SCELOPORUS-VARIABILIS-OLLOPORUS · · · · · · · · · · · · · ·	REPT SYST · · · · · ·	102793
SCENEDESMUS · · · · · · · · · · · · · · · ·	ECOL PLANT · · · · ·	104211
SCENEDESMUS-QUADRICAUDA · · · ·	PL PHOTOSYN· · · ·	113371
SCENEDESMUS-SUBSPICATA · · · · · ·	TOXIC GEN · · · · · ·	115918
SCHAFFERIA-NUBILUS · · · · · · · · · · ·	LEPIDOP SYST · · · ·	103000*GC

E. *Subject Index.* This index lists keywords from the title of the paper, plus other appropriate indexing terms supplied by the authors. These keywords are placed in the context of the wording of the title itself and the logical sequence of other terms:

Subject Context	▼ Keyword	Ref. No.
AL REJECTION/ PEDIATRIC	**LIVING-RELATED** AND CADAVERIC LIVER T	103388
MMUNOSUPPRESSANT-DRUG	DONOR CADAVERIC DON	116444
IC GRAFT LIVER SPITTING	DONOR XENOGRAFT/ AL	103263
L TRANSPLANTATION USING	HLA-IDENTICAL DONORS	116309
ULAR RECONSTRUCTION IN	LIVER TRANSPLANTATIO	99988
TO SALMONELLA SEROTYPE	**LIVINGSTONE** RESEARCH ARTICLE SALMON	108311
MISTRY RESEARCH ARTICLE	**LIZARD** ASPARTATE DOPAMINE GAMMA-AMI	115412
N/ AMEIVA EXSUL GROUND	BEHAVIOR NOTE AMEIVA-EXSUL SI	103667
E SCELOPORUS VARIABILIS	COMPLEX TAXONOMIC KEY SCELOPO	102793
INTS OF THE EURYTHERMIC	ELGARIA MULTICARINATA RESEARC	103735
F THE ADRENALECTOMIZED	MABUYA CARINATA SCHN. RESEARC	104870
OF THE ST. CROIX GROUND	ON GREEN CAY ST. CROIX U. S.	104594
PLEXIFORM LAYER OF THE	PODARCIS HISPANICA GOLGI AND	110972
IGH-ALTITUDE VIVIPAROUS	SCELOPORUS JARROVI RESEARCH A	103972
OD WEB RESEARCH ARTICLE	SPIDER SEA GRAPE SPECIES ABUN	103745
ION/ FOOD HABITS OF THE	TROPIDURUS ITAMBERE TROPIDURI	103718
DERMA HORRIDUM BEADED	VENOM RESEARCH ARTICLE HELODE	100792
LA MAMMALS FROGS BIRDS	**LIZARDS** CENTIPEDES BODY LENGTH STOMA	103664
NO ALDOSTERONE-TREATED	GALLOTIA GALLOTI RESEARCH AR	103454
BUTION/ STUDIES ON ROCK	LACERTA SAXICOLA COMPLEX REP	102785
L AND BISEXUAL WHIPTAIL	OF THE CNEMIDOPHORUS LEMNISC	102774

In the index, this paper is also listed alphabetically under each of the terms below that have the initial letter underlined:

<u>T</u>EMPERATURE <u>R</u>ELATIONSHIPS IN THE <u>H</u>IGH-ALTITUDE <u>V</u>IVIPAROUS <u>L</u>IZARD <u>S</u>CELOPORUS JARROVI <u>T</u>HERMAL <u>E</u>COLOGY <u>P</u>REGNANCY <u>A</u>RIZONA <u>U</u>SA

2. *Biological Abstracts/RRM (BA/RRM).*

   BA/RRM is the successor to *BioResearch Index* (as of 1980), and provides content summaries of over 192,000 books, book chapters, bibliographies, reports, reviews, meeting papers, research articles in minor and irregular serials, and biological software items. Complete abstracts are not given, but content is described by keywords, concept terms, and taxonomic categories, as indicated below:

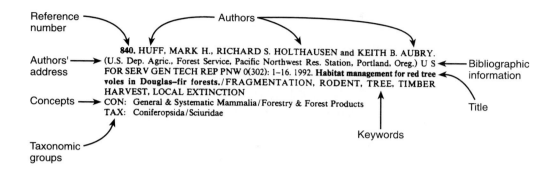

   BA/RRM contains a set of indexes identical to those for BA: author, biosystematic, generic, concept, and subject. Both BA and BA/RRM are part of a single *BIOSIS* (BioSciences Information Services) system, and do not overlap in content. BA/RRM is also issued twice monthly and in a biannual compilation.

3. *Science Citation Index.*

   This guide consists of four parts, the *Citation Index, Source Index, Permutation Subject Index,* and *Corporate Index.* It enables a researcher to determine if other workers have continued the studies reported in a particular paper. Suppose, for example, that you want to determine if any studies had followed up, and cited, the following paper:

   Hanski, I., and M. Gilpin 1991. Metapopulation dynamics: brief history and conceptual domain. Biological Journal of the Linnean Society 42:3–16.

   This requires examination of the *Citation Index* for each year since 1991. To do this, one looks up Hanski's name in the index. Under her name, one or more papers will be listed by year, journal, volume, and page number. These are Hanski's papers. Beneath each of these is a listing of papers that have cited Hanski. These are listed by author, year, journal, volume, and page. More information on the topics of each of these can be found by looking them up by author in the *Source Index* for the year of publication. Here, the titles are given, and you can determine more easily if they contain material of interest.

   In the *Permuterm Subject Index,* the significant words of titles of all papers in the *Source Index* are cross-indexed in pairs. By looking for pairs of keywords from the title of a useful reference, you can find other papers with titles containing the same keywords and published in the same year. The *Corporate Index,* issued once annually, lists organizations and institutions by geographical location. It gives the names of individuals that have published during the year, together with brief citations that can be cross-checked for more detail in the *Source Index.*

   Science Citation Index is published bimonthly, with annual and five-year cumulations.

4. *Ecological Abstracts.*

   This monthly index publishes abstracts of ecological papers, which are grouped under a detailed subject outline. A taxonomic index (genus and species) is provided at the end of each issue. A cumulative annual index is also available, providing access through subject keyword, author, and geographical indexes.

5. *Aquatic Sciences and Fisheries Abstracts.*

This index is compiled by the United Nations *Aquatic Sciences and Fisheries Information System (ASFA)*, which monitors about 5,000 serials. The index is organized into three series: *ASFA 1, Biological Sciences and Living Resources*; *ASFA 2, Ocean Technology, Policy and Non-living Resources*; and *ASFA 3, Aquatic Pollution and Environmental Quality*. Abstracts are published in monthly volumes. They are numbered and grouped under taxonomic and subject-matter headings and are indexed by author, subject, taxonomic (genus and species), and geographic indexes at the end of each volume. A cumulative annual index is also published. An example of an abstract is given below:

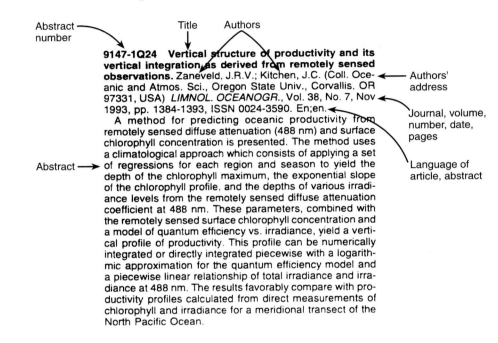

This database is also available on-line or in CD-ROM format at some institutions.

6. *Oceanic Abstracts.*

This index is published commercially, and consists of six bimonthly volumes and a year-end index volume. Numbered abstracts are grouped under major subject-matter headings. The following example comes from the December, 1991, issue (Vol. 28, No. 6):

91-097880 Age distribution of breeding female Antarctic fur seals in relation to changes in population growth rate. [En;en,fr] by I.L.Boyd, N.J.Lunn, P.Rothery, J.P.Croxall (Br. Antarctic Surv., Natl. Environ. Res. Counc., Madingley Rd., Cambridge CB3 0ET, UK) CAN. J. ZOOL./J. CAN. ZOOL., vol. 68, no. 10, Oct. 1990, pp. 2209-2213.

The age distribution of breeding female Antarctic fur seals at Bird Island, South Georgia, in 1988 was compared with the age distribution of a sample obtained in 1971-1973. The mean age in 1971-1973 was 7.41 (SE = 0.26) years and in 1988 it was 6.93 (SE = 0.20) years. After correction for age-dependent arrival time at the pupping beach in 1988, the mean age was 6.22 (SE = 0.14 years), which was significantly lower than in 1971-1973. Indicators of population size suggested that population growth at Bird Island had declined to below 3% annually by 1988 compared with rapid growth (17%) in 1958-1972. Exponential models fitted to the frequency distribution of age-classes greater than age 5 years and corrected for the rate of increase of the population gave adult survival rates of 0.66 (SE = 0.33) and 0.88 (SE = 0.02) for the 1988 and 1971-1973 samples, respectively. The reduced apparent adult survival rate in the 1988 sample was probably caused by emigration brought about by high densities of females on the pupping beaches. There are few signs from this analysis that the fur seal population at South Georgia is close to carrying capacity.

Each issue contains a subject index based on keywords, which are listed in their various permutations. Abstracts are referenced by number:

Agar,
   Seaweed products, Algae, ASW, Brazil 10478
Age composition,
   Females, Sexual reproduction, Growth, Arctocephalus
      gazella, PSW, South Georgia, Bird I. ....09788 ⟵
   Growth curves, Age determination, Mustelus californicus,
      Mustelus henlei, INE, USA, California 09690
   Growth, Stock assessment, Stock identification, Theragra
      chalcogramma, INE, Canada, British Columbia 11436
   Juveniles, Size distribution, Bottom trawls, Theragra
      chalcogramma, INE, Bering Sea 11486

Each issue also has an author index. The annual summary index combines the subject and author indexes of the bimonthly issues.

7. *Zoological Record.*

The *Zoological Record* is issued once a year. Only references, not abstracts, are given. The index appears as a set of separate sections covering about 20 specific subject areas and taxonomic groups.

References are listed alphabetically by the first author's name, and are numbered sequentially. Names of other authors are also listed and cross-referenced to the first author. The example below is taken from the *Comprehensive Zoology* volume for 1993:

Ylonen, H. & Magnhagen, C. [Eds]   (6260) ⟵——— Reference number
Predation risk and behavioural adaptations of prey: ecological and evolutionary
consequences. Symposium held at Konnevesi Research Station, 25-29 November
1991.
Ann Zool Fenn **29**(4) 1992: 177-320, illustr. [In English]
[Papers indexed separately]

Yoccoz, N.G., Engen, S. & Stenseth, N.C.   (6261)
Optimal foraging: the importance of environmental stochasticity and accuracy in
parameter estimation.
Am Nat **141**(1) 1993: 139-157, illustr. [In English]

Yockers, D.H.   (6262)
The WILD connection.
Natl Wildl Fed Sci Tech Ser No. 15 1991: 267-271. [In English]

Yoder, S.E. *see* Grifman, P.M. ⟵——————————————————— Second author
                                                            cross-reference
Yodzis, P. & Innes, S.   (6263)
Body size and consumer-resource dynamics.
Am Nat **139**(6) 1992: 1151-1175, illustr. [In English]

There are also subject, geographical, paleontological, and systematic indexes that list the first author's name and reference number for papers grouped under various headings.

Current Literature

The most current ecological literature has not yet been incorporated into these indexes. To check for very recent material, see if your library has the publication *Current Contents: Agriculture, Biology & Environmental Sciences,* which appears weekly. This publication gives tables of contents of the most recent issues of scientific journals, thus providing a source for references not yet included in the guides listed earlier. You also will find that it includes journals—perhaps a great many—that are not held by your library. In the back of each issue are keyword and author indexes. In addition, each issue lists the mailing addresses of the authors of all articles of the journal listed for that week. This enables you to write to an author and request a copy of an article that is of interest. *Current Contents* is usually shelved with current (unbound) periodicals or with the literature indexes.

Marine Science Contents Tables, issued bimonthly, reproduces the tables of contents of about 140 journals dealing with marine science.

Interlibrary Loan

Even relatively large libraries now find it impossible to maintain subscriptions to all, or even most, of the journals and other serials in science. You are very likely to find important references that are not available in your library. Almost all libraries, however, participate in systems of mutual exchange of specialized library materials. As you might expect, there is often a cost for such services, and it may take one or more weeks for particular items to be located and sent to your library. Nevertheless, interlibrary loan provides a valuable service to scientists who need to carry out a thorough survey of literature on a particular topic. Locate your interlibrary loan office, and inquire about the procedure and cost of the loan service.

COMPUTERIZED LITERATURE DATABASES

University libraries, colleges, or departments now provide access to various literature databases. Ascertain which of these are available, and familiarize yourself with the procedures for using them. Your instructor may request you to include information on the availability and use of these services in a lab report.

The Life Science Network

Biological Abstracts and *Biological Abstracts/RRM* are combined into a single electronic database, *BIOSIS Previews.* BIOSIS Previews and *Zoological Record Online* are included in the Life Science Network operated by BIOSIS (appendix B). Most libraries have access to this network or similar systems, and are able to conduct searches of these and other databases (e.g., Dissertation Abstracts Online, Pollution Abstracts). Searches of this type have a variable cost, depending on telecommunications time and number of citations and abstracts provided. The full texts of articles can also be ordered through a mail service, although not inexpensively.

Current Contents

The *Current Contents* listings are available on-line and on diskette, and many university libraries or academic branches subscribe to this service and make it available at library workstations or over their computer networks. You must check with your instructor about how to access this system. In using these versions of *Current Contents,* you must select one of the weekly issues for examination. A pull-down **Browse** menu allows you to look through the issue page by page, to view only the journals in a certain discipline, or to select specific journal titles for viewing. When you find an article of interest, you can click on an **Address** button to obtain the address of the author to whom a reprint request can be sent. The **Search** menu permits you to search an issue by several different criteria, such as keywords, disciplines, journal titles, or authors' names. You can create and save a search profile for future use. References can also be printed out or exported to certain bibliographic software programs (Pro-Cite and EndNote). Additional special capabilities of this system include ordering copies of articles and automatic preparation of reprint request cards.

UnCover

Many university libraries have computer workstations that allow one to access this commercial periodical database, which gives the tables of contents of about 14,000 periodicals, beginning in 1989. Use of the terminal is explained clearly on-screen. You can search this database by subject, journal name, or author's name. Although you cannot obtain authors' addresses directly from the database, you can obtain full bibliographic information that will enable you to find this same article in *Biological Abstracts,* where you can retrieve the author's mailing address. For some periodicals UnCover can FAX an article to you, for a service charge and copyright fee.

Local Library Databases

Many larger university libraries also have computer workstations that can access commercial databases covering selected periodicals and newspapers. These databases are typically supplied on tape or CD-ROM, and the particular ones available depend on the library's choice and budget. Two databases of particular interest to ecologists are (1) the *Expanded Academic Index* of magazines and journals, and (2) the *National Newspaper Index.* The Expanded Academic Index covers periodicals with general or broad content (e.g., *Science, Nature, Environment,* etc.). The National Newspaper Index covers articles in five important national newspapers. For both, the files cover only the past three years. Both indexes can be searched by topic or keyword. The references found, and in some cases an abstract, can be printed out. You may be able to obtain photocopies of the complete articles from the microforms collection of the library. The science-periodical coverage of the Expanded Academic Index is very similar to that of the *General Science Index* (table 2.1). *Newsbank* and *Newspaper Abstracts* are two other newspaper databases that may be available at library workstations.

Absearch

Absearch, a private company (see appendix B), markets read-only databases consisting of abstracts of articles from various groups of journals of interest to ecologists. Databases cover the journals of the *Ecological Society of America* (1945–93), *The Wildlife Society* (1937–92), and the *American Fisheries Society* (1945–93). Annual updates are provided. Other databases are under development. These databases can be searched by keyword and year. References can be exported to the Pro-Cite bibliographic software package.

✪ SUGGESTED ACTIVITIES

1. Browse a current issue of an ecological journal and find an article on a topic of interest to you. Write out a complete citation, using the format described for reference cards. Read the article, and in a few sentences summarize its content and major conclusions.

    Examine the bibliography of the article, and select *two* more references that are most similar in their content to the article itself. Write out complete citations for these articles. Examine *Science Citation Index* for recent papers that cite these references. Record complete citations for *two* articles that seem to be most similar to the current article you have just read.

    Using *Biological Abstracts* or *Biological Abstracts/RRM*, locate *two* additional papers by authors of the articles you have already listed above (including the one you read). Select keywords relating to the topic of the various papers for which you have citations. Using the subject index, find *one* additional reference. Write out citations for these articles.

    Use one of the other published indexes or computer databases to obtain *two* more citations from the recent literature. Counting the article you read, you now have 10 recent references on the topic.

2. As a class project, inventory the current periodicals that carry ecological articles. Each class member should examine a portion of the current periodical stacks and list the journals that contain at least some ecological articles. Remember that in addition to journals with "ecology" in their title, such periodicals might fall in the areas of general science, regional science, biology, botany, zoology, evolution, behavior, forestry, range management, wildlife, limnology, freshwater biology, marine biology, fisheries, conservation, agriculture, pollution, and environmental quality. Compile a complete list by combining the lists of different class members.

3. Examine recent issues of *Current Contents,* or scan the on-line version of this periodical. Review the tables of contents of ecological journals that are not in your library, and find two articles of interest to you. Retrieve the addresses of the corresponding authors of these articles. Prepare and mail reprint requests to these authors, using reprint request cards or a request format suggested by your instructor.

✪ QUESTIONS FOR DISCUSSION

1. What literature-search strategy would you use to obtain reference material on a current ecological issue? Select a topic that has been drawn to public attention in a newspaper or television report. Outline a search strategy for pulling together information to synthesize a report on the background and ecological nature of the topic.

2. What search strategy could you use to find information on the recent research activity of an ecologist and his or her colleagues at a particular university, so that you could evaluate the institution as a potential location for graduate study?

3. In what ways do you think that "scientific publication" will probably change in the next 10 years, based on emerging technologies of information transfer and retrieval?

SELECTED BIBLIOGRAPHY

Davis, E. B. 1981. Using the biological literature: a practical guide. Marcel Dekker, New York, New York, USA.

Kelly, M. C. 1991. Literature databases. Pages 955–965 *in* E. C. Dudley, editor. The unity of evolutionary biology. Vol. II. Dioscorides Press, Portland, Oregon, USA.

Malinowsky, H. R. 1994. Science and technology information sourcebook. Oryx Press, Phoenix, Arizona, USA.

Smith, R. C., W. M. Reid, and A. E. Luchsinger. 1980. Smith's guide to the literature of the life sciences. 9th ed. Burgess Publishing, Minneapolis, Minnesota, USA.

Wyatt, H. V. 1987. Information sources in the life sciences. Butterworth, London, England.

EXERCISE *3*

Experimental Design in Ecological Studies

INTRODUCTION

The conclusions that can be gained from an experiment depend primarily on its design. Unfortunately, many experiments are conducted with inadequate attention to design, so that enormous amounts of hard work often yield data that cannot be used to test the hypotheses that led to the experiment in the first place.

This exercise describes basic features of *manipulative experiments,* that is, experiments in which events affecting the experimental units are controlled by the investigator. These principles can be applied to field or laboratory studies in ecology.

EXPERIMENTAL DESIGN

Manipulative experiments involve one or more *treatments* that modify conditions in some or all of a set of *experimental units*. The effects of treatments are evaluated by measuring one or more *response variables*. In testing the influence of a fertilizer on productivity, for example, some plots—the experimental units—might be treated with a certain amount of fertilizer while others receive no treatment. Or, different sets of plots might be treated with different amounts of fertilizer. Features of the experimental units, such as the producer biomass or the rate of primary production, are then measured; these constitute the response variables.

The *design* of an experiment refers to the nature and number of experimental units, the types of treatments applied to them, the numbers of experimental units receiving each treatment, and how various treatments are applied to experimental units in space and time. Manipulative experiments are subject to several sources of *variability* that can make the results difficult or impossible to interpret. The confusion caused by these sources of variability (table 3.1) can be reduced by specific features of experimental design. In manipulative experiments, several critical aspects of design must be considered: replication, controls, randomization, and interspersion.

Replication

In manipulative experiments, experimental units chosen to be as similar as possible may exhibit differences in the response variables because the units were not truly identical. Likewise, differences can appear due to random errors in the measurement process itself. Of course, care in choosing the experimental units and in making measurements can minimize these sources of variability, but they cannot be eliminated completely. This variability, usually termed experimental error, can be measured if experimental units are *replicated*, that is, if several experimental units are assigned to each experimental treatment or condition.

Experimental units must be systems of the type to which the investigator intends to apply the conclusions of the experiment. For example, if one is interested in testing the hypothesis that addition of phosphorus fertilizer increases the primary productivity of ponds in general, the appropriate experimental units are *entire ponds*. Such an experiment, therefore, must consist of two or more ponds treated with fertilizer and compared to two or more untreated ponds of similar type. If only a single pond is fertilized, productivity measurements taken at several

Table 3.1. Sources of variability in experiments, and procedures for reducing confusion caused by such variability

Source of Variability	Reduction by Experimental Design
1. Variability among experimental units	Replication, interspersion, and simultaneous measurement
2. Random error in measurement of experimental results	Replication
3. Change in conditions through time	Controls
4. Unsuspected side effects of treatment procedures	Controls
5. Bias of investigator	Randomized assignment of treatments to experimental units
6. Chance influences on experiment in progress	Replication and interspersion

locations or at several times do not constitute replicate experimental units. The treatment of such measurements as replicate experimental units is termed *pseudoreplication* (Hurlbert 1984, Hurlbert and White 1993), and is a common error of design and analysis in ecological field studies.

Several types of pseudoreplication have been recognized. *Simple pseudoreplication* involves taking multiple measurements of response variables within individual experimental units and treating these measurements in statistical analyses as if they came from different experimental units. This is usually evidenced by the use in statistical tests of a number of degrees of freedom based on the total number of measurements rather than on the number of experimental units. For example, suppose that one pond is treated with fertilizer and another is untreated, and that 10 productivity measurements are subsequently made in each pond. A statistical test that purports to test whether fertilization affects pond productivity, and involves error degrees of freedom based on the 20 measurements (e.g., a *t* test of means with 20 − 2 degrees of freedom) is an example of simple pseudoreplication.

Temporal pseudoreplication occurs when measurements taken within the same experimental unit at different times are treated as coming from separate experimental units. If the 10 productivity measurements in the example above are taken on successive days, and 18 (20 − 2) error degrees of freedom are used in a comparison of the effect of fertilization on ponds, temporal pseudoreplication has been committed.

Sacrificial pseudoreplication occurs when the opportunity to distinguish variability of measurements within and between experimental units exists, but is ignored or "sacrificed." Suppose, for example, that three different ponds were fertilized and three were not fertilized, and that five productivity measurements subsequently were taken in each (a total of 30 measurements). A simple comparison of the effect of fertilization on pond productivity, using 28 (30 − 2) error degrees of freedom in a statistical test, is an example of sacrificial pseudoreplication. In this case, a nested analysis of variance (see exercise 8) could be used to partition variability in productivity values within ponds, among ponds of similar treatment, and between fertilized and unfertilized ponds. The appropriate test of the influence of fertilization on pond productivity would have 4 (6 − 2) error degrees of freedom.

Controls

In manipulative experiments, controls are experimental units that are identical to those receiving manipulative treatments except in the critical treatment factor. For example, if fertilizer is applied to a plot of grassland by spraying a solution of fertilizer in water, control plots should receive an identical spraying treatment, with water only. In this case, the control plot serves to reveal whether change in the plots receiving fertilizer is indeed due to the fertilizer, rather than to an accessory feature of the fertilizer treatment, such as addition of water or trampling by the experimenters. Controls can also reveal whether or not some change through time is tending to occur in the plots because of factors the experimenter cannot hold constant, such as seasonal change in daylength. Controls are essential for ecological field experiments, because it can rarely be assumed that conditions in nature will remain constant for any substantial time, and because almost any measurement or manipulation involves incidental impacts of the investigator on the natural system.

Randomization and Interspersion

In manipulative experiments, the investigator establishes replicated experimental units, some serving as controls and some receiving experimental treatments. These must be interspersed in space or time so that, on average, the different sets of units experience the same environmental conditions. Treatments (including controls) can be assigned to experimental units randomly. If the number of replicates is large, a random assignment procedure serves well, since it is very unlikely that, for instance, all controls would end up clustered in one location and all treatment units clustered in another.

In ecological field experiments, however, the number of replicates is often small, out of necessity. If an investigator sets up trapping grids to determine the effect of supplementary food on populations of rodents, the effort and large area required for each experimental unit probably limit the number of experimental units per treatment to three or four. If control or supplementary food status were assigned to grids randomly, it is very possible that control grids would by chance be clustered in one part of the study site and treatment grids in another. Such an arrangement opens the door to a chance systematic influence operating differentially on one part of the area in which the grids are located. Hurlbert (1984) terms such an influence a "non-demonic intrusion." For example, predators might enter the grid area more frequently from one side than from the other, or vegetation density might change gradually from one side to the other. Without interspersion, such differences might exert a biased effect on control or treatment plots.

Where such a possibility exists, semisystematic or systematic interspersion of experimental units is desirable. Control and treatment units can be alternated, or arranged in checkerboard fashion. A *randomized block design* (figure 3.1), which combines randomization and systematic interspersion, is a frequently used arrangement. In this procedure, the total number of experimental units is divided into sets known as blocks. Each block consists of a number of experimental units equal to the number of different types of treatments (including control). Within each block, treatments are then assigned randomly to the experimental units. A *Latin square design* (figure 3.1) is even more systematic in arrangement, having treatments assigned so that a given treatment occurs only once in each row or column.

UNREPLICATED PERTURBATIONS

In some cases, replication is impossible or impractical, as in the case of assessing the ecological impacts of a power plant, dam, or other large structure. Although these are, in a sense, giant experiments, in that they are planned ahead of time and their locations selected, they are single "treatments" that do not permit an experimental assessment of general treatment effects. It may be possible, however, to test whether or not the specific project causes a significant change in conditions that prevailed before it was constructed.

Suggested designs for evaluating such a perturbation are the *Before-After Control-Impact Pairs (BACIP) Method* (Stewart-Oaten et al. 1992) and a modification of BACIP involving the use of multiple control sites (Underwood 1994). These techniques involve simultaneously measuring response variables at impact (treatment) and control sites on a series of occasions before the structure is constructed and on a series of occasions afterwards. The control site or sites must be nearby and similar to the impact site, so that it can reasonably be assumed that without the impact the sites would change through time in parallel fashion (the validity of the technique depends on this assumption being fulfilled!). Intervals between measurements must also be long enough that the values

RANDOMIZED BLOCK DESIGN

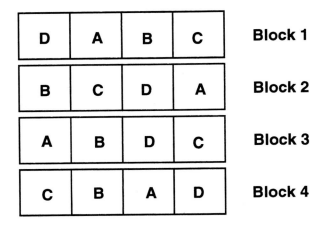

LATIN SQUARE DESIGN

Figure 3.1.
Schematic representation of plot layout for experiments with a
randomized block design and a Latin square design. The letters
A, B, C, and D indicate four different treatments.

obtained are statistically independent. The differences between measurements at control and impact sites can then be compared for the before and after periods by an appropriate statistical test.

EXPERTiMENTAL DESIGN

EXPERTiMENTAL DESIGN is a software package designed to help scientists and engineers select an appropriate experimental design for a particular study (appendix B). It is an expert systems program in which the user answers a set of questions about the nature of the study, including number of variables, whether or not the experiment involves screening, and other factors. The program then ranks the efficiency of some 17 experimental designs, including simple comparison of means, randomized block, and Latin square designs. The program also provides references to published descriptions of each design. This program is not directed specifically at ecological studies, and many of the designs are most appropriate for screening or testing in an industrial context.

✪ SUGGESTED ACTIVITIES

1. Examine the study outlines developed in exercise 1, and plan manipulative experiments that would be appropriate to the functional phase of the study, taking into consideration the use of replication, controls, and the randomization or interspersion of replicate experimental units.

2. Examine a published paper describing a manipulative field experiment. Review the methods and determine (1) whether or not adequate controls exist, (2) whether or not some form of pseudoreplication exists, and (3) whether or not interspersion problems exist for various treatment groups and controls.

3. Review issues of one or more regional natural history journals, and find a study that is largely descriptive. Outline a possible manipulative experiment to explore some aspect of functional significance, giving appropriate attention to the use of replication, controls, and the randomization or interspersion of replicate experimental units.

✪ QUESTIONS FOR DISCUSSION

1. What sorts of considerations might arise in a situation in which one must decide whether to use many small plots or fewer large plots in a field experiment? In a laboratory experiment in which experimental conditions are created in growth chambers?

2. What are some examples of ecological studies in which simple or temporal pseudoreplication is virtually impossible to avoid? Is the BACIP method applicable for these studies?

3. What sorts of "non-demonic intrusions" might operate in experiments (1) with potted plants in a greenhouse or growth chamber, (2) in an outdoor set of experimental ponds, or (3) in a set of experimental plots in a natural grassland ecosystem?

✪ SELECTED BIBLIOGRAPHY

Connell, J. H. 1974. Field experiments in marine ecology. Pages 21-54 *in* R. Mariscal, editor. Experimental marine biology. Academic Press, New York, New York, USA.

Dutilleul, P. 1993. Spatial heterogeneity and the design of ecological field experiments. Ecology 74:1646-1658.

Eberhardt, L. L., and J. M. Thomas. 1991. Designing environmental field studies. Ecological Monographs 61:53-73.

Hairston, N. G. 1989. Ecological experiments. Purpose, design, and execution. Cambridge University Press, Cambridge, England.

Hurlbert, S. H. 1984. Pseudoreplication and the design of ecological field experiments. Ecological Monographs 54:187-211.

———, and M. D. White. 1993. Experiments with freshwater invertebrate zooplanktivores: quality of statistical analyses. Bulletin of Marine Science 53:128-153.

Mead, R. 1988. The design of experiments. Cambridge University Press, Cambridge, England.

Potvin, C., and S. Tardif. 1988. Sources of variability and experimental designs in growth chambers. Functional Ecology 2:123-130.

Stewart-Oaten, A., J. R. Bence, and C. W. Osenberg. 1992. Assessing effects of unreplicated perturbations: no simple solutions. Ecology 73:1396-1404.

Underwood, A. J. 1981. Techniques of analysis of variance in experimental marine biology and ecology. Oceanography and Marine Biology Annual Reviews 19:513-605.

———. 1994. On beyond BACI: Sampling designs that might reliably detect environmental disturbances. Ecological Applications 4:3-15.

Zolman, J. F. 1993. Biostatistics: experimental design and statistical inference. Oxford University Press, New York, New York, USA.

Sampling Design in Ecological Studies

INTRODUCTION

Sampling activities serve a variety of purposes in both descriptive and functional studies in ecology. Sampling is intended to measure certain *variables,* that is, attributes of an ecological system that vary with environmental conditions and that can be described quantitatively. Except in rare cases in which all examples that show some attribute can be examined, ecologists must examine a *sample,* or subset, of all the examples available, which constitute the statistical *universe.* Sampling design relates to the types of variables that an investigator might choose to measure, together with the nature, spatial and temporal dispersion, and number of the samples used to measure these variables.

TYPES OF VARIABLES

Variables are of two rather different types: *continuous variables,* or characteristics that show gradual and incremental variation, and *discrete variables,* or characteristics that can be recorded as integers. The precision with which continuous variables, such as length, weight, salinity, temperature, and color, can be measured is limited only by the capabilities of the device with which the measurement is made. Discrete variables, exemplified by the numbers of individuals found in a sample or the number of items having one attribute as opposed to some other, are measured by counting, and consist only of whole numbers.

Continuous variables, such as color, can also be recorded as count data by dividing the range of possible variation into segments and tallying observations according to the segment in which they fall. In general, when this is done, information is lost, because the exact position of individual data values within the range divisions is not recorded.

Criteria for choosing variables to measure in a given situation include (1) a clear operational definition, (2) accuracy and precision of measurement, and (3) the effort required to obtain the measurement. An *operational definition* is a practical, working description of the variable in terms that can be understood and duplicated by other investigators. A variable such as "territory size" of a bird, for example, is nearly meaningless unless one can specify how many observations are used to define the area, how the boundaries of the area measured are defined from the observations, and whether all observations or just those involving territorial defense behaviors are utilized. The "foraging activity area" of a bird such as a swallow might be very interesting to know, but operationally impractical to define with existing technology. *Accuracy* is the degree to which the measured sample values correspond to the true values. *Precision* is the degree of correspondence among repeated measures of the same sample value. Both are of concern to ecologists. One can, for example, enumerate quite precisely the number of Cladocerans in a 1-liter water sample, but how accurately does the sampler trap the organisms in a 1-liter mass of lake water? The *effort* required to obtain measurements of one variable as opposed to a related variable is

Krebs (1994) 10:153–155   Smith (1990) 10:159–161

Table 4.1. Dimensions for quadrats of various areas and shapes

Area	Circular (Radius)	Square (Side)	Rectangular (1:2 Sides)	Rectangular (1:5 Sides)	Rectangular (1:10 Sides)
1	0.56	1.00	0.71 × 1.41	0.44 × 2.20	0.32 × 3.16
2	0.80	1.41	1.00 × 2.00	0.63 × 3.16	0.45 × 4.47
3	0.98	1.73	1.22 × 2.44	0.78 × 3.86	0.55 × 5.48
4	1.13	2.00	1.41 × 2.82	0.89 × 4.45	0.63 × 6.32
5	1.26	2.24	1.58 × 3.16	1.00 × 5.00	0.71 × 7.07
10	1.78	3.16	2.24 × 4.47	1.41 × 7.07	1.00 × 10.00
20	2.52	4.47	3.16 × 6.32	2.00 × 10.00	1.41 × 14.14
30	3.09	5.48	3.94 × 7.88	2.45 × 12.25	1.73 × 17.32
40	3.57	6.32	4.47 × 8.94	2.83 × 14.15	2.00 × 20.00
50	3.99	7.07	5.00 × 10.00	3.16 × 15.81	2.24 × 22.36
100	5.64	10.00	7.07 × 14.14	4.47 × 22.36	3.16 × 31.62
200	7.98	14.14	10.00 × 20.00	6.32 × 31.62	4.47 × 44.72
300	9.77	17.32	12.25 × 24.50	7.74 × 38.70	5.48 × 54.77
400	11.28	20.00	14.14 × 28.28	8.94 × 44.70	6.32 × 63.24
500	12.62	22.36	15.81 × 31.62	10.00 × 50.00	7.07 × 70.71
1000	17.84	31.62	22.36 × 44.72	14.14 × 70.71	10.00 × 100.00

frequently important to consider. A few detailed light-profile measurements made in different locations in a lake with a photometer, for example, might or might not be better than quick visual readings of transparency made with a Secchi disk.

SAMPLING UNITS

Very often, sampling requires the designation of area, volume, or time units within which variables are measured or counted. Such units are most often standardized, but sometimes they can be of variable size or duration. Fixed-area sampling units are usually termed *quadrats,* although such units need not be square, as suggested by the term.

The size of sampling units must be based on the size and density of the organisms being sampled. The area or volume sampled should be large enough to contain significant numbers of individuals, but small enough that they can be separated, counted, and measured without duplication or omission. For plant communities quadrats of 0.1-1.0 square meters for herbaceous vegetation, 10-20 square meters for shrubs or saplings up to about 3 meters in height, and 50-100 square meters for forest trees are commonly used.

The shape of quadrats can influence the ease, accuracy, and efficiency of quadrat sampling. In plant communities of low profile, for example, circular quadrats can be laid out very easily with a center pole and a freely rotating radius line. If the radius line is marked in appropriate distance units, quadrats of any desired size can be defined very quickly. Circular quadrats have the additional advantage of minimizing the quadrat edge, along which decisions often must be made about whether individual objects are ''in'' or ''out''—a type of decision that can sometimes be strongly biased. On the other hand, some studies suggest that elongate rectangular quadrats furnish a more accurate analysis of the composition of a stand of vegetation than an equal number of circular or square quadrats of the same area, especially when the long axis of the quadrat is oriented parallel to environmental gradients in the stand. Radii and rectangular dimensions for quadrats of various sizes and shapes are given in table 4.1.

The most effective size and shape of quadrats can be estimated using information (or best guesses) on the total time required to lay out and evaluate individual quadrats and the variability of the values obtained (Wiegert

1962). Effectiveness, E, is determined for whatever number of quadrats it takes to sample the same total area, for example, 5 quadrats of 1 m², 10 of 0.5 m², and 50 of 0.1 m². Effectiveness can be estimated from the relative time (RT) required to lay out and evaluate individual quadrats of various types and the relative variability (RV) of values obtained from the total set of quadrats of each type. The optimum quadrat size and/or shape is that with the highest value of E given by the equation

$$E = 1/(RT)\,(RV)$$

Relative time for a given quadrat type is simply the ratio of the time requirement, t, for a given quadrat type to that of the quadrat with the minimum time requirement (which will therefore have an RT of 1.00). Relative variability is the ratio of variability, v, determined for the set of quadrats of a given type to that of the quadrat type with the minimum variability (the RV for which is therefore 1.00). Variability can be estimated from the range of values observed or expected for the required number of quadrats of the various types. For this estimate, the range, w, must be multiplied by a conversion factor, c, related to the number of quadrats (figure 4.1). The estimated variability is then given by the equation

$$v = (wc)^2$$

For those already familiar with parametric statistics, the above estimate of variability corresponds to the sample variance, s^2, which can be used to calculate RV if it is available.

In some situations, particularly with vegetation analysis (see exercise 14), line transects or point-based sampling procedures are more practical to use than quadrats. Individual transects, transect sections, or points then constitute sampling units. Although it is not really practical to try to evaluate the effectiveness of size and shape features of such units as described above, the following discussions of sampling layout and sample number apply equally to quadrats, transects, and sampling points.

Sampling Layout

For us to make any valid inference from a sample, we must be assured that the sample constitutes an unbiased set of observations from the universe. Therefore, we must sample so that all portions of the universe to which we wish to generalize have an equal chance of being represented. A corollary of this requirement is that the selection of one part of the universe as a sample observation does not modify the probability that other specific parts of the universe will subsequently be selected.

In some cases unbiased sampling can be achieved with ease; in other cases it may be difficult or virtually impossible to attain. *Random sampling* is the procedure normally used to obtain an unbiased sample. When the universe consists of a limited number of individuals or items, these may be numbered. The sample can then be selected by using a set of random numbers. The simplest manner of doing this is to use a random numbers table (appendix A, table A.1). To use such a table, simply enter it in a manner that avoids bias in the selection of the first number. Once the starting point has been determined, random numbers can be obtained by reading in any consistent direction (down, up, across, etc.) from the point of entry.

A similar procedure can be utilized in sampling areas that can be subdivided by grid lines into quadrats. These units can be numbered, and a sample chosen by a set of random numbers. With areas too large to be covered by a grid, base lines at right angles to each other can be established along two sides or through two axes of the area in question. Sampling locations can then be selected by drawing pairs of random numbers to serve as co-ordinates for locating quadrats with reference to the base lines.

Volumetric samples can be taken in three-dimensional systems, such as lakes or soils, by similar procedures. Sampling also can be spread through time by random procedures.

Stratified random sampling involves subdividing the total space to be sampled into subsectors, within each of which sampling is carried out randomly. This assures that no major sector of the universe fails to be sampled. If these subsectors are of equal size, random sampling locations are then selected in the various subsectors in equal numbers. A large stand of vegetation, for example, might be divided into 10 strips, within each

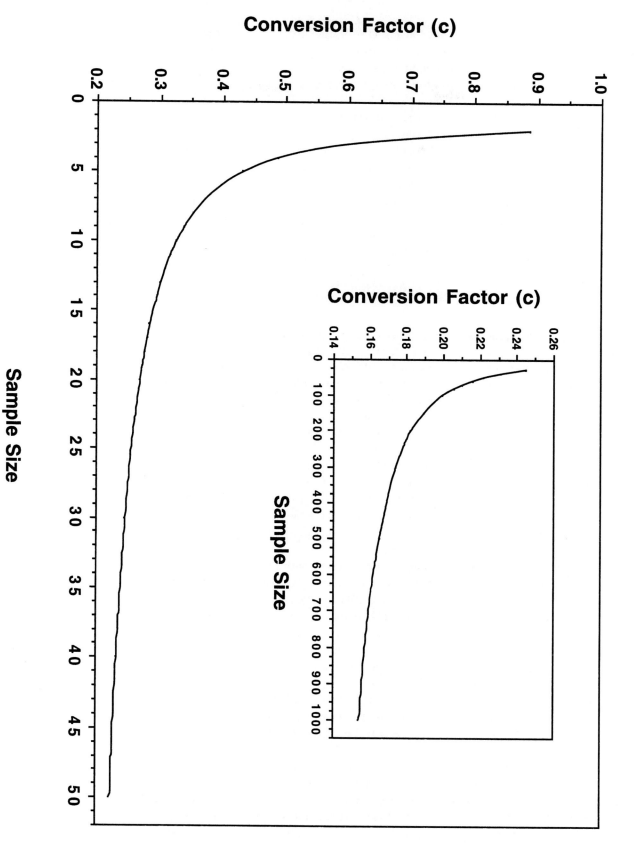

Figure 4.1.
Conversion factor, *c,* for adjusting the range, *w,* in calculations of sampling efficiency of quadrats of certain size and shape.

of which five quadrats could be placed at random. If the subsectors are of unequal size, random sampling sites can be assigned to subsectors in numbers proportional to area or volume. Sampling in a patchy habitat can be stratified so that the number of sampling locations in each patch is proportional to its area. Most statisticians concerned with ecological studies agree that stratified random sampling provides data amenable to statistical analysis.

Systematic sampling involves the uniform spacing of sampling activities in space or time. A systematic layout of plots is often relatively easy to create and is thought by some workers to give a closer approximation of the true composition of a population or community (Greig-Smith 1983). In principle, however, systematic samples are not independent, since the location of the first sampling location determines the location of all others. Thus, systematic sampling does not meet the basic requirement for statistical analysis. In practice, the characteristics of uniformly spaced sampling sites are rarely a function of the spacing pattern, but such a possibility cannot often be dismissed as impossible. Furthermore, few situations exist in which systematic sampling is much easier than random or stratified random sampling efforts that are planned ahead of time.

Modifications of these procedures may be appropriate in particular situations. In some cases, such as an analysis of museum specimens, random sampling is nearly impossible. In analyzing data from such sources, one must be especially careful to identify and evaluate potential sources of bias prior to any statistical analysis.

Number of Samples

The number of sample observations, and in the case of quadrats, their size, is important in relation to the precision of the estimates obtained from sampling data. In general, a large number of small or medium-sized quadrats is preferable to a small number of large ones, since the former provide a more detailed picture of the amount of variability in the universe. For most ecological work, 20–25 observations or quadrats are a minimum sample size; when the degree of variability is high, a much larger sample size may be necessary to give useful estimates.

If the desired precision, d, of an average value for some measurement obtained by sampling can be stated, an estimate of the needed number of samples, n, can be made. For this estimate, d must be stated as the greatest difference ($+$ or $-$) between the estimated and true average values that one wishes to attain, with a certain probability. Using the observed range, w, of values obtained in some preliminary sampling effort and the conversion factor, c, used previously (figure 4.1), and a probability coefficient, p, this estimate is given by the equation

$$n = (pwc/d)^2$$

In this equation, the probability value, p, can be taken roughly as 2.0 if one wishes to be 95% certain of the appropriate sample size, 2.6 if one wishes to be 99% certain, or 3.5 if a probability of 99.9% is desired. Bear in mind that these calculations are intended *only* to give preliminary estimates, and that refined techniques exist for estimating needed sampling effort. Krebs (1989) describes several of these techniques.

Software Useful in Sampling

Most programmable hand calculators now have a random number generator. Typically, this function yields random numbers, one at a time, between 0.0 and 1.0. To obtain random numbers between desired limits, one need only shift the decimal point to obtain the correct magnitude and ignore numbers outside the desired range.

E-Z Stat (appendix B), a basic statistical software package designed for ecologists, also contains a random number generator. This generator can be selected from the **DATA** pull-down menu, and one can command the creation of random numbers between specified limits, and with a specified number of decimal places. This software is available in versions designed for IBM PC–compatible and Macintosh microcomputers.

Table 4.2. Time (*t*), relative time (*RT*), variability (*v*), relative variability (*RV*), and overall effectiveness (*E*) of sampling with quadrats of certain size and shape, together with estimates of the number of quadrats required to yield a density estimate of certain precision (*d*) with a specified probability (*p*)

Size and Shape	d	p	t	RT	v	RV	E

SUGGESTED ACTIVITIES

1. Examine the study outlines developed in exercise 1, and plan a sampling strategy for some appropriate variable in the situation or situations outlined as the descriptive phase of the study. Set up a table of estimates of *t, RT, v, RV,* and *E* for quadrats of different size and shape. Define a certain desired precision, *d,* for average values of the variable and estimate the sample size, *n,* required to yield values of this precision with 95%, 99%, and 99.9% certainty.

2. Using the map of the desert plant community accompanying this manual, determine actual values of *t, RT, v, RV,* and *E* for sampling with quadrats of a range of sizes and shapes, using table 4.2 to summarize these data. Based on the observed range of values for quadrats of different types, estimate the sample size required to yield an average value within some predetermined precision, *d,* with 95%, 99%, and 99.9% certainty.

3. Using a simple habitat, such as an area of lawn on campus, sample some variable (e.g., number of dandelion plants) with quadrats of different size and shape, using table 4.2 to record *t, RT, v, RV,* and *E.* Based on the observed range of values for quadrats of different types, estimate the sample size required to yield an average value within some predetermined precision, *d,* with 95%, 99%, and 99.9% certainty.

QUESTIONS FOR DISCUSSION

1. In what kinds of ecological situations might an investigator be able to examine all examples of a statistical universe, rather than having to work with a sample?

2. Samples of soil are often taken by ecologists for laboratory analyses of texture, moisture, and nutrient content. Very often these are not taken randomly, but the results are often analyzed statistically. What sorts of biases might be introduced by this sort of practice?

3. How does the effort to specify the needed number of samples relate to the issue of type I and type II errors in testing ecological hypotheses (see exercise 7)?

⊘ SELECTED BIBLIOGRAPHY

Brummer, J. E., J. T. Nichols, R. K. Engel, and K. M. Eskridge. 1994. Efficiency of different quadrat sizes and shapes for sampling standing crop. Journal of Range Management 47:84–89.

Cochran, W. G. 1977. Sampling techniques. 3rd ed. John Wiley & Sons, New York, New York, USA.

Eberhardt, L. L., and J. M. Thomas. 1991. Designing environmental field studies. Ecological Monographs 61:53–73.

Green, R. H. 1979. Sampling design and statistical methods for environmental biologists. John Wiley & Sons, New York, New York, USA.

Grieg-Smith, P. 1983. Quantitative plant ecology. 3rd ed. Blackwell Scientific Publications, Oxford, England.

Krebs, C. J. 1989. Ecological methodology. Harper & Row, New York, New York, USA.

Wiegert, R. G. 1962. The selection of an optimum quadrat size for sampling the standing crop of grasses and forbs. Ecology 43:125–129.

Yates, F. 1981. Sampling methods for censuses and surveys. Charles Griffin and Co., London, England.

EXERCISE 5

Writing an Ecological Research Paper

INTRODUCTION

Sharing the results of scientific activity with other scientists is an integral part of the scientific process. The most effective way of doing this is through the publication of reports in scientific journals and other widely distributed publications.

A well-written scientific report must fulfill two objectives. First, it must describe clearly and completely the procedures that were followed and the results that were obtained. Second, it must place these results in perspective by relating them to the existing state of knowledge and by interpreting their significance for future study.

Clear and complete presentation of procedures and results is essential to one of the basic requirements of scientific research: repeatability. The reader of a report should be able to duplicate the study in its essential features, if desired. A full presentation of results, free of attached interpretations, is necessary to give the reader a chance to reach his or her own conclusions, whether or not they agree with those of the author. Learning to describe procedures and results clearly and completely is usually not the most difficult side of report preparation for the student, but it does require practice.

Learning how to place scientific research in perspective is usually more difficult; the work in question must be related in a meaningful way to the general body of knowledge on the topic. This requires that the scientist (1) trace the scientific origins of the research problem, (2) summarize the state of knowledge on the general topic, (3) state the critical hypotheses toward which the study itself is addressed, (4) interpret the results of the study in relation to these hypotheses and to the general state of knowledge, and (5) identify the scientific questions and procedural weaknesses that need to be addressed in the future.

Although it is necessary that the scientific report be complete, it should also be organized and concise. A general format for the structure of scientific reports in ecology has gradually evolved, and is now in wide use. This format is designed to help achieve clarity and conciseness, and consists of the organization of the text into four main parts: introduction, procedure, results, and discussion. Each of these is discussed in detail below. The manner of presentation of data in tables and figures, and the format for citation of other scientific work, also are described.

Basic principles of scientific writing are outlined in a number of guides to technical writing (e.g., Katz 1986, Alley 1987, McMillan 1988). The specific organization and style used in biological journals are covered thoroughly in the *CBE Style Manual (5th ed.)*, published in 1983, and in *Scientific Writing for Graduate Students*, published in 1986 by the Council of Biology Editors. For detailed examples of the application of these guidelines, the student should examine a current issue of the journal *Ecology*, which is available in almost every college or university library. The articles in this journal can be used as models for the preparation of scientific reports in ecology.

GENERAL ORGANIZATION

A scientific report in ecology is usually organized into the following sections:

Introduction
Procedure
Results
Discussion
Acknowledgments
Literature Cited

This arrangement, however, should not be considered an absolute pattern. When the topic being considered is complex, extensive, or unusual, subdivision or modification of these basic sections is often necessary. If the report is long, a *Summary* section following the *Discussion* may be appropriate. In articles in scientific journals, the function of the *Summary* is usually fulfilled by an *Abstract,* which is placed at the beginning of the article.

Introduction

This section should explain the objectives of the study, and why it is a worthwhile effort. An effective introduction should address the following questions:

- Why did the author undertake this study? The best initial statement is often a description of an observation in nature that stimulated the interest of the author, or of a hypothesis drawn from a survey of the literature.
- What is the existing state of knowledge about this topic? The author should synthesize information from the literature into an account that traces the development of knowledge of the problem and summarizes its current state. In particular, the gaps and inadequacies of current knowledge should be identified.
- What, specifically, is the author going to do? Here, the author must indicate the specific objectives or testable hypotheses that will be considered. These should be as specific as possible, and in the case of hypotheses, should be statements that are clearly capable of being either supported or refuted by the planned work.

A note of caution: the student should be careful to restrict background material to that which is directly pertinent to the problem at hand. The author of a scientific report should be prepared to answer the question, "Why did you include this material in the introduction to your paper?"

Procedure

This section must answer all the basic questions about the way the study was conducted: where, when, and how.

- Where: In the case of field studies, give the exact location. Describe relevant features of the study site, such as vegetation, climate, topography, and human disturbance. In the case of laboratory studies, give the location and identify the general facilities available to the research. In either case, explain why this was an appropriate site for the research.
- When: Give the time periods during which the work was done (year, month, day, time of day) whenever this is relevant to the type of data collected. Note the point within the seasonal (annual) cycle, or within other natural cycles (such as those of the tides), if appropriate. Describe special conditions of weather or other factors that prevailed during the study period.

- How: Explain the techniques of collecting data or conducting experiments, and describe the equipment used. Where these procedures and equipment are standard, do this by reference to published descriptions. Otherwise, describe them in enough detail that a reader can duplicate them. Indicate in an honest and accurate fashion how samples were obtained. For example, do not describe sampling as "random" unless some specific technique for obtaining randomness was employed. State the techniques used to record, summarize, and analyze the data. When statistical tests are utilized, cite a published source for them.

Results

This section should contain, in summarized form, the data and observations obtained. Do not present raw data; actual data sheets, if included in the report, should be placed in an appendix. Summarize the raw data in tables, figures, and the written text so that the reader can appreciate both the general patterns of these data and the degree of variability that they exhibit. Group individual data values into summary categories that can be appreciated easily by the reader. Present means and other descriptive statistics, and state the results of basic statistical analyses and comparisons.

In the presentation of results, begin with the most general features of the data and proceed toward the most specific. The written text should deal fully with the results and not merely refer to tables or figures in which the summarized data are presented. On the other hand, do not try to present tabular material in text format. Text, tables, and figures all have their unique capabilities for presentation of results. In general, the presentation of data as a table *or* figure is sufficient; the same data should not be presented in both ways. In achieving completeness of coverage in the text, concentrate on general patterns, trends, and differences in the results, and not on the details of the numbers themselves.

Combine statements about the significance of differences examined by statistical tests with a precise indication, commonly in parentheses, of the test used and the probability level chosen. For example, one might say, "The difference between means of the two samples was highly significant (paired t test; $t = 6.35$, $DF = 11$, $P < 0.01$; Zar 1984)." A parenthetical expression such as this might not need to include all of the information indicated, provided that this information is available elsewhere in the paper (e.g., an earlier reference to the published source of statistical procedures, or to the level of significance used consistently). It is essential, however, that the reader be able to ascertain all of the above for each test utilized.

Finally, the *Results* section should be free of interpretation of the data. This is frequently difficult for the student, since it is always tempting to tell why the data turned out as they did, or what they mean, when they are first presented. Nevertheless, this is not the function of this section, and the distinction required in writing here is useful in forcing the scientist to maintain the critical distinction between results and their interpretation in his or her own mind.

Discussion

In this section, the author must interpret the data in relation to the original objectives or hypotheses, and relate these interpretations to the present state of knowledge and future needs for research. Care must be taken to make this section genuinely interpretive and not just a rehash of results at some higher level of generality.

The author should attempt to include several of the following tasks in a good discussion section:

- Reach conclusions about the initial hypotheses
- Compare conclusions to those of others
- Identify sources of error and basic inadequacies of technique
- Speculate on broader meanings of the conclusions reached

- Identify needed next steps in research on the problem
- Suggest improvements of methods

Acknowledgments

In this section, give credit to those who helped in the study by contributing work, advice, permission, technical assistance, funds for conducting the actual work, and help with preparation of the manuscript.

Literature Cited

This section should contain, in alphabetical order, only those items specifically referred to in the text. In spite of considerable effort, the format for references has not become fully standardized in the scientific literature. The format recommended by the *CBE Style Manual* is followed by many American journals, but not by most book publishers. Here, we shall recommend the format adopted by the journal *Ecology*. It is perhaps the simplest format that provides all the essential information. The format for three types of common references is given below:

> Journal article:
> Tilman, D. 1994. Competition and biodiversity in spatially structured habitats. Ecology 75:2–16.
> (This reference includes the name of the author or authors, the year of publication, the article title, the fully spelled journal name, and the volume and page numbers of the article itself.)
> Article in edited volume:
> Root, T. L. 1993. Effects of global climate change on North American birds and their communities. Pages 280–292 *in* P. M. Kareiva, J. G. Kingsolver, and R. B. Huey, editors. Biotic interactions and global change. Sinauer Associates, Sunderland, Massachusetts, USA.
> (This reference includes the article author, date of publication, article title, pages in volume, names of volume editors, volume title, and publisher's name and location.)
> Book:
> Louw, G. N. 1993. Physiological animal ecology. Wiley & Sons, New York, New York, USA.
> (This reference gives the author, date of publication, title, and publisher's name and location.)

For the format for citing other types of publications, see either the *CBE Style Manual* or a current issue of *Ecology*.

GENERAL CONSIDERATIONS

Text Citations

Do not use footnotes. Instead, use parenthetical references that give names and/or dates keyed to the *Literature Cited* section. This may be done in several ways, as indicated below:

> "Tilman (1994) found that. . . ."
> "The spatial competition hypothesis (Tilman 1994). . . ."
> "At Cedar Creek Natural History Area (Inouye et al. 1987), plants compete. . . ."
> (This last form is used when more than two authors exist. The abbreviation *et al.* is for the Latin phrase meaning "and others.")

Reading a paper or two in *Ecology* will give the student insight into how other types of material are referred to in the text.

Quotations

Do not use extensive quotations in scientific writing. Usually, the *only* reasons for using the exact words of another author are (1) because it is this exact wording that is the focus of discussion, or (2) because the wording carries the intended meaning in so striking a fashion that it cannot easily be paraphrased.

Remember also that unacknowledged use of the exact wording of other authors, in length greater than phrases, is plagiarism, and is a violation of copyright law. Plagiarism in the preparation of papers and reports by students carries very severe penalties at most universities.

Tables and Figures

Prepare tables and figures according to the format shown in a journal such as *Ecology*. In a manuscript, place them on pages separate from the text, one table or figure per page. Useful guidelines for figure preparation are given by Allen (1977), CBE Scientific Illustration Committee (1988), Papp (1976), and Tufte (1983).

Number tables (Arabic numerals) and title them fully in a legend placed above the table itself. Within the body of the table, use headings to identify clearly the nature and units of the data given. Keep vertical lines to a minimum in tables, or better, do not use them at all. Number figures (Arabic, again) and provide a full legend, which in this case is placed below the figure. The axes of the figure (if it is a graphical presentation of data) should be specified clearly and the units identified and scaled appropriately. Legends for both tables and figures should give enough information about what these items contain, so that they can be understood without reference to the text.

Leave adequate margins on pages with tables and figures, just as for text pages. Refer to all tables and figures by parenthetical citations, e.g. (table 1) or (figure 3), at an appropriate point or points in the text discussion. When making these references, do not say simply, "Table 1 contains data on leaf sizes in the two areas." Instead, say something significant about these data, such as "Leaf sizes were much more variable in area A than in area B (table 1)."

Other Points

The following points should also be observed:

- Number all pages.
- Italicize all scientific names (indicate by underlining in typed manuscript when italic type is unavailable).
- Use the metric system and other international units whenever possible (see American Society for Testing and Materials 1976, Pennycuick 1974).
- Write numbers as numerals whenever they are associated with measurement units (e.g., 3 meters) or are parts of dates or mathematical expressions. In other cases, spell them out for numbers less than 10 (e.g., five rabbits), and give them as numerals for larger values (e.g., 14 rabbits).

COMPUTER SOFTWARE

Microcomputers and word-processing software are replacing typewriters for the preparation of reports and manuscripts. The student should gain access to a word processor and become familiar with one of the several excellent word-processing programs now available. A good word-processing program should allow control over font styles and sizes, page composition, manipulation of blocks of text, and construction of tables. It should have a spelling check, a thesaurus, and a function for exporting or importing text and graphics from other files.

✪ SUGGESTED ACTIVITIES

1. Review the outline you prepared for an ecological study that progresses from descriptive through functional stages. Make a topic outline of the kinds of things you would present in each of the sections of a scientific report, as outlined above.

2. Select for analysis an article from a recent issue of *Ecology* or another ecological journal. Does the article follow the outline presented above? Is all methodology placed in a "methods" section? Are results and discussion rigorously separated? Does the discussion section effectively interpret and compare the results?

3. As a class project, review ecological articles or reports for the following, each student examining one publication. Summarize results for the class, and determine the frequency of various faults.

    - International units not used consistently
    - Inappropriate quoted material included in text
    - Some tables and figures not cited in text
    - Sentences citing tables and figures lack content
    - Table and figure legends not fully explanatory
    - Figures or tables missing labels or units
    - No reference given to source of statistical tests
    - Results of statistical tests incompletely stated
    - Some references in literature cited section not cited in text
    - Some articles cited in text not in literature cited section
    - Incorrect authorship or date for articles cited in text
    - Some references in literature cited section incomplete

✪ QUESTIONS FOR DISCUSSION

1. Why is it important to be rigorous in separating methods, results, and discussion?
2. What constitutes appropriate use of quoted or reproduced material, the source of which is acknowledged?
3. This exercise addresses many of the mechanical considerations in writing ecological reports. What are some of the stylistic considerations that also are important?

✪ SELECTED BIBLIOGRAPHY

Allen, A. 1977. Steps toward better scientific illustrations. 2nd ed. Allen Press, Lawrence, Kansas, USA.

Alley, M. 1987. The craft of scientific writing. Prentice-Hall, Englewood Cliffs, New Jersey, USA.

American Society for Testing and Materials. 1976. Standard for metric practice, ANSI/ASTM 380-76. American Society for Testing and Materials, Philadelphia, Pennsylvania, USA.

CBE Scientific Illustration Committee. 1988. Illustrating science: standards for publication. Council of Biology Editors, Bethesda, Maryland, USA.

CBE Style Manual Committee. 1983. CBE style manual. 5th ed. Council of Biology Editors, Bethesda, Maryland, USA.

CBE Writing Committee. 1986. Scientific writing for graduate students. Council of Biology Editors, Bethesda, Maryland, USA.

Katz, M. J. 1986. Elements of the scientific paper. Yale University Press, New Haven, Connecticut, USA.

Mack, R. N. 1986. Writing with precision, clarity, and economy. Bulletin of the Ecological Society of America 67:31-35.

McMillan, V. E. 1988. Writing papers in the biological sciences. St. Martin's Press, New York, New York, USA.

Papp, C. S. 1976. A manual of scientific illustration. American Visual Aid Books, Sacramento, California, USA.

Pennycuick, C. J. 1974. Conversion factors. SI units and many others. University of Chicago Press, Chicago, Illinois, USA.

Tufte, E. R. 1983. The visual display of quantitative information. Graphics Press, Cheshire, Connecticut, USA.

University of Chicago Press. 1993. The Chicago manual of style. 14th ed. University of Chicago Press, Chicago, Illinois, USA.

Zweifel, F. W. 1988. A handbook of biological illustration. 2nd ed. University of Chicago Press, Chicago, Illinois, USA.

Quantitative Description of Ecological Samples

INTRODUCTION

Ecological research requires the sampling of many attributes of individuals, populations, communities, and ecosystems. Sampling data typically exhibit two features of major interest: a certain central tendency and a certain pattern of variability. The *central tendency* is the value or condition that best typifies the attribute examined. For example, *most* flowers in the sample were red, their *average* diameter was 2 centimeters, and the plant *midway* between the tallest and shortest was 22 centimeters tall. *Variability* is the spread of individual values about this central tendency. For example, flowers *ranged* from 1 to 3 centimeters in diameter, but 95% were *within* 0.1 centimeter of the average. Ecologists use statistical techniques to describe the central tendency, express the variability within the sample data, and state the degree of confidence that can be placed on estimates of central tendency. This exercise describes the basic techniques used for these purposes.

Parameters and Statistics

In our discussions of sampling relationships, we must be careful to distinguish whether we are referring to the sample or to its universe when we speak of measures of central tendency and variability. When such values refer to the universe, we term them *parameters* and designate them with Greek letters. The symbol μ, for example, designates the *mean,* or arithmetic average, of values for a universe. Values obtained from samples, on the other hand, are termed *statistics* and are designated by Roman letters, such as $\overline{X}$, the mean calculated for a sample from some universe. In fact, our objective in obtaining and analyzing sample statistics is to infer something about the universe from which the sample came. Our purpose is to analyze sample values in a manner that enables us to make quantitative statements about the universe and at the same time specify the risk of these statements being false.

Parametric and Nonparametric Procedures

Many statistical procedures assume that the frequency distribution of the sampling data conforms to a certain mathematical distribution. The mathematical characteristics of this distribution can then be utilized in the description of sample data. Statistical techniques of this type are termed *parametric.* The mathematical distribution most commonly utilized in the analysis of sampling data is the *normal distribution*. In the normal distribution, values are symmetrically distributed about the mean, forming a bell-shaped frequency curve. Other mathematical distributions, such as binomial and Poisson distributions, exist, however, and important parametric procedures are based on them.

39

DESCRIPTIVE STATISTICS

Measures of Central Tendency

For continuous data (exercise 4) three measures of central tendency are commonly used. For a sample, the individual values are designated by the symbol X and the number of values by n. The *sample mean*, $\overline{X}$, is the arithmetic average of individual X values. It is an estimate of the *universe mean*, μ. When individual observations are grouped into classes, each of which contains observations with a certain value or range of values, the sample mean is obtained by multiplying the class value, or midpoint of the class range, by the number of observations in the class, summing these products for all classes, and dividing by the total number of values, n. For characteristics that conform to a normal distribution, the mean is the most useful measure of central tendency. In many cases, however, the distribution of individual values may be highly asymmetrical. In these cases, the most useful measures of central tendency might be the *mode,* or class of values with the greatest frequency, or the *median,* the value midway between one extreme and another.

For discrete data (exercise 4) the mode is most useful for describing the central tendency of discrete variables that are assessed by counting. The class of values containing the observation midway between extremes is also the median class.

Measures of Variability

Variability also can be described in several ways. The *range* is simply a statement of the greatest extremes shown by a set of values, e.g., the tallest and shortest plants or the extremes of flower color in a universe or sample. Variability can also be described by stating the percent of values that fall within specified distances of a mean, mode, or median.

For continuous sampling data that show a normal distribution, however, the most useful descriptors of variability are the *standard deviation* and the *variance.* The standard deviation, designated by σ for the universe and by s for a sample, is one of the mathematical terms defining the normal distribution. No matter how tall and compact or broad and flattened a normal curve might look, the values are distributed symmetrically about the mean so that 95% of them lie within 1.96 standard deviations and 99% within 2.58 standard deviations on either side of the mean. The sample variance, s^2, is a statistic that describes the variability within the sample in terms of the deviation (squared) of individual observations from their sample mean. The sample variance is equal to the sum of the squared deviations of individual observations from the mean, divided by one less than the total number of observations, n. The variance, s^2, is normally computed in the process of determining the standard deviation, using the deviations of individual sample values, X, from their mean, $\overline{X}$, by the equation

$$s^2 = \frac{\Sigma(X - \overline{X})^2}{n - 1}$$

With grouped data, the variance is obtained by squaring the deviation of the class value or class midpoint from the mean, multiplying this by the class frequency, summing these products, and dividing by $n - 1$. When calculations are done by calculators capable of accumulating products, an easier equation is

$$s^2 = \frac{\Sigma X^2 - \dfrac{(\Sigma X)^2}{n}}{n - 1}$$

The standard deviation is the square root of the variance.

Testing for Normality of Distribution

A variable tends to be normally distributed if it is influenced by many independent factors, each having a small effect. Many types of ecological data, particularly measurement data, fit a normal distribution. Before this assumption is made, however, the sample values should be tested for normality by plotting them and examining the form of the distribution. If the plotted data show a single frequency peak and symmetrical form, the normal distribution usually can be assumed, and various parametric procedures followed. Alternatively, formal tests for normality of distribution can be conducted (see Bliss 1967, pp. 101-112). If the graphed distribution possesses several peaks or is markedly asymmetrical, so that many values lie close to one end of the distribution, or if a formal test for normality indicates otherwise, analyses assuming normality should not be used.

Transforming Data to Create Normality

Data values that are not normal in distribution can sometimes be made normal by arithmetic modification. Four kinds of transformations are commonly used:

1. Logarithmic. In this case, values are converted to the logarithm (base 10 or any other base) of the value + 1 (to allow transformation of values of 0). This transformation is often used when sample standard deviations are proportional to the means in value.
2. Square root. Here, the values are converted to the square root of their raw value or the raw value + 0.5. This transformation is often used when sample variances are proportional to the means in value.
3. Reciprocal. The transformed values in this case are equal to the inverse of the raw value. This transformation is often used when sample standard deviations are proportional to the square of their means.
4. Arcsine. In this instance, the transformed value is the sine of the square root of the raw value. This transformation is almost always used when the raw values are percentages.

Inferences about the Mean of the Universe

The sample mean, $\overline{X}$, is only an estimate of the true mean, μ, of the universe. The accuracy of the sample mean depends on the number and variability of the sample values. These two factors are incorporated into a statistic termed the standard error of the mean, $s_{\overline{x}}$, which is calculated by the equation

$$ s_{\overline{x}} = \sqrt{s^2/n} \quad \text{or} \quad s/\sqrt{n} $$

The standard error of the mean permits one to calculate limiting values, known as the confidence limits of the sample mean, between which, with a certain probability, the true mean of the universe lies. These limits are calculated by the expression

$$ \overline{X} \pm t \, (s_{\overline{x}}) $$

In this expression, t is a value obtained from a table of critical values of the t distribution (appendix A, table A.2) for the desired confidence level and $n - 1$ degrees of freedom. The level of confidence refers to the probability, selected by the investigator, that the universe mean will be included in the calculated limits. For most ecological work, 95% or 99% confidence limits ($\alpha = 0.05$ or 0.01, respectively) are considered appropriate. The statistical concept of degrees of freedom refers to the number of items that can vary independently within a data set. Once the value of the mean has been fixed, only $n - 1$ sample values can vary independently. Once these values are specified, the last one is also determined, since it must be a value that will produce the observed mean.

COMPUTER SOFTWARE FOR SAMPLE DESCRIPTION

E-Z Stat (appendix B), a basic statistical software package designed for ecologists, contains a descriptive statistics module that features pull-down menus that enable a user to enter data, conduct the analyses described and produce plots or graphs of the data. This software is available in versions designed for IBM-compatible and Macintosh microcomputers.

The IBM PC software package *Ecological Analysis-PC Vol. 1,* distributed by Oakleaf Systems (appendix B), also contains a descriptive statistics module.

SUGGESTED ACTIVITIES

1. Obtain a set of measurement data by random sampling, such as the lengths of leaves obtained from a randomly selected set of plants. Calculate the various measures of central tendency and variability. Examine the data for normality and determine the standard error and the 95% and 99% confidence limits on the mean. Reorder the same data by tallying frequencies in an appropriate set of range classes. Repeat the calculations and compare the results with the previous analysis.

2. At a location on campus, or using the desert plant community map, sample numbers of plants or total crown cover with randomly placed quadrats (exercise 4). Determine the mean, mode, and median for the sample. Test for normality, and if the data conform to a normal distribution, determine the variance, standard deviation, and standard error of the mean for 95% and 99% levels of confidence. Reorder the data by tallying frequencies of quadrats with different numbers of plants or ranges of cover. Repeat the calculations and compare the results with the previous analysis.

3. Examine a set of published data from *Ecology, Ecological Monographs,* or another ecological journal. Test for normality, and if the data conform to a normal distribution, determine the variance, standard deviation, standard error of the mean, and 95% and 99% confidence limits. Determine whether your results agree with those presented in the publication.

QUESTIONS FOR DISCUSSION

1. What sorts of data would you expect to show a normal distribution? What sorts of data probably would not show a normal distribution?
2. What relation should exist for the mean, median, and mode if a data set shows a normal distribution? How might these statistics differ for data sets that are not normal?
3. Should a mean be presented without an indication of its standard deviation or standard error? Explain.

SELECTED BIBLIOGRAPHY

Bliss, C. I. 1967. Statistics in biology. Vol. 1. McGraw-Hill Book Co., New York, New York, USA.

Cox, C. P. 1987. A handbook of introductory statistical methods. John Wiley & Sons, New York, New York, USA.

Hampton, R. E. 1994. Introductory biological statistics. Wm. C. Brown, Dubuque, Iowa, USA.

Hassard, T. H. 1991. Understanding biostatistics. Mosby-Year Book, St. Louis, Missouri, USA.

Norman, G. R., and D. L. Streiner. 1994. Biostatistics: the bare essentials. Mosby-Year Book, St. Louis, Missouri, USA.

Pagano, M., and K. Gauvreau. 1993. Principles of biostatistics. Wadsworth, Belmont, California, USA.

Sokal, R. R., and F. J. Rohlf. 1980. Biometry. 2nd ed. W. H. Freeman Co., San Francisco, California, USA.

———. 1987. Introduction to biostatistics. 2nd ed. W. H. Freeman Co., San Francisco, California, USA.

Williams, B. 1993. Biostatistics. Chapman & Hall, New York, New York, USA.

Zar, J. H. 1984. Biostatistical analysis. 2nd ed. Prentice-Hall, Inc., Englewood Cliffs, New Jersey, USA.

EXERCISE 7

Testing Basic Ecological Hypotheses about Samples

INTRODUCTION

Having considered sampling and experimental procedures, together with basic statistical techniques of describing sampling data, we can now consider how to test hypotheses. In this exercise we describe basic tests of hypotheses about the central tendencies and patterns of variability of one or two samples.

Type I and Type II Errors

Testing a quantitative hypothesis involves rejection or acceptance of the null hypothesis (see exercise 1). It has been said that science advances by rejection of null hypotheses, since rejection leads to acceptance of an alternate hypothesis that expresses a new or different idea about some relationship.

Testing a null hypothesis is an exercise in probability, and is subject to two types of errors. The *type I* error is rejection of the null hypothesis when it is actually true. This error is the conclusion that a certain relationship exists, when, in fact, it does not. For example, in an experiment in which predators were removed from some plots and not others to see if prey populations are reduced by predation, rejection of the null hypothesis leads to the conclusion that predators *do* affect prey abundance. If the prey populations really were not affected, a type I error was committed. The *type II* error is acceptance of the null hypothesis when it is actually false. In the above experiment, failure to recognize that predator removal affected prey abundance when it really *did* exemplifies a type II error. Tests of hypotheses must therefore be designed to keep the chance of both types of errors low, but type I errors are of particular concern, since they wrongly suggest that new knowledge has been gained.

In hypothesis testing, the probability of a type I error, α, is usually held to 5% ($\alpha = 0.05$) or less. Tables of critical statistical values are typically structured to give values for critical α levels of 0.05, 0.01, and 0.001, and it is traditional to report that the null hypothesis has been rejected with an α less than one of these levels. Modern statistical software now allows exact α values to be determined for many tests, so that nowadays null hypotheses are often reported as being rejected at an exact α level. By general consensus, however, ecological null hypotheses are rarely considered to be rejected if α is greater than 0.05.

One-Tailed and Two-Tailed Tests

Tests of hypotheses often can be nondirectional or directional. *Two-tailed tests* are nondirectional tests in which the null hypothesis is one of "no difference," and in which deviation can occur in two directions. *One-tailed tests* state a null hypothesis that is subject to rejection by deviation in only one direction. A two-tailed test using a critical α of 0.05 assumes a rejection zone of 0.025 at one end of the normal curve and 0.025 at the other, meaning that the absolute value of a test statistic must fall at one of these points or beyond. A one-tailed test concentrates all of the rejection zone (0.05) in one tail of the curve. The value of a test statistic (given that it has the right sign) thus need not be as great to lead to rejection.

One-tailed tests have been misused frequently in ecology (Hurlbert and White 1993, Lombardi and Hurlbert *in press*). A one-tailed test is not appropriate simply because the investigator suspects that a difference will probably appear in a certain direction. Rather, a one-tailed test is appropriate only when a difference in one direction *only* is relevant to the hypothesis at hand.

Parametric and Nonparametric Tests

Like the procedures for describing variance and standard deviation (exercise 6), *parametric tests* are based on certain mathematical distributions. These tests, including those based on the normal distribution, require that the data sets do not deviate from the assumed distribution.

Nonparametric tests make no specific assumptions about the form of the sampling distribution—only that the data are unbiased. Thus, they can be used either for measurement or count data, and for data that do not show (or cannot be assumed to show) a normal distribution. The power of many of these techniques—that is, their ability to identify real differences as being statistically significant—is not much less than that of comparable parametric tests. Since many types of ecological data tend to deviate from normality, nonparametric techniques are very useful to the ecologist.

PARAMETRIC TESTS OF SAMPLE AND UNIVERSE MEANS

Comparison of Sample and Universe Means

A null hypotheses often tested by parametric procedures is that of no difference between an observed sample mean and that of some universe. This requires that the mean, μ, and in some cases the standard deviation, σ, of the universe be known. If both parameters of the universe are known, the difference is evaluated with a z test:

$$z = \frac{\overline{X} - \mu}{\sigma_{\overline{x}}} \qquad \qquad \text{Where: } \sigma_{\overline{x}} = \sigma/\sqrt{n}$$

Thus, z is the difference between sample and universe means in standard errors of the universe mean, σ_{μ}. The question in this case is whether or not the difference is so great that the sample mean is unlikely to have come from that universe. The critical values of z are 1.96 for the 0.05 α level and 2.58 for the 0.01 level. Calculated z values greater than these indicate that the observed difference could only occur by chance (if the sample actually came from the universe in question) in 5% or 1% of all samplings, respectively. Since these are very low probabilities, the null hypothesis is rejected with these chances of a type I error.

If only the mean of the universe is known, a t test must be used instead. In this test the sample standard deviation, s, is the only available estimate of variation of values about the mean. The calculated t value must be compared to a critical value for the desired level of significance and $n - 1$ degrees of freedom. The t value is calculated as follows:

$$t = \frac{\overline{X} - \mu}{s_{\overline{x}}} \qquad DF = n - 1$$

The calculated t value is compared to critical values for different α levels (appendix A, table A.2). The null hypothesis is rejected if the calculated t exceeds the critical t at the chosen α level.

Comparison of Variances and Means for Two Samples

Comparing statistics for two samples evaluates the null hypothesis that the difference is no greater than expected for two samples from the same universe. This hypothesis is either accepted, meaning that no significant difference between the two samples exists, or is rejected, indicating that the two samples differ significantly in value.

Unpaired Samples. If the two samples consist of measurements on different objects, an unpaired t test is appropriate. In this test the two sample variances are first compared for homogeneity. This comparison determines whether the two samples show the same degree of variability or come from universes with different degrees of variability. The test statistic, F, is the ratio of the sample variances, the larger variance forming the numerator. Since there is no a priori reason for one sample having the larger variance, this is a two-tailed test of whether the first sample has a variance significantly greater than the second, or the second greater than the first. The calculated F value is compared to a tabled critical value (appendix A, table A.3) for a selected α level (e.g., 0.05 or 0.01) and $n - 1$ degrees of freedom for each sample. A calculated F in excess of the tabled value rejects the null hypothesis, meaning that the two samples differ significantly in variability. If this is true, the means may not be compared with a t test (below). In this case, the data values in the two samples may be transformed (exercise 6) and a new F test done to see if the variances are equalized enough to permit a t test. Even if the variances are not homogeneous, the central tendencies of the samples may be compared by the nonparametric Mann-Whitney test (described later).

If the calculated F value is less than the critical tabled value, the two variances can be regarded as estimates of a common value. They then can be combined and used in a t test of difference between the sample means. The difference between the means is compared to the standard error of the difference, $s_{\bar{x}_1 - \bar{x}_2}$, which is calculated by the equation

$$s_{\bar{x}_1 - \bar{x}_2} = \sqrt{\frac{(n_1 - 1)s_1^2 + (n_2 - 1)s_2^2}{n_1 + n_2 - 2}\left(\frac{1}{n_1} + \frac{1}{n_2}\right)}$$

Where: $\bar{x}_1, s_1^2, n_1$ = Values for one sample

$\bar{x}_2, s_2^2, n_2$ = Values for second sample

A t value is then calculated by the equation

$$t = \frac{\bar{x}_1 - \bar{x}_2}{s_{\bar{x}_1 - \bar{x}_2}}$$

The calculated t value is compared to a tabled critical value (appendix A, table A.2) for the desired α level and $n_1 + n_2 - 2$ degrees of freedom. A calculated value in excess of the tabled value rejects the null hypothesis. The means differ more than expected by chance, and thus probably came from two different universes. An example of a unpaired t test is given in table 7.1.

Paired Samples. If the two samples are actually two sets of measurements on the same objects (i.e., organisms, populations, ecosystems), a simpler version of the t test can be used. In this case, the differences, d, between the measurements are determined by subtracting values for the second set from those for the first set (keeping track of whether they are + or −). The mean and standard error of the mean of the d's are calculated (exercise 6). A t value is then obtained by the equation

$$t = \frac{\bar{d}}{s_{\bar{d}}} \qquad \text{Where: } s_{\bar{d}} = \text{Standard error of } d\text{'s}$$

This value has a number of degrees of freedom one less than the number of paired comparisons. Otherwise, its significance is judged as for the unpaired t test.

Table 7.1. Example of parametric comparison of data for two samples (weights of acorns from oak trees on north and south slopes)

Acorn Weights (Grams)

Sample #1 (N slope)	Sample #2 (S slope)
2.33	2.02
2.51	1.90
2.12	2.13
2.70	2.50
2.00	2.30
2.42	2.21
2.54	2.21
2.60	1.80
2.44	2.64
2.53	2.14

$n_1 = 10$

$\Sigma X = 24.19$
$\bar{X} = 2.42$

$n_2 = 10$

$\Sigma X = 21.85$
$\bar{X} = 2.18$

$$s_1^2 = \frac{58.9359 - \dfrac{(24.19)^2}{10}}{9}$$

$s_1^2 = 0.0467$
$s_1 = 0.2161$
$s_{\bar{x}_1} = \sqrt{0.0467/10}$
$s_{\bar{x}_1} = 0.0683$

$$s_2^2 = \frac{48.3247 - \dfrac{(21.85)^2}{10}}{9}$$

$s_2^2 = 0.0647$
$s_2 = 0.2544$
$s_{\bar{x}_2} = \sqrt{0.0647/10}$
$s_{\bar{x}_2} = 0.0804$

95% Confidence Limits for Sample Means

$t_{0.05,9} = 2.262$

Sample #1
$2.42 \pm 2.262\,(0.0683)$
95% CL $= 2.27 - 2.57$

Sample #2
$2.18 \pm 2.262\,(0.0804)$
95% CL $= 2.00 - 2.36$

Tests of Sample Means Against Mean of Universe ($\mu = 2.50,\ \sigma = 0.20$)

Sample #1
$$z = \frac{|2.42 - 2.50|}{0.20/\sqrt{10}} = 1.26$$

$$t = \frac{2.42 - 2.50}{0.0683} = 1.07$$

Sample #2
$$z = \frac{|2.18 - 2.50|}{0.20/\sqrt{10}} = 5.06$$

$$t = \frac{2.18 - 2.50}{0.0804} = 3.98$$

$t_{0.05,9} = 2.262$

F test of Homogeneity of Variances of Samples

$$F = \frac{0.0647}{0.0467} = 1.38 \quad DF = 9, 9 \quad F_{0.05;9,9} = 4.03$$

t test of Difference Between Sample Means

$$s_{\bar{x}_1 - \bar{x}_2} = \sqrt{\frac{9(0.0647) + 9(0.0647)}{10 + 10 - 2}\left(\frac{1}{10} + \frac{1}{10}\right)} = 0.1055$$

$$t = \frac{2.42 - 2.18}{0.1055} = 2.274 \quad DF = 18 \quad t_{0.05,18} = 2.101$$

NONPARAMETRIC COMPARISON OF SAMPLES

Unpaired Samples. The nonparametric *Mann-Whitney test* compares the central tendencies of two samples of unpaired values, regardless of the distribution form of the sample data. The test may be used with measurement data, whether normally distributed or otherwise, or with appropriate types of count data. It is especially useful for comparing sets of population estimates of plants or animals. For normally distributed data it is almost as powerful as the *t* test.

To perform this test, the values from both samples are first arranged in a single sequence of increasing size. Each value in this sequence is then assigned a rank, beginning with 1 for the smallest value. For ties (observations having the same value), each of the tied values is assigned the average of the ranks for which it is tied. The rank of the largest observation is thus equal to the total number of values for the two samples.

The ranks are then summed separately for the two samples. For samples containing 20 or fewer observations, either of the rank-sums (W) can be used to calculate U and U' values by the equations

$$U = n_1 n_2 + \frac{n_1(n_1 + 1)}{2} - W_1$$

$$U' = n_1 n_2 - U$$

Where: n_1 = Number of observations in sample 1
n_2 = Number of observations in sample 2
W_1 = Rank-sum for sample 1

The smaller of the values U or U' is then compared to critical values for the 0.05 and 0.01 α levels (appendix A, Table A.4). If the U value is equal to or less than the tabled value for the corresponding sample sizes, the sample central tendencies are different at the selected probability level.

For samples with more than 20 values, rank-sums for samples from a single universe show a normal distribution about a certain mean. The mean and standard deviation of this universe depend only on the sample size. The deviation of an observed rank-sum value from its theoretical mean thus can be compared to the standard deviation for the theoretical rank-sum distribution, and the probability of a deviation that large determined. This is a form of z test, carried out by the equation

$$z = \frac{W_1 - \frac{n_1(n_{1+2} + 1)}{2}}{\sigma}$$

Where: W_1 = Rank-sum for sample 1 (sample with lower value)
n_1 = Number of observations in sample 1
n_2 = Number of observations in sample 2
$\dfrac{n_1(n_{1+2} + 1)}{2}$ = Theoretical mean rank-sum
σ = Standard deviation of theoretical rank-sum distribution

If no ties exist in the ranked values, the standard deviation is calculated by the equation

$$\sigma = \sqrt{\frac{n_1 n_2 (n_{1+2} + 1)}{12}}$$

If ties do exist, the following equation must be used:

$$\sigma = \sqrt{\frac{n_1 n_2 [n_{1+2}(n_{1+2}^2 - 1) - \Sigma T]}{12 n_{1+2} (n_{1+2} - 1)}}$$

Where: T = Correction value for each set of ties calculated by the equation
$T = t(t - 1)(t + 1)$
in which t = number of tied values in group

A calculated z value exceeding critical values (see above) for the chosen level of confidence indicates that the null hypothesis of no difference between the central tendencies of the two samples must be rejected. Table 7.2 shows an example of sample comparison using the Mann-Whitney test.

Paired Samples. The *Wilcoxon signed-rank test* involves sets of paired measurements, like those in the paired *t* test. In this test the differences, *d,* between the measurements are first determined (keeping track of

Table 7.2. Example of nonparametric Mann-Whitney test of difference between the central tendencies of two samples

A. Samples of n = 20 or smaller (acorn weights from table 7.1)

Rank	Sample #1	Sample #2	Rank
3	2.00	1.80	1
5	2.12	1.90	2
11	2.33	2.02	4
12	2.42	2.13	6
13	2.44	2.14	7
15	2.51	2.21	8.5
16	2.53	2.21	8.5
17	2.54	2.30	10
18	2.60	2.50	14
20	2.70	2.64	19

$W_1 = 130$ $\qquad$ $W_2 = 80$

$$U = (10)(10)\frac{10(11)}{2} - 130 = 25$$

$$U' = (10)(10) - 25 = 75$$

$$U_{0.05;\ 10,\ 10} = 23$$

B. Samples larger than n = 20

Sample #1	Sample #2
$n_1 = 23$	$n_2 = 28$
$W_1 = 387$	$W_2 = 939$

$$z = \frac{\left| 387 - \dfrac{23(52)}{2} \right|}{\sqrt{\dfrac{(23)(28)(52)}{12}}} = 3.99$$

$$z_{0.05} = 1.96$$

C. Samples larger than n = 20 and containing ties (data as in B)

3 sets of ties involving 3 items

$$T = 3(2)(4) = 24$$

5 sets of ties involving 2 items

$$T = 2(1)(3) = 6$$

$$\Sigma T = 3(24) + 5(6) = 102$$

$$z = \frac{\left| 387 - \dfrac{23(52)}{2} \right|}{\sqrt{\dfrac{(13)(28)[51(51^2 - 1) - 102]}{12(51)(50)}}}$$

$$z = 4.00$$

$$z_{0.05} = 1.96$$

whether they are + or −). Ignoring differences of 0, the absolute values of differences are ranked from smallest to largest, with tied values assigned the average of the tied ranks. The sum of ranks, T, for the differences with the least frequent sign (+ or −) is used to calculate T' by the equation

$$T' = \frac{n(n + 1)}{2} - T$$

T and T' are compared to critical values for the 0.05 and 0.01 α levels (appendix A, table A.5). If either T or T' is equal to or less than the tabled value for the corresponding sample sizes, the sample central tendencies are different at that probability level.

NONPARAMETRIC GOODNESS-OF-FIT TESTS

The *chi-square goodness-of-fit test* may be used whenever a theoretical expectation of the frequency of sample observations among two or more categories exists. The null hypothesis is that observed frequencies do not differ from those theoretically expected. Count data are thus required for this test. Typically, such data concern the numbers of objects or individuals that have various characteristics, or the numbers of individuals that show various responses. Data for continuous variables, such as color, weight, or length, can be analyzed only if they can be grouped in categories for which some theoretical expectation of frequency exists.

Based on the number of observations, and the theoretical expectation of their distribution among categories, expected values summing to the same total as that for the observed values are first calculated. How closely the observed and expected sets correspond is then tested by calculating chi-square:

$$\text{Chi-Square} = \Sigma \frac{(\text{Observed} - \text{Expected})^2}{\text{Expected}}$$

Suppose, for example, that we determine preferences of 90 individuals of some animal species by placing them in a chamber having equal areas of three substrate types. We can test the observed responses against an expectation of no preference (and thus, equal numbers on each substrate) as follows:

Substrate	Expected	Observed	O − E	$\frac{(O - E)^2}{E}$
Rock	30	50	20	13.3
Sand	30	25	5	0.8
Mud	30	15	15	7.5
	Total 90	90		21.6 = Chi-Square

Chi-square goodness-of-fit values have 1 fewer degrees of freedom than the number of observed-expected categories (two in the above example). The calculated chi-square is compared to tabled critical values for the corresponding number of degrees of freedom and chosen probability level (appendix A, table A.6). A calculated chi-square that exceeds the tabled critical value rejects the null hypothesis of no difference between observed and expected sets.

Chi-square testing requires a certain minimum quantity of data. It is usually recommended that no expected category have a value less than 1.0, and that no more than 20% have values less than 5.0. Furthermore, in tests with 1 degree of freedom, the absolute value of the difference between observed and expected is usually reduced by 0.5. This adjustment, known as Yates' correction, takes into account the fact that expected values can be fractional and observed values cannot, thus creating deviations that are simply artifacts. The 20% requirement is now known to be somewhat too severe, however, and Yates' correction somewhat too great.

NONPARAMETRIC CONTINGENCY ANALYSES

This is one of the most useful tests for ecologists. It is employed with count data grouped in a matrix defined by two variables. For example, in a substrate preference test, we may record the preferences in individuals for different substrates in the light, on one hand, and in the dark, on the other. Results of such tests might be grouped as follows:

	Substrate		
	Sand	Mud	Total
Light	35 (a)	25 (b)	60 (a+b)
Dark	15 (c)	25 (d)	40 (c+d)
Total	50 (a+c)	50 (b+d)	100 (a+b+c+d = T)

Contingency analysis is simply a test of interaction between the two variables, substrate and light in the above example. The null hypothesis is that the effect of one variable is not influenced by the other. In the example, this means that selection of substrate is not influenced by, or contingent upon, light conditions.

For 2×2 contingency tables, like the above, a chi-square value can be calculated in either of two ways. Expected values for each cell can be computed and the chi-square value calculated for observed-expected differences as for the goodness-of-fit test. In this case, expected frequencies for each cell are given by the following expressions:

$$\text{Cell a: } \frac{(a+b)(a+c)}{T}$$

$$\text{Cell b: } \frac{(a+b)(b+d)}{T}$$

$$\text{Cell c: } \frac{(c+d)(a+c)}{T}$$

$$\text{Cell d: } \frac{(c+d)(b+d)}{T}$$

Alternatively, the following calculation, which already includes Yates' correction, can be used:

$$\text{Chi-Square} = \frac{(|ad - bc| - 0.5T)^2(T)}{(a+b)(a+c)(b+d)(c+d)}$$

Contingency analyses can be used in situations having more than two expressions of each variable, that is, for contingency tables with more than two rows and/or columns. In such cases the calculation of separate expected frequencies for each cell in the table is unavoidable. The expected frequency for a particular cell is given by the appropriate form of the expression

$$\frac{(\text{row total})(\text{column total})}{(\text{grand total})}$$

Expected and observed frequencies are then compared and a chi-square value calculated as for a goodness-of-fit test.

In all chi-square contingency analyses, the number of degrees of freedom is equal to the number of rows minus one multiplied by the number of columns minus one. In the case of a 2×2 table, of course, this gives 1 degree of freedom. Thus, for 2×2 tables, Yates' correction should be applied. A contingency table with 4 rows and 3 columns has 6 degrees of freedom.

Calculated chi-square values are compared with tabled critical values (appendix A, Table A.6) for the corresponding number of degrees of freedom and selected level of confidence. A calculated value exceeding the tabled value leads to rejection of the null hypothesis, and thus to the conclusion that a significant interaction occurs between the two variables.

Contingency chi-square tests are also subject to the recommendation that no cell have an expected frequency less than 1.0, and no more than 20% of the cells have expected frequencies less than 5.0.

COMPUTER SOFTWARE FOR TESTING HYPOTHESES ABOUT SAMPLES

E-Z Stat (Appendix B), a basic statistical software package designed for ecologists, contains statistics modules that feature pull-down menus that enable a user to enter data, produce plots or graphs of the data, and carry out most of the tests described above. This software is available in versions designed for IBM-compatible and Macintosh microcomputers.

SUGGESTED ACTIVITIES

1. Collect cones of pines from slopes of different aspect, steepness, or other habitat difference. Select a measurement variable (e.g., height, mass) for analysis. Compare the samples with the non-paired *t* test and with the Mann-Whitney test.
2. Measure the length of stem growth increments for two consecutive years (the same years!) on individuals of a woody plant species (e.g., pine, ocotillo). Compare the central tendencies of the samples by a paired *t* test and by the Wilcoxon paired-sample test.
3. Collect a sample of individuals of an animal species (e.g., crayfish, shore crab) for which the sexes can be distinguished. Sex the animals and test the observed ratio against a theoretical 1:1 sex ratio.
4. Collect leaves from sun-exposed and shaded portions of the canopy of a tree or shrub. Identify a characteristic by which the leaves can be categorized (e.g., with/without galls, with/without herbivore damage). Use a chi-square contingency analysis to test whether or not sun exposure and leaf characteristics are independent.

QUESTIONS FOR DISCUSSION

1. How does the number of values in a sample, together with the variability among these values, affect statistics such as the variance, the standard deviation, and the standard error of the mean?
2. How will such factors affect the size of the difference required for rejection of the null hypothesis in two-sample tests?
3. What sorts of ecological data are likely to demand analysis by nonparametric tests?
4. Although nonparametric tests do not assume normal distribution of sampling data, what assumptions must be met for these tests to be valid?

SELECTED BIBLIOGRAPHY

Cox, C. P. 1987. A handbook of introductory statistical methods. John Wiley & Sons, New York, New York, USA.

Hampton, R. E. 1994. Introductory biological statistics. Wm. C. Brown, Dubuque, Iowa, USA.

Hassard, T. H. 1991. Understanding biostatistics. Mosby–Year Book, St. Louis, Missouri, USA.

Hurlbert, S. H., and M. D. White. 1993. Experiments with freshwater invertebrate zooplanktivores: quality of statistical analyses. Bulletin of Marine Science 53:128–152.

Lombardi, C. M., and S. H. Hurlbert. 1994. Misprescription and misuse of one-tailed tests. Animal Behaviour, in press.

Norman, G. R., and D. L. Streiner. 1994. Biostatistics: the bare essentials. Mosby–Year Book, St. Louis, Missouri, USA.

Pagano, M., and K. Gauvreau. 1993. Principles of biostatistics. Wadsworth, Belmont, California, USA.

Sokal, R. R., and F. J. Rohlf. 1980. Biometry. 2nd ed. W. H. Freeman Co., San Francisco, California, USA.

———. 1987. Introduction to biostatistics. 2nd ed. W. H. Freeman Co., San Francisco, California, USA.

Williams, B. 1993. Biostatistics. Chapman & Hall, New York, New York, USA.

Zar, J. H. 1984. Biostatistical analysis. 2nd ed. Prentice-Hall, Inc., Englewood Cliffs, New Jersey, USA.

EXERCISE 8

Regression, Correlation, and Analysis of Variance

INTRODUCTION

Regression, correlation, and analysis of variance are used extensively in ecology. Regression and correlation determine whether the values of two or more variables are significantly related. Analysis of variance tests whether or not the variation in a response variable can be attributed to different levels of influence of one or more other variables.

In *regression analysis,* a single *dependent variable, Y,* is considered to be a function of one or more *independent variables, X1, X2,* etc. The values of the independent variables are regarded as being determined in an error-free manner. Those of the dependent variables are influenced to a greater or lesser extent by those of the independent variables, and are also subject to random variation. In a *simple regression* analysis, only one independent variable is considered; in *multiple regression* analysis there are two or more independent variables. *Parametric regression* analysis assumes that for any given value of the independent variable, values of the dependent variable vary normally about some mean. In the simplest form of regression analysis, the relationship between dependent and independent variables is assumed to be linear.

In *correlation analysis,* two or more variables are assumed to be related in some fashion, but each is subject to random variation. Again, parametric techniques of correlation analysis assume that under a given set of conditions, variation in each of the variables follows a normal curve. In correlation analysis, the letters *X* and *Y* are often used to designate different variables, but this does not imply that the variable called *Y* is dependent on the variable called *X*.

Analysis of variance (ANOVA) allows the simultaneous testing of several samples to see if variation is significantly related to one or more variables. In single-factor ANOVA, the samples differ in only one variable, such as the time at which they were taken, the location from which they came, or, in an experiment, the amount of fertilizer applied. Single-factor ANOVA tests whether or not significant variation exists in the means of the samples. In two-factor ANOVA, the samples are grouped according to two variables, such as time and location of collection, or fertilizer and irrigation treatment. Two-factor ANOVA tests whether or not each variable significantly affects the sample mean. When each sample contains replicates, two-factor ANOVA can also determine if there is a significant interaction of the two variables, i.e., if the effect of one variable is significantly influenced by the level of the other variable.

In this exercise we shall consider simple linear regression analysis and multiple regression, using parametric procedures. Correlation and ANOVA can be carried out by both parametric and nonparametric techniques. We shall describe simple parametric and nonparametric forms of correlation and one-factor ANOVA. Last, we shall describe general features of multifactor ANOVA.

REGRESSION ANALYSIS

Simple Linear Regression

Simple linear regression derives an equation for a straight line that expresses the relationship between dependent and independent variables. This equation has the form

$$Y = a + bX$$

Where: Y = Value of dependent variable
a = Intercept of regression line on y-axis
b = Slope of regression line
X = Value of independent variable

To conduct a linear regression analysis, the following values must be calculated (most electronic calculators can determine these values in a single operation sequence):

1. The sum (ΣY) and mean ($\overline{Y}$) of the dependent variable values.
2. The sum (ΣX) and mean ($\overline{X}$) of the independent variable values.
3. The sum of squares of deviations of each value of the dependent variable from the mean (SSY). This value can be calculated in either of two ways, the second being simpler on most electronic calculators:

$$SSY = \Sigma (Y - \overline{Y})^2 \quad \text{or} \quad SSY = \Sigma Y^2 - (\Sigma Y)^2/n$$
Where: n = Number of Y or X values

4. The sum of squares of deviations of each value of the independent variable from the mean (SSX). This value can be calculated in either of two ways:

$$SSX = \Sigma(X - \overline{X})^2 \quad \text{or} \quad SSX = \Sigma X^2 - (\Sigma X)^2/n$$

5. The sum of cross-products of corresponding X and Y values from their means (SCP). This value can be calculated in either of two ways:

$$SCP = \Sigma(X - \overline{X})(Y - \overline{Y}) \quad \text{or} \quad SCP = \Sigma XY - \Sigma X \Sigma Y/n$$

The unknowns b and a then may be calculated by the equations

$$b = SCP/SSX$$
$$a = \overline{Y} - b\overline{X}$$

Note that the sign of b depends only on the sign of SCP.

These calculations are facilitated by setting up a table like table 8.1, in which an example of simple linear regression is worked out. In this example, the number of land bird species on certain West Indian islands is considered to be a value, Y, dependent upon the area, X, of the island (expressed as a logarithm).

Whether or not the calculated relationship is statistically significant can be determined by an F test. In this test, the portion of the variation of the dependent variable that is accounted for by the independent variable is compared to that not accounted for by calculating an F ratio (table 8.2). This F value is compared to a critical F value (from appendix A, Table A.3) for 1 and $n - 2$ degrees of freedom. The critical F value is that for a one-tailed test, since we are interested only in whether the variation explained exceeds that not explained, and not the opposite.

The calculated value of b also can be tested against various hypothesized values, such as zero. To do this, the standard error of the regression slope, s_b, is calculated:

$$s_b = \sqrt{\frac{\text{residual MS}}{SSX}}$$

Table 8.1 Simple linear regression analysis of data on island area and number of terrestrial bird species for selected islands of the West Indies

Island	Y Number of Land Bird Species	Y − Ȳ	X Land Area (Log 10 km²)	X − X̄
Hispaniola	69	33.7	4.884	2.236
Puerto Rico	50	14.7	3.949	1.301
St. Lucia	41	5.7	2.790	0.142
St. Vincent	36	0.7	2.590	−0.058
St. Kitts	20	−15.3	2.255	−0.393
Saba	18	−17.3	1.114	−1.534
Anguilla	13	−22.3	0.954	−1.694

$\Sigma Y = 247$ $\bar{Y} = 35.3$ $\Sigma X = 18.536$ $\bar{X} = 2.648$

$\Sigma(X - \bar{X})^2 = 12.093$ $\Sigma(X - \bar{X})(Y - \bar{Y}) = 165.574$

$b = \dfrac{165.574}{12.093} = 13.692$ $a = 35.3 - 13.692(2.648) = -0.956$

Table 8.2 Testing the significance of a simple linear regression

Line	Variance Component	Sum of Squares	DF	Mean Square	F
1.	Accounted for by independent variable	$b[\Sigma(Y - \bar{Y})(X - \bar{X})]$	1	Same as sum of squares	$\dfrac{\text{Line 1 mean square}}{\text{Line 2 mean square}}$
2.	Residual variance	Line 3 − line 1	n − 2	$\dfrac{\text{Sum of squares}}{\text{DF}}$	- -
3.	Total variance	$\Sigma(Y - \bar{Y})^2$	n − 1		- -

Test of Regression Data from Table 6.1

Line		Sum of Squares	DF	Mean Square	F
1.	Explained	2267.04	1	2267.04	76.38
2.	Residual	148.39	5	29.68	
3.	Total	2415.43	6		

$F_{(0.05;\ 1,5)} = 6.61$

The calculated slope, b, can then be tested against any hypothesized value, c (including 0), by a t test (see exercise 7) with $n - 2$ degrees of freedom:

$$t = \frac{b - c}{s_b}$$

Slope estimates for different regressions can be compared in a similar fashion, but for this the standard error of the difference between slopes, s_d, must be used:

$$s_d = \sqrt{\frac{(\text{residual MS})_p}{(\text{SSX})_1} + \frac{(\text{residual MS})_p}{(\text{SSX})_2}}$$

Where: $(\text{residual MS})_p = \dfrac{(\text{residual sum of squares})_1 + (\text{residual sum of squares})_2}{(\text{residual DF})_1 + (\text{residual DF})_2}$

A t test is then conducted:

$$t = \frac{b_1 - b_2}{s_d} \quad \text{with } n_1 + n_2 - 4 \text{ degrees of freedom}$$

Multiple Regression

Multiple regression analysis is useful in identifying significant relationships in ecological situations in which it is difficult to carry out controlled experiments or single-variable manipulations. Multiple regression is designed to derive an equation that describes the combined relationship of a single dependent variable to several independent variables. The general form of such an equation is

$$Y = a + b_1X_1 + b_2X_2 + b_3X_3. \ldots \ldots + b_kX_k \quad \text{Where: } k = \text{Number of independent variables}$$

Manual calculation of multiple regression equations, even for only a few independent variables, is very tedious, because all possible cross-product combinations for deviations of these values from their means must be computed and manipulated in a complex fashion. Fortunately, however, sophisticated computer programs for such analyses are available for personal computers (see below).

CORRELATION ANALYSIS

Simple Linear Correlation

This procedure calculates a coefficient of correlation, r, using the same values computed for linear regression analysis:

$$r = \frac{SCP}{\sqrt{(SSX)(SSY)}}$$

Values of r vary from -1.0 (perfect negative correlation) to $+1.0$ (perfect positive correlation). A t test is used to determine if a calculated r is significantly different from zero (no correlation). A t value is calculated by the following expression, the denominator of which is the standard error of the coefficient:

$$t = \frac{r}{\sqrt{(1 - r^2)/(n - 2)}}$$

This calculated t value is then compared to the critical t for a selected significance level and $n - 2$ degrees of freedom (appendix A, table A.2). A calculated value exceeding the critical value indicates a correlation significant at the selected alpha level.

To test the significance of differences between correlation coefficients (r_1 and r_2), the coefficients are transformed to z' values:

$$z' = 0.5 \ln \frac{1 + r}{1 - r}$$

The z' values are then compared by a z test. Here, it is important to keep in mind the difference between the test statistic, z, and the transformed correlation coefficients, z'. The z statistic is calculated:

$$z = \frac{z'_1 - z'_2}{\sigma_d}$$

Where: σ_d = Standard error of difference

$$\sigma_d = \sqrt{\frac{1}{n_1 - 3} + \frac{1}{n_2 - 3}}$$

Table 8.3 Spearman rank correlation analysis on the number of resident land birds and lizard species on selected islands of the West Indies

Island	X Birds	X Rank	Y Lizards	Y Rank	d_i	d_i^2
Hispaniola	69	7	54	7	0	0
Puerto Rico	50	6	18	6	0	0
St. Lucia	41	5	9	5	0	0
St. Vincent	36	4	7	3.5	0.5	0.25
St. Kitts	26	3	7	3.5	0.5	0.25
Saba	18	2	4	1	1.0	1.0
Anguilla	13	1	5	2	1.0	1.0
						2.50 $= \Sigma d_i^2$

$$r_s = 1 - \frac{6\,(2.50)}{7^3 - 7} = 1 - \frac{15.0}{336} = 0.955$$

r_s (critical) for $n = 7$, $\alpha = 0.05$ (2-tailed) $= 0.786$

Nonparametric Spearman Rank Correlation

The Spearman rank correlation coefficient, r_s, is determined by ranking X and Y values from smallest to largest (assigning each a rank number, or the average of the ranks for which they are tied), determining the difference, d, for ranks of each XY combination, and calculating the sum of the squared d's. The coefficient is then calculated:

$$r_s = 1 - \frac{6\Sigma d_i^2}{n^3 - n}$$

Critical r_s values are given in appendix A, table A.7. Table 8.3 illustrates the calculation of r_s for bird and lizard diversity on some West Indian islands.

ANALYSIS OF VARIANCE

Parametric Single-Factor ANOVA

Single-factor ANOVA is illustrated in table 8.4, part I, in which plant growth at four watering levels, each with four replicates, is considered. The null hypothesis is that mean growth does not differ under the four watering levels. ANOVA is a technique of partitioning the total variation in the response variable into explained and unexplained components. The procedure requires the calculation of various sums of squares (SS), degrees of freedom (DF), mean squares (MS), and an F ratio. The F ratio compares the fraction of the variation attributable to the grouping factor (watering) to that (within-groups, or "error") of replicates receiving the same treatment.

The *total SS* is calculated as the sum of squares of all the n individual X values (16, in this example) minus the total of the Xs squared and divided by n:

$$\text{Total SS} = \Sigma X^2 - (\Sigma X)^2/16$$

The *groups SS* is calculated by taking the sum of values for each water group, squaring and dividing by the number of values (4), summing these values for the four groups, and subtracting the total of all values squared and divided by n (16):

$$\text{Groups SS} = (\Sigma Xa)^2/4 + (\Sigma Xb)^2/4 + (\Sigma Xc)^2/4 + (\Sigma Xd)^2/4 - (\Sigma X)^2/16$$

Table 8.4. Single-factor parametric ANOVA and Kruskal-Wallace nonparametric ANOVA for data on growth of plants under four levels of watering

Part I. Data (Values and Ranks)

		Watering Levels		
n	*a*	*b*	*c*	*d*
1	4.36 (1)	5.06 (5)	6.28 (11.5)	7.22 (16)
2	5.07 (6)	6.13 (10)	6.47 (13)	6.83 (15)
3	4.91 (3)	5.20 (7)	5.44 (8)	6.49 (14)
4	4.54 (2)	4.96 (4)	6.05 (9)	6.28 (11.5)
	$\Sigma Xa = 18.88$	$\Sigma Xb = 21.35$	$\Sigma Xc = 24.24$	$\Sigma Xd = 26.82$
	$\bar{X}a = 4.72$	$\bar{X}b = 5.34$	$\bar{X}c = 6.06$	$\bar{X}d = 6.70$
	$\Sigma Ra = 12.0$	$\Sigma Rb = 26.0$	$\Sigma Rc = 41.5$	$\Sigma Rd = 56.5$

$\Sigma Xt = 91.29$
$\bar{X}t = 5.70$

Part II. Single-Factor Parametric ANOVA

Variance Component	SS	DF	MS	F
Groups (Water)	8.92	3	2.98	15.68
Error	2.30	12	0.19	
Total	11.22	15		

Critical F (0.001) for 3, 12 DF = 10.8

Part III. Kruskall-Wallace Nonparametric ANOVA

$$H = \frac{12}{16(17)} \left(\frac{12^2}{4} + \frac{26^2}{4} + \frac{41.5^2}{4} + \frac{56.5^2}{4} \right) - 3(17) = 12.248$$

Critical H (0.001) for $n_{4,4,4,4} = 11.338$

The *within-groups SS,* a measure of the variation among samples receiving identical treatment, is simply the difference between the total and groups *SS*s:

$$\text{Within-Groups SS} = \text{Total SS} - \text{Groups SS}$$

The degrees of freedom for the total *SS* are the total number of values (*n*) minus 1 (i.e., 15), for the groups *SS* the total number of groups minus 1 (i.e., 3), and for the within-groups *SS* the difference between these two values (12). To obtain the mean squares, the sums of squares are divided by the degrees of freedom.

To test whether the variance explained by the watering group variable is significant, an *F* ratio is computed by dividing the groups *MS* by the within-groups *MS*. This *F* value is then compared to critical values for the corresponding number of degrees of freedom for a one-tailed test (appendix A, table A.3). In the example, the result of the test shows a significant influence of watering treatments on growth.

Kruskal-Wallace Nonparametric ANOVA

In this test, the *X* values are ranked from smallest to largest (table 8.4, part III), and the rank values summed for the treatment groups (watering levels). Tied values are assigned the mean value of the ranks involved. Using the total number of *X*s (n_t), the number of *X*s in each group ($n_a \ldots n_d$), and the sums of the ranks ($R_a \ldots R_d$), the Kruskal-Wallace *H* Statistic is calculated:

$$H = \frac{12}{n_t(n_t - 1)} \Sigma \frac{R^2}{n} - 3(n_t - 1)$$

(This value is compared to critical values from appendix A, table A.8). A calculated *H* exceeding the critical value indicates a significant influence of the grouping factor at the chosen alpha level.

Multifactor ANOVAs

ANOVA with two or more variables of classification of independent variables become complicated in design and analysis. The basic principle of these analyses is similar: the variance attributable to each variable is determined, and an *F* ratio is calculated with this component as numerator and an appropriate component of unexplained variance as denominator. Inasmuch as hand calculation of multifactor ANOVAs can be very tedious (and subject to arithmetic errors!), these analyses are now done by statistical software. In using such software, however, it is essential that the investigator understand ANOVA design, and how to select the version of ANOVA that the software is commanded to perform.

In multifactor ANOVA, the distinction between the effects of *fixed* variables and *random* variables becomes important. Fixed variables are those imposed experimentally, as in the case of levels of watering used in an experiment. Random variables are those chosen for comparison, but not fully under experimental control and therefore subject to random variation. For example, an experiment might be done on plots randomly located on soils of different types (e.g., heavy clay, loam, and sandy types). In a two-variable experiment, watering levels (a fixed variable) could be examined on different soil types (a random variable). Thus, a two-factor ANOVA could involve two fixed variables, two random variables, or one fixed and one random variable. The structure of the *F* ratios for testing independent variables differs in each of these cases (Zar 1984).

Multifactor ANOVAs can also differ in whether variables are *crossed* or *nested*. A crossed design is one in which all levels of one variable are combined with all levels of the other(s). An example of a crossed design with watering levels and soil types is the examination of all watering levels on all soil types. If small mammal populations were sampled on four plots in each of three forest types, but six persons did the sampling, each studying two plots in one of the forest types, the ANOVA design would be nested. The "persons" variable is nested within that of forest types.

The design of multifactor ANOVAs can thus become quite complicated as a result of the number of variables, random and fixed effects, and crossed or nested design. Whether the design is *balanced* or *unbalanced*, meaning that each combination of variables has the same or a different number of replicates, also complicates calculations.

In multifactor ANOVA in which each combination of variables is replicated, *F* tests can be conducted to evaluate the overall effect of each variable, that is, its effect across all levels of the complementary variable(s). In addition, tests can be done to determine if a significant *interaction* exists among pairs of variables. A significant interaction implies that the value of one variable changes with that of the other.

SOFTWARE FOR REGRESSION, CORRELATION, AND ANOVA ANALYSES

EcoStat (appendix B), a basic statistical software package designed for ecologists, contains statistics modules that feature pull-down menus that enable a user to enter and graph data, and to carry out simple linear regression and correlation, Spearman rank correlation, single-factor ANOVA, and Kruskal-Wallace nonparametric ANOVA. This software is available in versions designed for IBM PC (DOS) and Macintosh microcomputers. The Ecological Analysis–PC statistical package (appendix B) also includes modules for parametric regression and correlation analyses.

Analysis of multifactor ANOVAs is best done with one of the more sophisticated statistical software packages now available commercially for DOS, Windows, and Macintosh microcomputers. These include Minitab, Statistica, Statistix, Systat, SigmaStat, Strata, Crunch, S-Plus, Gauss, BioΣtat, ABstat, SAS, SPSS, and BMDP. These software packages differ greatly in user-interface features, capabilities, and price, all of which should be considered carefully before one invests in them. The *Bulletin of the Ecological Society of America* contains a "Technological Tools" section that presents reviews of these and other software packages of particular interest to ecologists (see, for example, Ellison 1992 and later issues).

☯ Suggested Activities

1. Measure the length of last year's growth on a sample of a woody plant species (e.g., pine, ocotillo), noting also some measure of maximum size. Compute a simple linear regression of growth (Y) as a function of size (X).

2. Examine a set of data on the number of plant species and the number of bird or other animal species on islands in the Galápagos (e.g., Connor and Simberloff 1978). Use simple linear correlation and Spearman rank correlation to test whether these variables are correlated. Use simple linear regression to test whether these variables are dependent on features such as area, maximum elevation, area of nearest neighboring island, and other factors.

3. Examine a publication or thesis that gives a set of data on which a single ANOVA was performed. Repeat the analysis with the Kruskal-Wallace nonparametric ANOVA. Would the investigator have reached the same conclusion if using the parametric ANOVA had been deemed inappropriate?

☯ Questions for Discussion

1. Is it necessarily the case that nonparametric tests will be less powerful, i.e., less able to reject a null hypothesis when it is false, than the corresponding parametric test?

2. Suppose that the height of a single plant is measured at 3-day intervals for a month during a period of active growth. Is it legitimate to do a single-factor regression of height (Y) vs. time (X)?

3. If a correlation is found between two variables, what are the possible categories of explanations for it?

☯ Selected Bibliography

Bennington, C. C., and W. T. Thayne. 1994. Use and misuse of mixed model analysis of variance in ecological studies. Ecology 75:717–722.

Connor, E. F., and D. Simberloff. 1978. Species number and compositional similarity of the Galapagos flora and avifauna. Ecological Monographs 48:219–248.

Dixon, W. J., and F. J. Massey, Jr. 1983. Introduction to statistical analysis. 4th ed. McGraw-Hill Book Co., New York, New York, USA.

Ellison, A. M. 1992. Statistics for PCs. ESA Bulletin 73:74–87.

Montgomery, D. C., and E. A. Peck. 1982. Introduction to linear regression analysis. John Wiley & Sons, New York, New York, USA.

Myers, R. H. 1986. Classical and modern regression with applications. Duxbury Press, Boston, Massachusetts, USA.

Wonnacott, T. H., and R. J. Wonnacott. 1981. Regression: a second course in statistics. John Wiley & Sons, New York, New York, USA.

Zar, J. H. 1984. Biostatistical analysis. 2nd ed. Prentice-Hall, Inc., Englewood Cliffs, New Jersey, USA.

Soil Arthropod Sampling

INTRODUCTION

Soil and leaf litter invertebrates are difficult to sample due to their small size and secretiveness, together with the nature of the matrix in which they occur. Various techniques, however, have been developed to extract such forms from soil and litter samples (Kevan 1955, Macfadyen 1962, Murphy 1962, Crossley et al. 1991, Gorny and Grum 1993). Some of these techniques are strictly mechanical and separate organisms from matrix on the basis of size, specific gravity, or body surface properties. Other techniques are dynamic in nature and involve the active movement of organisms toward some attractant or away from some unfavorable environmental condition. Dynamic extraction techniques vary in the attractant or repellent used, and in whether the extraction process is carried out on a dry or a wet sample. The Tullgren funnel technique is a dry-sample, dynamic extraction technique that employs heat and desiccation to drive organisms downward from the sample through a funnel apparatus and into a vial of preservative. The value of this technique in general ecological studies lies in the simplicity of its equipment and procedure, and in its applicability to a wide variety of terrestrial arthropods.

TULLGREN FUNNEL PROCEDURE

The soil arthropod sampling procedure outlined in this exercise involves three main steps: (1) collection of samples, (2) Tullgren funnel extraction, and (3) identification and counting of the extracted organisms.

Collection of Samples

The procedure for collecting samples should be designed to allow standardization of the surface area and depth of the soil or leaf litter samples collected. For general ecological work, this requirement can usually be satisfied through the use of some type of cylindrical metal coring device. For sampling in porous soils or sands, a tin can with the top and bottom removed may be satisfactory. For more compact soils, a section of metal pipe with a machined cutting edge at one end may be necessary. Directions for the construction of more refined sampling devices are given by Murphy (1962, p. 161).

The sample volume should be kept small, since the efficiency of extraction decreases with increased volume. In most situations, samples with a cross-sectional area of about 25 square centimeters are adequate. To obtain the sample, the corer is forced into the soil or leaf litter to the desired depth and dug out with a trowel. The sample should be removed from the corer with as little disturbance as possible and placed in a container such as a wide-mouth jar, aluminum tin, or cylindrical pint ice-cream carton.

Colinvaux (1993): 25:526–539   Smith (1990): 9:179–189, 29:730–740
Odum (1993): 5:137–139   Smith (1992): 9:137–149
Ricklefs (1993): 8:143–145

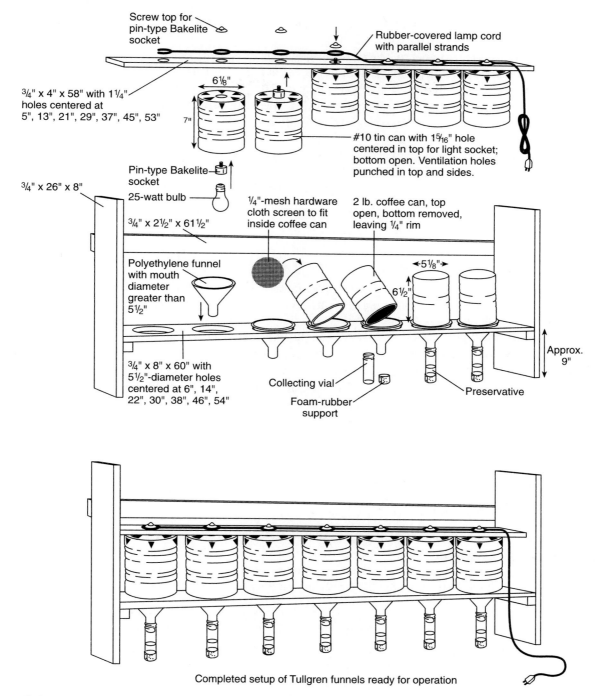

Screw top for pin-type Bakelite socket

Rubber-covered lamp cord with parallel strands

3/4" x 4" x 58" with 1 1/4" holes centered at 5", 13", 21", 29", 37", 45", 53"

6 1/8"

7"

#10 tin can with 1 5/16" hole centered in top for light socket; bottom open. Ventilation holes punched in top and sides.

3/4" x 26" x 8"

Pin-type Bakelite socket

25-watt bulb

3/4" x 2 1/2" x 61 1/2"

1/4"-mesh hardware cloth screen to fit inside coffee can

2 lb. coffee can, top open, bottom removed, leaving 1/4" rim

5 1/8"

6 1/2"

Polyethylene funnel with mouth diameter greater than 5 1/2"

Approx. 9"

3/4" x 8" x 60" with 5 1/2"-diameter holes centered at 6", 14", 22", 30", 38", 46", 54"

Collecting vial

Foam-rubber support

Preservative

Completed setup of Tullgren funnels ready for operation

Figure 9.1.
Construction plans for an inexpensive and easily constructed Tullgren funnel system (courtesy of K. K. Bohnsack, San Diego State University)

Extraction Procedure

The Tullgren funnel employs electric lights to heat and dry the samples. Tullgren funnel systems vary greatly in degree of complexity, however. A plan for construction of a simple, inexpensive set of funnels with materials readily available in most areas is given in figure 9.1. Another simple design is given by Crossley et al.

(1991, p. 189). Descriptions of more refined Tullgren funnel extractors are given by Auerbach and Crossley (1969), Macfadyen (1961), and Murphy (1962, pp. 174–178).

The soil or leaf litter material to be extracted should be wrapped in one to two layers of cheesecloth and placed on the hardware-cloth screen above the funnel. Extraction is usually most efficient when the sample is kept intact and placed upside down in the sample chamber. The cheesecloth functions to reduce the amount of soil sifting down into the specimen vial as the sample dries. This sifting may be further reduced by use of a device termed an autosegregator, which catches soil particles dropping into the funnel but allows animals to pass through freely (Newell 1955).

For most purposes, 80% ethanol is a satisfactory preservative. Ethanol acts as a repellent for some organisms, however, so for careful studies of some groups other preservatives may be desirable (Macfadyen 1962). The time required for extraction varies with the size of the sample. For samples of normal size, extraction should be nearly complete after 48 hours.

Identification and Counting

Animals may be identified and counted by placing the sample in a flat dish and examining it with a dissecting microscope. Plastic Petri dishes having the bottom marked with a Quebec-style counting grid are especially useful for this type of examination.

Most animals can be identified to order or family with the aid of standard reference books (see bibliography). For most soil arthropod groups, however, identification beyond this level is very difficult. The large number of species and small body size of many groups mean that careful anatomical study is required for complete identification. The taxonomy of many groups is still incomplete. Furthermore, many of the individuals encountered in samples are immature forms for which descriptions are generally unavailable. For certain types of ecological studies, identification to major group is adequate. For other purposes, it may be possible to distinguish individual species, but designate them only with code numbers or letters. Numbers of individuals of different taxa obtained in Tullgren funnel extraction studies can be recorded in table 9.1.

🌀 Suggested Activities

The Tullgren funnel extraction technique can be used in connection with a variety of ecological studies. For example, data obtained from Tullgren funnel extraction studies can be used in laboratory exercises dealing with community structure and diversity. Data on the composition of soil arthropod communities from different situations also can be used as the basis for the laboratory exercise dealing with community ordination.

1. Compare abundance and diversity of soil arthropods in communities of different stages of biotic succession. Or compare samples taken from different habitats, at different soil depths, or at different seasons.
2. Compare soil arthropod populations of mull and mor litter horizons in broad-leafed and coniferous forests from the same general region.
3. Compare the efficiency of extraction of soil arthropods by extracting samples of different sizes taken from the same habitat.

🌀 Questions for Discussion

1. What conditions probably influence the effectiveness of the Tullgren funnel technique in extracting different groups of arthropods from soil and litter samples?

Table 9.1 Numbers of soil arthropods of various taxa obtained by Tullgren funnel extraction of soil or litter samples

Taxon	Sample Number																								

2. Do differences in density and taxonomic composition of the soil arthropod populations correlate with differences in habitat conditions such as litter depth, soil humus content, soil moisture, or pH?

3. Based on information in reference books, what are the feeding niches of the various organisms obtained? Can the various groups of animals be linked together into a logical food web?

SELECTED BIBLIOGRAPHY

Identification

Arnett, R. H., Jr., N. M. Downie, and H. E. Jacques. 1980. How to know the beetles. Wm. C. Brown Co., Dubuque, Iowa, USA.

Baker, E. W., J. H. Camin, F. Cunliffe, T. A. Woolley, and C. E. Yunker. 1958. Guide to the families of mites. Contribution No. 3, Institute of Acarology, Department of Zoology, University of Maryland, College Park, Maryland, USA.

———, and C. W. Wharton. 1952. An introduction to acarology. Macmillan, Inc., New York, New York, USA.

Balogh, J. 1972. The Oribatid genera of the world. Akademiai Kiado, Budapest, Hungary.

Borror, D. J., D. M. Delong, and R. E. White. 1989. An introduction to the study of insects. 7th ed. Holt, Rinehart & Winston, New York, New York, USA.

————, and R. E. White. 1970. A field guide to insects. Houghton Mifflin Co., Boston, Massachusetts, USA.

Chu, H. F. 1949. How to know the immature insects. Wm. C. Brown Co., Dubuque, Iowa, USA.

Dindal, D. E. 1990. Soil biology guide. John Wiley & Sons, New York, New York, USA.

Essig, E. O. 1958. Insects and mites of western North America. Macmillan, Inc., New York, New York, USA.

Goodey, T. 1951. Soil and freshwater nematodes. John Wiley & Sons, New York, New York, USA.

Hoff, C. C. 1949. The pseudoscorpions of Illinois. Illinois Natural History Survey Bulletin 24(4):411–498.

Kaston, B. I., and E. Kaston. 1953. How to know the spiders. Wm. C. Brown Co., Dubuque, Iowa, USA.

Krantz, G. W. 1970. A manual of acarology. Oregon State University, Corvallis, Oregon, USA.

McDaniel, B. 1979. How to know the mites and ticks. Wm. C. Brown Co., Dubuque, Iowa, USA.

Sasser, I. N., and W. R. Jenkins, editors. 1960. Nematology. University of North Carolina Press, Chapel Hill, North Carolina, USA.

White, R. E. 1983. A field guide to the beetles of North America. Houghton Mifflin Co., Boston, Massachusetts, USA.

Wooley, T. A. 1988. Acarology. Mites and human welfare. John Wiley & Sons, New York, New York, USA.

Extraction Procedures

Burges, A., and F. Raw, editors. 1967. Soil biology. Academic Press, New York, New York, USA.

Crossley, D. A., Jr., D. C. Coleman, P. F. Hendrix, W. Cheng, D. H. Wright, M. H. Beare, and C. A. Edwards, editors. 1991. Modern techniques in soil ecology. Agriculture, Ecosystems and Environment 34(1–4):1–510. (Also published separately by Elsevier, New York, New York, USA.)

Gorny, M., and L. Grum. 1993. Methods in soil zoology. Elsevier, Amsterdam, The Netherlands.

Kevan, D. K., editor. 1955. Soil zoology. Academic Press, New York, New York, USA.

Kunnelt, W. 1961. Soil biology. Faber and Faber, London, England.

Macfadyen, A. 1961. Improved funnel-type extractors for soil arthropods. Journal of Animal Ecology 30:171–184.

————. 1962. Soil arthropod sampling. Advances in Ecological Research 1:1–34.

Marshall, V. C. 1972. Comparison of two methods of faunal extractors for soil microarthropods. Soil Biology and Biochemistry 4:417–426.

Murphy, P. W., editor. 1962. Progress in soil zoology. Butterworth and Co., London, England.

Newell, I. M. 1955. An auto-segregator for use in collecting soil-inhabiting arthropods. Transactions of the American Microscopical Society 74:3898–392.

Phillipson, J., editor. 1971. Methods of study in quantitative soil ecology. Blackwell Scientific, Oxford, England.

Schouten, A. J., and K. K. M. Arp. 1991. A comparative study on the efficiency of extraction methods for nematodes from different forest litters. Pedobiologia 35:393–400.

Southwood, T. R. E. 1975. Ecological methods with particular reference to the study of insect populations. Chapman & Hall, London, England.

Ecology

Anderson, J. M. 1975. Succession, diversity and trophic relations of some soil animals in decomposing leaf litter. Journal of Animal Ecology 44:475–495.

————, and A. Macfadyen, editor. 1976. The role of terrestrial and aquatic organisms in decomposition processes. Blackwell Scientific Publications, Oxford, England.

Fitter, A. H., D. Atkinson, D. Read, and M. B. Usher, editors. 1985. Ecological interactions in soil: plants, microbes and animals. Blackwell Scientific Publications, Oxford, England.

Lohm, U., and T. Person, editors. 1977. Soil organisms as components of ecosystems. Ecological Bulletins (Stockholm) 25:1–614.

Peterson, H., and M. Luxton. 1982. A comparative analysis of soil fauna populations and their role in decomposition processes. Oikos 39:287–388.

Price, P. Insect ecology. 2nd ed. John Wiley & Sons, New York, New York, USA.

Santos, P. F., J. Phillips, and W. G. Whitford. 1981. The role of mites and nematodes in early stages of buried litter decomposition in a desert. Ecology 62:654–663.

Teuben, A., and G. R. B. Smidt. 1992. Soil arthropod numbers and biomass in two pine forests on different soils, related to functional groups. Pedobiologia 36:79–89.

Wallwork, J. A. 1970. Ecology of soil animals. McGraw-Hill, London, England.

———. 1976. The distribution and diversity of soil fauna. Academic Press, London, England.

Mark and Recapture Population Estimates

INTRODUCTION

For many animals, estimates of population size by direct counts are impractical. For mobile or secretive forms, it is difficult to obtain direct estimates of population density, even in areas of very small size. Often, however, estimates of population size can be obtained by marking a segment of the population on one occasion and sampling the numbers of marked and unmarked animals on one or more later occasions. These techniques are discussed in detail by Seber (1986), Krebs (1989), and Pollock et al. (1990).

SINGLE MARKING AND SINGLE RECAPTURE, CLOSED POPULATION

The Lincoln-Petersen technique involves a single marking and single recapture in a closed population: one in which no immigration, emigration, or recruitment occurs. Although sometimes known as the Lincoln or Petersen index, it is not an index of relative population size, but an estimate of the number of individuals in the area sampled.

In this technique, a sample of individuals is captured, marked, and released into the population, thus creating a certain ratio of marked to total animals. After an interval sufficient to allow dispersal of the marked animals through the population, a second sampling is carried out to estimate this ratio (Pollock et al. 1990). From the size of the sample marked (n_1), the total in the second sample (n_2), and the number of marked animals recaptured in the second sample (m_2), the size of the population ($\hat{N}$) is given by the equation

$$\hat{N} = n_1 n_2 / m_2,$$

or, in a form somewhat less biased when numbers are small,

$$\hat{N} = \frac{(n_1 + 1)(n_2 + 1)}{(m_2 + 1)} - 1$$

The approximate variance, s^2, of this estimate is

$$s^2 = \frac{(n_1 + 1)(n_2 + 1)(n_1 - m_2)(n_2 - m_2)}{(m_2 + 1)^2(m_2 + 2)}$$

With the standard deviation, s, 95% and 99% confidence limits on the estimate are given by

$\hat{N} \pm 1.96(s)$ (95% confidence limits)

$\hat{N} \pm 2.58(s)$ (99% confidence limits)

Krebs (1989) presents techniques that may be more appropriate for determining confidence limits, especially when the number or ratio of marked animals in the recovery sample is small.

Krebs (1994): 10:155–159, Appendix II:698–699
Ricklefs (1993): 14:251

The population estimate is valid for the time of release of the marked animals into the population, *not* the recapture time. Mortality may cause the population to decline substantially after the release of marked animals without changing the ratio of marked to total animals. This technique therefore allows for normal mortality in the population between marking and recapture.

Several conditions must be met for the estimate to be valid. First, all individuals must be equally catchable. The ratio of marked to total animals must not change between release and recapture, meaning that no recruitment of unmarked animals can occur, either by reproduction or by immigration. Likewise, there must be no loss of marks and no differential loss of marked and unmarked animals by death or emigration. Lastly, the recapture sample must be an unbiased estimate of the ratio of marked to total animals. This can best be assured by carrying out either the release or the recapture sampling randomly, and by allowing an adequate interval for dispersal of the marked animals through the population.

If several (k) recapture samples are taken from a population, each yielding estimates of the population, $\hat{N}$, more precise estimates of the population and its confidence limits are possible. The various values of $\hat{N}$ are averaged, giving a mean estimate, $\overline{\hat{N}}$. The approximate standard error of this estimate, $s_{\overline{\hat{N}}}$, is given by the equation:

$$s_{\overline{\hat{N}}} = \sqrt{\frac{1}{k(k-1)} \Sigma (\hat{N} - \overline{N})^2}$$

Confidence limits for the mean estimate, $\overline{\hat{N}}$, are then calculated as follows:

$\overline{\hat{N}} \pm 1.96(s_{\overline{\hat{N}}})$ (95% confidence limits)

$\overline{\hat{N}} \pm 2.58(s_{\overline{\hat{N}}})$ (99% confidence limits)

REPEATED MARKING AND RECAPTURE, CLOSED POPULATION

The Schnabel technique allows marking and recaptures on several occasions, not necessarily separated by equal intervals. The population must remain constant throughout the sampling period, and immigration, reproduction, differential mortality, and other factors must not change the ratio of marked to total animals created by the investigator.

This method is often used for animals that are difficult to capture and can be obtained only in small numbers. The results give a series of population estimates and standard errors of increasing reliability, and the investigator can essentially continue the study until satisfied with the estimates.

On the first date, a sample of animals is captured, marked, and released. On each subsequent date, a sample, n_i, is captured. The number of marked recaptures, m_i, is noted and the remaining animals are marked and returned to the population. Thus, the total number of marked animals, M_i, increases through time. Note that M_i is the number of marked animals in the population just *before* the sample on date i is taken.

Population estimates, $\hat{N}$, are calculated for each recapture date (table 10.1):

$$\hat{N} = \frac{\Sigma n_i M_i}{\Sigma m_i}$$

The standard error of $1/\hat{N}$ can be calculated as follows:

$$s_{1/N} = \Sigma m_i / (\Sigma n_i M_i)^2$$

With this standard error, confidence limits can be calculated by the following expressions:

$\hat{N} = 1/1.96(s_{1/\hat{N}})(1/N)$ (95% confidence limit)

$\hat{N} = 1/2.58(s_{1/\hat{N}})(1/N)$ (99% confidence limit)

Table 10.1. Population estimates by the Schnabel method of repeated marking and recapture

Date (i)	Marked Animals in Population (M_i)	Captured n_i	$M_i n_i$	Recaptures m_i	Σm_i	Newly Marked and Released	$\hat{N}_i$	$s_{\hat{N}}$	___% Confidence Limits

SINGLE MARKING WITH MULTIPLE RECAPTURES, OPEN OR CLOSED POPULATION

The Lincoln-Petersen technique requires that the ratio of marked to total animals in the population stay constant after release of the marked animals. If this ratio does change, but at a relatively constant rate, the rate of change can be estimated and projected back to estimate the ratio at the instant the marked animals were released. This procedure, developed by Jackson (1939), requires an initial marking of n_0 animals, and at least three or four recapture samples at equal intervals, in which numbers of total animals, n_i, and marked animals, m_i, are recorded.

The numbers of marked recaptures, m_i, are first corrected to values designated y_i, the number per 100 animals marked and 100 recaptured, by the equation

$$y_i = (m_i/1)(100/n_0)(100/n_i)$$

Next, a weighted ratio ($r+$) showing the rate of decrease of the corrected recapture values, y_i, over the t recapture times is calculated by one of the equations below:

$$r+ = \frac{y_2 + y_3 + \ldots \ldots y_t}{y_1 + y_2 + \ldots \ldots y_{t-1}}$$

or, if many recapture values are available,

$$r+ = \frac{y_3 + y_4 + \ldots \ldots y_t}{y_1 + y_2 + \ldots \ldots y_{t-2}}$$

With this weighted ratio, the theoretical number of recaptures that would have been obtained at the time of release (y_0) is calculated by the equation

$$y_0 = (y_1 + y_2 + \ldots .y_{t-1})/r+ - (y_1 + y_2 + \ldots y_{t-2})$$

With y_0, the theoretical recapture value for time zero, an estimate of the population size is obtained by substituting this value in the following equation:

$$\hat{N} = 10,000/y_0$$

This method can be combined with the single recapture method simply by taking recapture samples at several times subsequent to the release of marked animals. Calculations of corrected recapture values, the weighted ratio $r+$, the theoretical recapture value at the time zero, and the resulting population estimate can be recorded in table 10.2. The procedure for calculating confidence limits for this estimate is involved. This procedure, with a table of values required in the computation, is given by Jackson (1939).

SINGLE MARKING AND MULTIPLE RECAPTURES WITH UNEQUAL CATCHABILITY IN A CLOSED POPULATION

Often it is unrealistic to assume that individuals are equally catchable. With multiple recaptures, however, some indication is given of differential catchability by the numbers of individuals that are caught once, twice, three times, and so on. Chao (1987) has suggested a procedure for estimating population size, $\hat{N}$, when individuals are marked and recaptured on several occasions, as in the Schnabel procedure. In such an effort, a certain total number of different individuals, S, will be obtained. This total will be made up of individuals caught once, f_1, those caught twice, f_2, those caught three times, f_3, and so on, up to f_k, those caught in all of the k capturing efforts (including the first).

From these summary values, the population estimate is obtained by the equation

$$\hat{N} = S + f_1^2/2f_2$$

Table 10.2. Calculations for multiple-recapture population estimates

Date of marking and release Number marked and released (M)				
Corrected recapture values	y_1 y_2 y_3 y_4 y_5 y_6 y_7 y_8 y_9 y_{10}			
Weighted ratio (r+)				
y_0 Population estimate (N̂)				

The variance of this estimate, s^2, is given by the equation

$$s^2 = f_2 \left[0.25(f_1/f_2)^4 + (f_1/f_2)^3 + 0.5(f_1/f_2)^2 \right]$$

With s, the square root of the variance, s^2, 95% and 99% confidence limits are calculated by the expressions

$\hat{N} \pm 1.96s$ (95% confidence limits)
$\hat{N} \pm 2.58s$ (99% confidence limits)

THE JOLLY-SEBER METHOD, OPEN OR CLOSED POPULATION

In many, if not most, populations some immigration, emigration, and recruitment are likely during a mark-recapture study. If it is assumed that every marked animal on one sampling date has the same probability of surviving to the next (together with equal probability of capture on any given date and of not losing its mark), entry and departure of individuals can be estimated, and population estimates derived. The Jolly-Seber method involves estimating the "survival" rate of individuals in the population, along with the rates of entry (by both immigration and reproduction) and departure (by both death and emigration). If the population is closed, however, entry and departure rates equal recruitment and mortality rates.

The Jolly-Seber method requires the individual marking of animals, and the recording of which individuals are recaptured on each sampling date. Animals may be marked on each sampling date, and estimates thus obtained for different dates. A raw data sheet must therefore be kept, like that illustrated in table 10.3.

Computation of Jolly-Seber estimates is involved, and is discussed in detail in Seber (1986), Krebs (1989), and Pollock et al. (1990). Several versions of the Jolly-Seber model have been formulated, differing in whether they assume that survival and capture probabilities differ for the various intervals between captures (Model A), that survival is constant but capture probabilities differ (Model B), that survival varies but capture probability is constant (Model C), or that both capture and survival probabilities are constant (Model D). Computer software programs (below) are available for computation of Jolly-Seber estimates.

Table 10.3. Example of a data sheet (a) for Jolly-Seber population estimates and format (b) for B-TABLE input file to program JOLLY

a. Data sheet

		Sampling Dates 1	2	3	4	5	6	7	8
		Numbers Last Captured on Each Earlier Date							
Sampling	1	0	3	2	1	0	0	0	0
Dates	2		0	2	1	1	0	0	0
	3			0	3	2	1	0	0
	4				0	2	2	2	0
	5					0	3	3	1
	6						0	4	1
	7							0	2
	8								0
Total Caught		12	11	13	14	13	12	16	10
Total Released*		12	11	12	14	12	12	15	10

b. B-TABLE format

```
TITLE = POCKET GOPHERS
PERIODS = 8
TYPE = B-TABLE
FORMAT = (8F4.0)
    0    3    2    1    0    0    0    0
         0    2    1    1    0    0    0
              0    3    2    1    0    0
                   0    2    2    2    0
                        0    3    3    1
                             0    4    1
                                  0    2
                                       0
   12   11   13   14   13   12   16   10
   12   11   12   14   12   12   15   10
```

*Total released might be fewer than total caught due to mortality during capture process.

SOFTWARE FOR CAPTURE-RECAPTURE ANALYSES

EcoStat, a Trinity Software package (appendix B), contains easy-to-use programs for IBM-compatible and Macintosh microcomputers for the Lincoln-Petersen, Schnabel, and Jolly-Seber (Model D) analyses. The Jolly-Seber analysis requires that data be entered on (1) numbers of animals captured on each date, (2) numbers newly marked and released on each date, and (3) numbers recaptured on each date of those marked and released on each previous date.

The FORTRAN computer program JOLLY is available for IBM PCs to perform Jolly-Seber computations for Models A–D (appendix B). The files for this analysis should be loaded into a directory on the hard disk of an IBM PC or compatible. A DOS data file should then be created, using the DOS editor (e.g., edlin). The input data required for this file, which must be derived from the raw data sheet (table 10.3a), are

1. The number of animals captured on each date
2. The number of animals released on each date (the number captured minus any dying or not released for other reasons)
3. The number of animals captured on each given date that were *last* captured on each earlier date (e.g., on sampling date 4, the numbers of marked animals last captured on dates 1, 2, and 3)

These data can be structured in various formats, the easiest being the "B-TABLE" format illustrated in table 10.3b.

The program output includes estimates of population size (with 95% confidence intervals), survival rate, and capture probability, and tests of differences between estimates using the different models.

SUGGESTED ACTIVITIES

1. Carry out mark-recapture studies on field populations of an invertebrate or small vertebrate that can be marked easily (e.g., snails, isopods, or insects by quick-drying paint; crayfish by clipping notches in tail plates; amphibians by clipping toes).

2. Conduct mark-recapture experiments in a laboratory colony of flour beetles or mealworms, which can be marked with quick-drying paint. Simulate open populations by removal and addition of animals from other colonies.

3. Using beans of several colors, simulate capture-recapture studies of the types outlined above. Marking can be simulated by replacing beans of one color (in "capture" samples) by those of another color. The dynamics of open populations can be simulated by removing numbers of beans ("unmarked" and "marked") and replacing them with those of the designated "unmarked" type.

QUESTIONS FOR DISCUSSION

1. List a variety of animals in your area for which it would be interesting to obtain population estimates. How could samples from these populations be captured and marked? Are the populations open or closed? Are mortality, recruitment, and recapture probability important or variable? What mark-recapture technique would be most appropriate for each?

2. How does the fraction of the population marked affect the confidence limits for the single recapture estimate? How does the size of the recapture sample (n_2) affect these confidence limits?

3. What assumption is made in all techniques about loss of marks? About chance of mortality, or emigration of marked individuals?

SELECTED BIBLIOGRAPHY

Begon, M. 1979. Investigating animal abundance. University Park Press, Baltimore, Maryland, USA.

Blower, J. G., L. M. Cook, and J. A. Bishop. 1981. Estimating the size of animal populations. Allen and Unwin, London, England.

Caughley, G. 1977. Analysis of vertebrate populations. John Wiley & Sons, New York, New York, USA.

Chao, A. 1987. Estimating the population size for capture-recapture data with unequal catchability. Biometrics 43:783–791.

Cormack, R. M. 1993. Variances of mark-recapture estimates. Biometrics 49:1188–1193.

————, P. P. Ganapati, and D. S. Robson. 1979. Sampling biological populations. International Co-operative Publishing House, Fairland, Maryland, USA.

Jackson, C. H. N. 1939. The analysis of an animal population. Journal of Animal Ecology 8:238–246.

Krebs, C. J. 1989. Ecological methodology. Harper & Row, New York, New York, USA.

Montgomery, W. I. 1987. The application of capture-mark-recapture methods to enumeration of small mammal populations. Symposium of the Zoological Society of London 58:25–57.

Pollock, K. H. 1981. Capture-recapture models: a review of current methods, assumptions, and experimental design. Studies in Avian Biology 6:426–435.

————, J. D. Nichols, C. Brownie, and J. E. Hines. 1990. Statistical inference for capture-recapture experiments. Wildlife Monographs No. 107.

Seber, G. A. F. 1982. The estimation of animal abundance and related parameters. 2nd ed. Charles Griffin, London, England.

————. 1986. A review of estimating animal abundance. Biometrics 42:267–292.

Wilbur, H. M., and J. M. Landwehr. 1974. The estimation of population size with equal and unequal rates of capture. Ecology 55:1339–1348.

Catch per Unit Effort Population Estimates

INTRODUCTION

Catch per unit effort techniques use the principle of diminishing returns to estimate population size when it is not practical to obtain exact counts of individuals in unit-area or unit-volume samples. If a series of samples is taken successively from a population, and the individuals are not returned, a decrease in numbers captured per unit effort is usually noted in the later samples. If the rate of decrease is constant, it can be measured and used to estimate the total population size.

This exercise describes the calculation of catch per unit effort population estimates together with their confidence limits.

SAMPLING PROCEDURE

The population selected for this study must be small enough for sampling to remove a significant fraction of the population. If the population is too large, little reduction in numbers caught per unit effort is likely. Best results are usually obtained when the total number captured is at least one-third of the total population.

The sampling procedure requires that equal-effort samples be drawn from the population on successive occasions. Equal-effort sampling can be accomplished in several ways. In many situations, equal numbers of sub-samples can be taken on each occasion. For animals such as fish or mammals, equal numbers of traps operating for standard time periods may yield equal-effort samples. In areas of low or open vegetation, equal searching time by the same observers may constitute a satisfactory equal-effort technique for populations of large invertebrates such as snails.

Sampling data can be recorded in table 11.1. The decrease in catch per unit effort can be shown graphically by plotting catch per unit effort against cumulative catch. For this, and for the subsequent analysis, the cumulative catch corresponding to the first sampling period should be taken as 0. When data are plotted, it can be seen that if a line through the plotted points is extended until it strikes the x-axis, where per-unit effort is 0, an estimate of the total population size is obtained.

LINEAR REGRESSION ESTIMATE

One method for locating and projecting the catch per unit effort trend to obtain a population estimate is the simple linear regression technique described in exercise 8. In this analysis, catch per unit effort is the dependent variable, *Y*, and prior cumulative catch the independent variable, *X*. For the first sampling period the

Krebs (1994):10:152–160

Table 11.1. Least-squares linear regression analysis of catch per unit effort and cumulative catch data

Sampling Period	Catch per Unit Effort Y	Prior Cumulative Catch X	X²	XY
1		0		
2				
3				
4				
5				
6				
7				
8				
9				
10				
Total				
Average				

Linear Regression

$b =$ _____ $b_u =$ _____ $b_l =$ _____

$a =$ _____ $a_u =$ _____ $a_l =$ _____

$\hat{N} =$ _____ $\hat{N}_l =$ _____ $\hat{N}_u =$ _____

Maximum Likelihood

$t_2 =$ _____ $\hat{N}_l =$ _____ $\hat{N}_u =$ _____

$q =$ _____

$\hat{N} =$ _____

$s_{\hat{N}} =$ _____

prior cumulative catch is, of course, zero. Linear regression analysis can be carried out in table 11.1. When the intercept, a, and the slope, b, have been determined, an estimate of the total population is given by the equation

$$N = -a/b$$

Confidence limits on N are calculated by determining the standard error of the regression slope, s_b. For this, critical 2-tailed t values for $\alpha = 0.05$ (95% confidence limit) or 0.01 (99% confidence limit) and $k-2$ degrees of freedom should be read from table A.2, appendix A. In reading these values, k is the number of catch per unit effort samples. The upper and lower confidence limits on b are then calculated by the equations

$b_u = b + t(s_b)$ (upper limit)

$b_l = b - t(s_b)$ (lower limit)

In turn, b_u and b_l are used to calculate corresponding values for the intercept, a (a_u and a_l, respectively), using the same means for X and Y that were used to calculate b and a. Upper and lower confidence limits on N are then obtained by the equations

$$N_u = -a_l/b_l$$
$$N_l = -a_u/b_u$$

MAXIMUM LIKELIHOOD ESTIMATE

As described by Skalski and Robson (1982), the population can also be estimated by the equation

$$\hat{N} = \frac{\Sigma Y}{1 - q^k}$$

Where: q = Probability of an individual avoiding capture
k = Number of equal-effort samples taken

To estimate q, one must calculate the expression

$$t_2 = \sum_{i=1}^{k} (i - 1)Y_i$$

The probability of capture avoidance, q, is then obtained by iteration of the expression

$$\frac{t_2}{\Sigma Y} = \frac{q}{1 - q} - \frac{kq^k}{1 - q^k}$$

In doing this, values of q are selected and used to calculate the right side of the expression until a value equalling the left side is obtained. Alternatively, q can be read from table 3 in Robson and Chapman (1961). In this table, the value of $t_2/\Sigma Y$ is found in the table body in the column with "number of classes" corresponding to the number of equal-effort samples taken. The value of q is read along the left margin of the table (where it is designated s).

The standard deviation of this estimate, s_N, is given by the equation

$$s_N = \sqrt{\frac{\hat{N}(1 - q^k)q^k}{(1 - q^k) - (pk)^2(q^{k-1})}}$$

Where: p = 1 − q

Confidence limits are calculated for 95% and 99% levels by the following expressions:

N ± 1.96s_N (95% confidence limits)

N ± 2.58s_N (99% confidence limits)

Skalski et al. (1984) describe a procedure for comparing catch per unit effort estimates obtained from different populations by the same method.

SOFTWARE FOR CATCH PER UNIT EFFORT ESTIMATES

MicroFish is an IBM PC–compatible software package for maximum likelihood analysis of catch per unit effort data by either interactive data input or the preparation of a data file (Van Deventer and Platt 1985). The interactive program requests catch per unit effort values for a single species and yields the population estimate with 95% confidence limits. By creating a named data file, numbers of several species from several locations, together with their weights and lengths, can be submitted, and a comprehensive analysis obtained for data sets such as those obtained in a series of electrofishing samples. MicroFish also includes a sample-size module that determines the number of unit-effort samples required to obtain a population estimate with a specified precision.

To use this program, you must review sections of the user's guide that are basic to the execution of a simple analysis. The program creates an output file that can be printed after the program itself has been exited.

🌐 SUGGESTED ACTIVITIES

1. Sample a field population of invertebrates, such as snails, by a series of equal-effort searches (equal number of persons searching for a specified period). It may be appropriate to give the search personnel experience in finding the target animals by first having them search an area of habitat similar to that of the study area.

2. Sample a colony of flour beetles or mealworms by taking a standard number of subsamples of the colony matrix and sieving to remove the animals. Replace the matrix material removed with an equal amount of new material. Between samplings, mix the colony matrix or allow sufficient time for dispersal of animals into the new material, so that subsequent sampling reflects the altered population density.

3. Use equal-effort searching to sample an artificial population of known size, such as a certain number of marbles scattered in a lawn area, or a certain number of snails introduced into an area of snail-free habitat. Evaluate the accuracy of the catch per unit effort technique in estimating the true population size.

⊘ QUESTIONS FOR DISCUSSION

1. What assumptions are made about immigration, reproduction, emigration, and death in the population for the purposes of these techniques? What assumptions are made about the probability of capture of different individuals?

2. What would be the significance of the plotted points forming a concave or convex curve rather than a straight line? What procedure could you use if the catch per unit effort values do not fall into a straight line?

3. For what sorts of animal populations is this technique best suited?

⊘ SELECTED BIBLIOGRAPHY

Hammill, M. O., and T. G. Smith. 1990. Application of removal sampling to estimate the density of ringed seals (*Phoca hispida*) in Barrow Strait, Northwest Territories. Canadian Journal of Fisheries and Aquatic Sciences 47:244-250.

Harding, C. M., A. W. Heathwood, R. G. Hunt, and K. L. Q. Read. 1984. The estimation of animal population size by the removal method. Applied Statistics 33:196-202.

Otis, D. L., K. P. Burnham, G. C. White, and D. R. Anderson. 1978. Statistical inference for capture data on closed animal populations. Wildlife Monographs 62:1-135.

Peterson, J., M. Taylor, and A. Hanson. 1980. Leslie population estimate for a large lake. Transactions of the American Fisheries Society 109:329-331.

Robson, D. S., and D. G. Chapman. 1961. Catch curves and mortality rates. Transactions of the American Fisheries Society 90:181-189.

Skalski, J. R., and D. S. Robson. 1982. A mark and removal field procedure for estimating population abundance. Journal of Wildlife Management 46:741-751.

———, M. A. Simmons, and D. S. Robson. 1984. The use of removal sampling in comparative censuses. Ecology 65:1006-1015.

Soms, A. P. 1985. Simplified point and interval estimation for removal trapping. Biometrics 41:663-668.

Van Deventer, J. S., and W. S. Platts. 1983. Sampling and estimating fish populations from streams. Transactions of the North American Wildlife and Natural Resources Conference 48:349-354.

———. 1985. A computer software system for entering, managing, and analyzing fish capture data from streams. USDA Forest Service Research Note INT-352.

Distance Sampling of Animal Populations

INTRODUCTION

In recent years, new census techniques have been developed for animals that can be observed along transect routes (Emlen 1977, Burnham et al. 1980) or at a series of census points (Buckland 1987). These techniques assume that all individuals of the target species are observed at or very close to the transect line or station, and that detection declines with increasing distance. They provide estimates of absolute density (as opposed to indices of relative abundance) from counts of individuals and data on their perpendicular distance from the observation line or point. The techniques also equate estimates for species with different detectabilities. Several specific techniques exist, some more sophisticated than others. One variant, the variable-radius circular plot technique, involves data collection at specific stations rather than along transects.

Distance-sampling techniques have been used extensively to sample populations of small land birds, but are also applicable to surveys of other vertebrates, such as whales and dolphins at sea or large mammals in grasslands or tundra. In this exercise we shall describe the technique as applied to censusing of birds and other conspicuous small animals.

Land birds have been a popular subject for population studies because of their diurnal activity, conspicuousness, and interesting behavior. Accurate censuses of land birds, however, are difficult to obtain. Some species, although quite vocal, are very secretive, and in habitats with dense vegetation the activities of many species, particularly those of the high canopy, are difficult to observe. Techniques for censusing land birds have until recently been very time consuming. Censuses of breeding birds, based on mapping techniques, required many visits to a small area of uniform vegetation, and the plotting of locations of territorial song or aggressive encounters of all birds detected. In the non-breeding season, censuses were simply the average of repeated counts taken along transects through an area of habitat, the assumption being that all birds were observed. Despite the benefits of multiple visits to census areas, neither method provides a reliable accounting of the population present.

We shall describe two simple versions of the distance-sampling procedure, as applied to small land birds. These techniques are suitable for birds that do not travel in large flocks. They do not provide good estimates for raptors or for aerial feeders such as swallows or swifts.

EMLEN STRIP-TRANSECT ANALYSIS

In this procedure, the observer follows a linear transect route and tallies birds detected by song or sight on either side, noting their perpendicular distance from the transect line. It is assumed that an alert observer will detect all active birds in the immediate vicinity of the transect line, and that away from the line detection becomes less and less complete. Most songbirds can be detected to a distance of 125 m to either side, a distance equivalent

Begon et al. (1990):4:129–131
Krebs (1994): 10:153–160
Ricklefs (1990): 15:287–288

to a census area of 25 hectares for each kilometer of transect. Loud or especially conspicuous species can often be detected to 250 m on either side, a distance equivalent to an area of 50 hectares for each kilometer of transect.

For non–breeding-season censuses, when territorial song is infrequent, no distinction need be made between birds detected by song, call notes, or sight. For breeding-season censuses, however, when males engage in frequent territorial song, detections by territorial song should be recorded separately from those involving other cues (sight or call notes). Separate estimates of populations should be made for song-cue detections and for total detections (for calculating the population from song cues, the number of tallies in the full-detection strip should be doubled on the assumption that each territorial male is mated). The larger of the two estimates is then taken as the best estimate of the true population.

Census Procedure

1. Select an area of habitat with conditions that are uniform for 125 m or more on either side of a transect route.
2. Mark the transect route at 100 m (or shorter) intervals for a distance of 1–2 km.
3. Determine an appropriate width of the total strip transect for the species in question, based on conspicuousness (e.g., 125 m or 250 m to either side of the transect route).
4. Select and practice methods for measuring, pacing, or estimating perpendicular distances from the transect line to a distance of 35 m, and for using a range-finder to estimate points at greater distances.
5. Conduct the census by walking slowly (1–2 km/hr in woodland, 2–3 km/hr in open country) along the transect route, pausing now and then to look and listen. At this speed one should pass 100-m points each 3–6 minutes in woodland or each 2–3 minutes in open country. In a notebook, record each bird detection by species and by right-angle distance of the spot of first detection from the transect route. You may or may not want to use "squeek" or "psssh" sounds to call birds into view; these are often regarded as introducing an undesirable variable into census procedure. Ask your instructor to demonstrate these sounds, and discuss their use. Note the time at which you pass each 100 m-point so that you can keep a relatively uniform pace. Avoid long stops, since these may allow birds already tallied to move ahead of you, leading to double counting. Note the type of cue by which each individual was detected (S = song, C = call note, V = vision). Tally song and nonsong detections by distance from the transect line in table 12.1.

Emlen Data Analysis

The width of the full-detection strip is first estimated, and the count within this strip is multiplied by the ratio of the total census strip width to the full-detection strip width to estimate the true population of the entire strip transect. Determine the width of the full-detection strip as illustrated in figure 12.1. In practice, the number of detections in the first or second strip adjacent to the transect line is often smaller than in the next strip or two because birds move away from the projected route of the observer in advance of his approach. Determine the strip for which the number of detections drops more than 20% below the average number per 5 m strip for the region closer to the transect line. The area up to, but not including, this 20% decline strip is defined as the full-detection strip. Multiply the number of detections in the full-detection strip (or 2 × the number for territorial song detections) by the ratio of strip-transect width to full-detection strip width (see figure 12.1) to obtain the population estimate for the total strip width.

The total number of detections across the full strip transect can be divided by the true population estimate to give a coefficient of detectability (CD). This CD is specific for the species, the strip-transect width, the habitat, and the time of year (perhaps for other variables, as well!). If it is determined that censuses

Table 12.1. Strip-transect census results and calculation of total bird populations for the census area

Species	Detection Cue	5 m Strips							30 m Strips			125 m Strip	Total
		0–5	5–10	10–15	15–20	20–25	25–30	30–35	35–65	65–95	95–125	125–250	
	Song												
	Other												
	Total												
	Song												
	Other												
	Total												
	Song												
	Other												
	Total												

Species	Census Transect Width	Full-Detection Distance	Calculated Population Number/_____ Ha	Coefficient of Detection

in a particular type of habitat at a given season show a nearly constant CD, the species in question can be censused with considerable accuracy without the need to make detailed estimates of the distance of all detections from a transect line. One simply needs to tally all detections within the defined limits of the strip transect and divide by the species' CD to obtain the population estimate.

FOURIER ANALYSIS

A more rigorous approach to estimating population density involves use of the Fourier series, a mathematical expansion series, to fit the distribution of numbers of observations versus distance from the transect line (Burnham et al. 1980). With the mathematical relationship underlying this series, data on the perpendicular distances, x, of the total number, n, of observations, together with the effective census-strip half-width, w, (for simplicity taken as equal to the largest observed value of x) can be used to estimate $f(0)$, a quantity related to the probability of detection of individuals:

$$\hat{f}(0) = \frac{1}{w} + \sum_{k=1}^{m} a_k$$

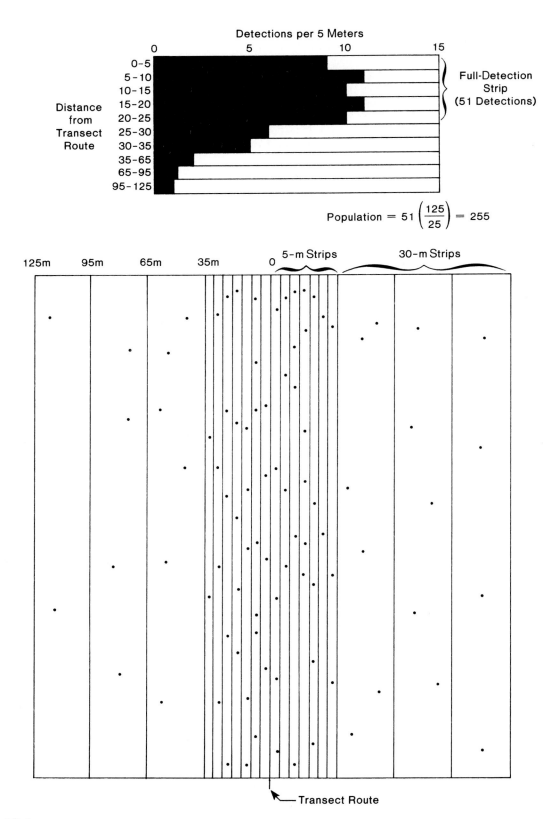

Population $= 51 \left(\dfrac{125}{25} \right) = 255$

Figure 12.1.
Diagram of a strip-transect census area illustrating the pattern of detections that might be obtained for a hypothetical species. For this example, 51 individuals were detected in the strips from 0 to 25 m (the full-detection strip), yielding a population estimate of 255 individuals. Since 38 detections occurred beyond the full-detection strip, the coefficient of detection (CD) for the entire strip-transect area is 89/255, or 0.349

Values of a_k in the summation above are obtained for values of $k = 1$ up to $k = m$ by the equation

$$\hat{a}_k = \frac{2}{nw} \left[\sum_{i=1}^{n} \cos \left(\frac{k \pi x}{w} \right) \right]$$

In this calculation, the cosine should be calculated in radians, not degrees. A value of m equal to 6 or less should be chosen (values of a_k will become smaller as k increases, and the sequence of calculations should be terminated when a_k becomes very small).

Given the value of $\hat{f}(0)$, the estimated density, $\hat{D}$, of individuals along a transect of length L is given by the equation

$$\hat{D} = \frac{n \, \hat{f}(0)}{2L}$$

This density will be for one square unit of distance measure (e.g., 1 m² if distances are measured in m), and, thus, conversion to a larger area (e.g., an acre or hectare) may be desirable. As for the Emlen technique, this density value should be doubled if the individuals tallied were singing males.

Burnham et al. (1980), Buckland (1985, 1987), and Buckland et al. (1993) describe this and related mathematical approaches in detail, including the application of this technique to observations at individual points rather than along transects.

PROGRAM *DISTANCE*

The computer program DISTANCE has been developed for the IBM PC microcomputer and compatibles to analyze sampling data of the type described in this exercise (appendix B). This program incorporates advanced methodology for analysis of distance sampling data, as described in Buckland et al. (1993). Use of this program is described fully in a user's guide (Laake et al. 1993). To use this program, you must review sections of the user's guide that are basic to the execution of a simple analysis.

In DISTANCE, data are entered by a simple input program that defines the transect characteristics (OPTIONS), the numbers of individuals detected in each belt (DATA), and the nature of the analysis desired (ESTIMATE). The program is capable of applying various mathematical relationships to the curve of change in detections with distance from the transect line, and choosing the best one for density estimation. Using accessory commands permits multiple data sets to be analyzed and more sophisticated analyses to be carried out. Data collected at points (variable-radius circular plot technique) can also be analyzed by this program. The program creates an output file that can be printed after the program itself has been exited.

✷ SUGGESTED ACTIVITIES

1. Choose an extensive area of relatively homogeneous habitat, such as an area of grassland, chaparral, or deciduous forest. Select one or a few species for study, as outlined in the exercise above. Collect data and analyze them by the Emlen method, by Fourier analysis, and using the DISTANCE program. Compare population estimates and CDs for different observers, for the same bird species at different times of day or in different habitats, and for different bird species.

2. Select a grassland, meadow, or old-field habitat and designate a transect route through it. Collect data for one or more species of butterflies, recording the perpendicular distance of each individual from the transect line when first observed. Analyze the data as in #1 above.

3. Sample the desert plant community accompanying this manual. Select a species for data collection along line transects, which can be designated by string and thumb tacks. Assume that only individuals that can be reached by an unblocked perpendicular line from the transect are "detectible." Record the numbers of such individuals in belts of appropriate, uniform width paralleling the transect. Analyze the data as in #1 above.

✪ QUESTIONS FOR DISCUSSION

1. What are the effects of bird inactivity, such as tight-sitting of incubating birds, on the results of this census technique? In what situations might it be unrealistic to assume that an alert observer can see all birds near the transect line (hint: think tall!)?
2. How would the presence of a considerable number of "floaters" (nonterritorial individuals living inconspicuously in an area fully divided up by territories of other individuals) affect results of a breeding census?
3. What assumption about the length of the observation period at a given point along a transect, or at a given variable-radius observation station, is made in this technique?

✪ SELECTED BIBLIOGRAPHY

Bibby, C. J., N. D. Burgess, and D. A. Hill. 1992. Bird census techniques. Academic Press, London, England.

Buckland, S. T. 1985. Perpendicular distance models for line transect sampling. Biometrics 41:177–195.

———. 1987. On the variable circular plot method of estimating animal density. Biometrics 43:363–384.

———, D. R. Anderson, K. P Burnham, and J. L. Laake. 1993. Distance sampling. Chapman & Hall, London, England.

Burnham, K. P., D. R. Anderson, and J. L. Laake. 1980. Estimation of density from line transect sampling of biological populations. Wildlife Monographs No. 72, 202 pages.

Emlen, J. T. 1977. Estimating breeding season bird densities from transect counts. Auk 94:455–468.

Laake, J. L., S. T. Buckland, D. R. Anderson, and K. P. Burnham. 1993. DISTANCE user's guide. Colorado Cooperative Fish and Wildlife Research Unit, Colorado State University, Fort Collins, Colorado, USA.

Ralph, C. J., and J. M. Scott, editors. 1981. Estimating numbers of terrestrial birds. Studies in Avian Biology 6:1–630. (This volume contains 80 individual papers on various aspects of bird censusing.)

Ratti, J. T., L. M. Smith, J. W. Hupp, and J. L. Laake. 1983. Line transect estimates of density and the winter mortality of gray partridge. Journal of Wildlife Management 47:1088–1095.

Waide, R. B., and P. M. Narins. 1988. Tropical forest bird counts and the effect of sound attenuation. Auk 105:296–302.

Mapping Home Ranges and Territories

INTRODUCTION

Most animals restrict their activity to areas with which they are familiar and have formed a behavioral attachment. Presumably, familiarity with an area enhances their ability to forage, avoid enemies, and reproduce. The total area utilized by an individual, pair, or social group is termed the *home range*. A home range or portion thereof that is actively defended against other individuals, pairs, or groups is termed a *territory*.

This exercise describes techniques for estimating the area of a home range or territory from a series of observed locations of individuals, their signs, or their territorial defense behaviors. Such observations are obviously a sample of activity locations, and the statistical objective of various techniques is thus to estimate the total activity area from the sample.

TECHNIQUES OF DELIMITING ACTIVITY AREAS

Several procedures have been used commonly to estimate the area of a home range or territory. These techniques use spatial coordinates, such as latitude and longitude values or positions on x- and y-axes, of activity locations in two-dimensional space. The complexity of calculations for most techniques requires that computer software be used for this exercise.

Minimum Convex Polygon

This simple procedure, recommended by Mohr (1947), has been used widely in bird and mammal studies. The method is simply to connect the outermost observation points to form a polygon with no internal angles greater than 180 degrees. Although the method is effective for compact, sharply defined home ranges or territories, the area obtained is strongly influenced by the location of the outermost observations, and is essentially unresponsive to the degree of concentration of observations in the central region. Furthermore, if the activity area is elongate and either curved or bent, a large area with no actual occurrences can become included in the convex polygon.

Bivariate Normal Estimate

Jennrich and Taylor (1969) proposed this parametric technique, which is based on the normal curve. It assumes that occurrences are normally distributed along two axes, which intersect at some center of activity. The area defined thus ranges from a circle, if the axes are of equal length, to an ellipse, if they are unequal. Since the

Brewer (1994): 4:109–111   Odum (1993): 6:161–162   Smith (1992): 13:197–201 method assumes that the probability of occurrences declines outward from the center of activity, the calculated area must be determined as that within which a specified percentage, e.g., 95%, of occurrences fall. Weaknesses of the method are obviously the assumptions that the area must have a simple circular or elliptical shape and that activity is normally distributed with respect to a single point. One consequence of these assumptions might be that if an animal's activity area were U-shaped, the calculated range would be centered on a point where the animal was never found.

Harmonic Mean Estimate

Dixon and Chapman (1980) devised a nonparametric method using radial distances from locations of occurrence to all points in a rectangular grid superimposed on the area in question. The average values of such measurements (or their reciprocals) are termed *harmonic means.* For example, for each grid point, a harmonic mean can be calculated as the average of reciprocals of the radial distances to all occurrence locations. One or more grid points are identified as centers of activity by high values of the harmonic mean, which must always lie within areas where occurrences are concentrated. Isolines can then be calculated to enclose a specified percentage of occurrences. One weakness of the technique is that calculations of the harmonic mean are distorted if a grid point falls on or very close to an actual occurrence site, so that the grid must be positioned over the area so as to avoid such proximity.

Fourier Transform Estimate

Anderson (1982) suggested a nonparametric method of defining the activity range by smoothing the two-dimensional histogram of observed locations of activity. This technique is the Fourier transform, which uses trigonometric functions to create a smoothly contoured frequency "landscape." For this landscape, areas with specified percentages of the smoothed frequency distribution can then be determined. Worton (1989) notes that this procedure can yield negative density estimates in some situations, and that the procedure limits the determined boundaries of the range to those of the originally specified two-dimensional frequency plane.

Adaptive Kernel Estimate

Worton (1989) suggested another nonparametric estimator of home range or territory size. This procedure assumes that each occurrence site is, in effect, the center of localized activity within the home range or territory. Mathematically, a bivariate normal distribution, or *kernel,* is superimposed on each occurrence site to represent such activity. The overlapping distributions of kernels are then aggregated to create an overall activity distribution with one or more peaks where occurrences are concentrated. In the *adaptive kernel* procedure, the surface of this overall distribution is smoothed by use of a parameter that varies in value, lightly smoothing central areas of high frequency of activity but strongly smoothing the fringing areas of low activity.

COMPUTER SOFTWARE FOR HOME RANGE AND TERRITORY ANALYSIS

The following two software packages for IBM PCs and compatible microcomputers provide home range and territory analyses by most of the above procedures. Larkin and Halkin (1994) review and provide source information on several other programs for computing areas of territories and home ranges.

Program CALHOME

The computer program CALHOME (version 1.0) (Kie et al. 1994) has been developed by the Pacific Southwest Research Station of the U.S. Forest Service and the California Department of Fish and Game to determine the areas of home ranges or territories by all methods listed above except the Fourier transform (see appendix B for information on program acquisition). The program is designed for IBM-compatible computers with a hard disk, 640K of RAM, and a VGA color monitor. Printed and graphical output can be sent to an HP Plotter, an HP Laserjet printer (II, III, or IV), or Epson-compatible dot-matrix printers.

Data Input

CALHOME uses x- and y-coordinate data on animal occurrences. These data can be in feet, meters, 10-meter units, or 100-meter units. Data must be recorded as columns of x (first column) and y (second column) values. Although a FORTRAN format statement can be used to specify the exact spaces occupied by x and y values, the default format is "free," indicated by an asterisk symbol in the format menu. A "free" format means that the x and y values need only be separated by one or more spaces, the program thereby reading the first number as an x value and the second as a y value. The data file should be given a simple DOS name, such as **DATA.001.**

Analysis

Open CALHOME from its directory by typing **CALHOME** and touching the **Return** key. View the introductory screens, and touch the **Return** key until you have reached the program window, which lists several menus across the top.

The **Files** menu allows you to specify the name of your data file, or to select the example data file, **DATA001,** included with the program. Following this you are automatically asked to indicate the data file format (see above) and to specify the data units. You must also provide a name (without a three-character extension, e.g., **OUTPUT1**) for a results file.

The **Methods** menu allows you to choose one or more of the four methods of area determination. For the methods you select, you must also choose one or more "utilization percentages," which are contour lines enclosing specified percentages of the maximum area defined by the points according to that method. Up to four such percentages can be specified. For example, if you specify **50, 95, 100,** you will obtain estimates of the area within which 50%, 95%, and 100% of all observations fall, or in the case of the bivariate normal method, the areas defining the respective fraction of the total spatial distribution defined. For the bivariate normal method, a specification of 100% cannot truly be determined; if this percentage is specified the program determines the area for 99.9% of the distribution.

Since the harmonic mean and adaptive kernel procedures use grid or scaled kernel units, you must also specify a grid cell size, or approve a default size suggested by the program.

The **Analyze** menu enables you to command program execution, and the **Output** menu enables you to direct the results to the monitor or to a printer or plotter. If you specify a hard copy output, you must also indicate the computer port to which the device is linked (check with your instructor). Hard copy is not printed automatically because of these operations, however (see below). These files are stored in a DOS file named **QUEUE.PRT,** which can be accessed after exiting CALHOME. The **Other** menu allows you to modify features of data plotting, output scaling, computation units, and other features. The **Quit** menu enables you to leave the program once analyses have been completed.

Obtaining Hard Copy

After you have exited CALHOME, type **HOMEPRT.** This will cause the files in QUEUE.PRT to be printed or plotted, after which this file will be erased.

Program McPAAL

Program McPAAL (Version 1.22) (Stuwe and Blohowiak 1992) includes calculations of all the above procedures except the adaptive kernel estimate. This program is also developed for IBM PCs and compatibles with CGA or Hercules graphic capabilities (see appendix B for source information). Textual and graphical screen output can be printed.

Data Input

McPAAL requires a DOS input file giving x- and y-coordinate data on animal occurrences. These data can be in any units. Data must be recorded as columns of x- and y-coordinates. The data file should be given a simple DOS name, such as **DATA.001,** and placed in the directory with the McPAAL program files.

Analysis

Open McPAAL from its directory by typing **McPAAL** and touching the return key. The program presents a main menu that contains a module entitled **Format X** and **Y Data.** Selecting this item opens a menu for specifying the name of the data input file, for which you may select your file or an existing sample data file. You must then specify the name of the formatted output file (with the extension **.MCP**) that will be used in the program's home range analyses. In this same menu, you must indicate the range of columns in which x and y values lie by keying in the first and last column numbers.

On returning to the main menu, activity range analyses can be carried out by the various procedures. When a specific procedure is selected, a menu is opened that requests your choice on various analysis and display items. You must (choice 8) specify your map scale in terms of the distance in meters between coordinate units. You must also (choice 7) specify width of a grid scale in meters, which determines the accuracy of the calculation in some procedures. This value should essentially be the value to which your map locations are accurate. When you have completed the choices, you can command the program to process the data, giving the calculated activity area and a graph of the activity-area shape.

McPAAL has other features, such as a module for placing a map of the study area into the program, that can be activated as described in the program manual.

SUGGESTED ACTIVITIES

1. Map the territory of one or more dragonflies in the marshy border zone of a stream or lake. Use small plastic flags to record locations. After at least 25 locations have been marked, lay out base lines at right angles to each other with 50-m measuring tapes. Record x- and y-coordinates for each point. Prepare a computer data file and carry out territory analysis with CALHOME or McPAAL.

2. Using a map with a scale of 1 inch to 200 feet, map the movements, singing posts, and territorial encounters with neighbors for a selected bird. Obtain at least 25 locations. With a clear plastic grid overlay, obtain x- and y-coordinates for each point. Prepare a computer data file and carry out territory analysis with CALHOME or McPAAL.

3. Locate and mark soil heaps and tunnel openings of a pocket gopher or mole, using small plastic flags. Use 50-m tapes to define base lines at right angles to each other, and determine x- and y-coordinates for each heap or tunnel opening. Prepare a computer data file and carry out territory analysis with CALHOME or McPAAL.

QUESTIONS FOR DISCUSSION

1. Do home ranges and territories exist only in two-dimensional space? In what situations might they be three-dimensional? How could such activity areas be measured?

2. How are activity areas or territories of various animals likely to change with seasonal change in habitat conditions? With stage in an animal's breeding cycle?

3. How is the size of home ranges or territories likely to vary with the trophic level of an animal? With its body size? With its pattern of body temperature regulation?

🌓 SELECTED BIBLIOGRAPHY

Anderson, D. J. 1982. The home range: a new nonparametric estimation technique. Ecology 63:103–112.

Calhoun, J. B., and J. U. Casby. 1958. Calculation of home range and population density of small mammals. Public Health Monograph Number 55. United States Department of Health, Education, and Welfare, Washington, D.C., USA

Dixon, K. R., and J. A. Chapman. 1980. Harmonic mean measure of animal activity areas. Ecology 61:1040–1044.

Jennrich, R. I., and F. B. Turner. 1969. Measurement of non-circular home range. Journal of Theoretical Biology 22:227–237.

Kie, J. G., J. A. Baldwin, and C. J. Evans. 1994. User's manual for Program CALHOME. U.S. Forest Service General Technical Report, in press.

Larkin, R. P., and D. Halkin. 1994. A review of software packages for estimating animal home ranges. Wildlife Society Bulletin, 22:274–287.

Mohr, C. O. 1947. Table of equivalent populations of North American mammals. American Midland Naturalist 37:223–249.

Spencer, W. D., and R. H. Barrett. 1984. An evaluation of the harmonic mean measure for defining carnivore activity areas. Acta Zoologica Fennica 171:255–259.

Stuwe, M., and C. E. Blohowiak. 1992. Manual for McPAAL: micro-computer programs for the analysis of animal locations. Conservation and Research Center, Front Royal, Virginia, USA.

Worton, B. J. 1987. A review of models of home range for animal movement. Ecological Modelling 38:277–298.

———. 1989. Kernel methods for estimating the utilization distribution in home-range studies. Ecology 70:164–168.

EXERCISE 14

Vegetation Analysis

INTRODUCTION

The most widely used sampling techniques for vegetation analysis involve (1) quadrats in which individuals can be counted and measured, (2) line transects along which the intercepted individuals are measured, and (3) points at which measurements are made of distance to and attributes of the nearest individuals. These techniques can be adapted for almost all plant communities. Many communities of sessile or sedentary animals (e.g., intertidal invertebrates) can also be sampled by these techniques.

The quadrat technique is easy to use in communities with a profile low enough that a frame can be placed over the area to be sampled, or a radius line rotated around a central point. It is also appropriate for forest communities that are open enough to allow plot boundaries to be marked easily and accurately with a tape measure. The line-transect technique is effective for low, dense communities in which quadrats are difficult to place on the surface. It is also the easiest technique to use when individual organisms are difficult to distinguish from one another, or are irregular in shape. Point-based techniques are best when individuals are widely spaced or when the dominant organisms are large and dense enough that quadrat boundaries or straight transect lines are difficult to lay out.

This exercise describes the use of quadrat, line-intercept, and point-quarter techniques of community analysis. These techniques are applied to communities of plants or animals occupying areas ("stands") of two-dimensional habitat space (e.g., terrestrial plant communities, sessile rocky intertidal invertebrates, and the like).

GENERAL SAMPLING CONSIDERATIONS

Considerations of the size, shape, and number of sampling units, and whether they should be spaced systematically, randomly, or in a stratified random manner, were discussed in exercise 4. This discussion should be reviewed and the procedures to be used for this exercise chosen.

For any sampling procedure, it is usually most convenient to establish a baseline along one side of the stand or through the stand center. Quadrats, transects, or sampling points can be located randomly, in a stratified random manner, or systematically with reference to this baseline. Quadrats or sampling points can be located randomly by drawing pairs of random numbers to serve as coordinates of sampling locations, one number indicating the distance along the baseline, the other the distance into the area. Awbrey (1977) describes another technique for locating random points. Transects for line-intercept measurements can be initiated at randomly chosen points along the baseline. For stratified random sampling, the entire stand can be either divided into equal subareas or transected by sampling lines at regular intervals. Equal numbers of quadrats, transects, or sampling points can then located randomly within the subareas or along the transect lines. For line-intercept sampling in a rectangular area, one randomly selected point can be used to define the locations of equally spaced transects crossing the plot. This procedure is systematic in design but unbiased in estimates of density, cover, and frequency (Butler and McDonald 1983).

Table 14.1. Data sheet for recording species, numbers of individuals, and dominance values in quadrat sampling

Date _____ Locality _____ Stand (Number or Type) _____
Observer Name _____ Quadrat Size _____

Species	Quadrat #1	Quadrat #2	Quadrat #3	Quadrat #4	Quadrat #5

QUADRAT SAMPLING

Once the quadrat sampling design has been defined, the types of quantitative data to be obtained should be determined. In sampling communities of plants or sessile animals, the species, the number of individuals, the space they cover, and the frequency with which they occur are usually the most important characteristics.

In counting the individuals of some plants and animals, such as multistemmed shrubs or colonial corals, an arbitrary definition of what constitutes an individual will be needed. Criteria for inclusion or exclusion of individuals occurring on the edge of a quadrat may also be necessary. For example, plants with rooted bases lying more than halfway inside the boundary might be counted and measured as if they lay completely inside, and plants lying more than halfway outside completely excluded.

The area covered by individuals can be measured in several ways. For large woody plants, the diameter or circumference of the trunk can be measured and the *basal area* (cross-sectional area of the trunk) determined with a hand calculator. For smaller plants, the diameter of the crown foliage may be measured and the areal cover of the crown calculated. Individual basal area or cover values can be recorded by quadrat number and species in a notebook or data sheet (table 14.1), where they serve to indicate both the number and the size of the individuals.

Often, the measurement of cover values for individual organisms is impossible or impractical. Cover can be recorded, however, as estimates of the percent of the quadrat covered by each species. Usually this is done by assigning cover class estimates (Goldsmith et al. 1986). Several scales for ranking cover have been suggested, one of the most common being the following:

Cover Class	Range of Percent Cover	Midpoint
1	0–1%	0.5%
2	1–5%	3.0%
3	5–25%	15.0%
4	25–50%	37.5%
5	50–75%	62.5%
6	75–100%	87.5%

Considering the difficulty of obtaining accurate measurements of the diameters of irregular canopies of many plants, cover class estimates are often as accurate as more time-consuming direct measurements.

The frequency of occurrence of a species is provided by the fraction of quadrats that contain the species. In addition to number of individuals, basal area or cover, and frequency, one may wish to record other quantitative values, such as biomass, plant height, number of tillers (grasses) or stems, or numbers of flowers and fruits.

In summarizing quadrat data, density, dominance, and frequency values can be determined for each species. *Density* refers to the number of individuals per unit area, *dominance* to the basal area or crown cover per unit area, and *frequency* to the fraction of sample plots containing the species. For a particular species, these values may be expressed either in an absolute form or as *relative density, relative dominance,* and *relative frequency,* which show the percentage of the individual species' value with respect to the total for all species. Relative values for density, dominance, and frequency then can be combined into a single *importance value,* which combines these three somewhat different measures of the importance of the species in the community. These values are calculated by the following equations:

$$\text{density} = \frac{\text{number of individuals}}{\text{area sampled}}$$

$$\text{relative density} = \frac{\text{density for a species}}{\text{total density for all species}} \times 100$$

$$\text{dominance} = \frac{\text{total of basal area or areal coverage values}}{\text{area sampled}}$$

$$\text{relative dominance} = \frac{\text{dominance for a species}}{\text{total dominance for all species}} \times 100$$

$$\text{frequency} = \frac{\text{number of plots in which species occurs}}{\text{total number of plots sampled}}$$

$$\text{relative frequency} = \frac{\text{frequency value for a species}}{\text{total of frequency values for all species}} \times 100$$

$$\text{importance value} = \text{relative density} + \text{relative dominance} + \text{relative frequency}$$

Two of the above relative values, rather than all three, may sometimes be used to obtain an importance value, so that some ecologists prefer to calculate the importance value as the average, instead of the total, of the available values.

LINE-INTERCEPT ANALYSIS

Line-intercept analysis can yield quantitative estimates of density of individuals, areal cover, and frequency comparable to those obtained by quadrat sampling. The most satisfactory device for laying out a line transect is a measuring tape 30–50 meters in length. With the aid of the tape scale, the transect line can be subdivided into sections of any desired length, which will allow for the determination of frequency. The tape-measure scale also makes it easy to measure the length of the segments of transect line intercepted by individuals, an essential value for calculation of areal cover. In the absence of a long tape measure, a transect line of string, rope, or wire, marked off into regular intervals for the recording of frequency, can be used. Meter sticks or short tape measures are also needed for measuring other characteristics of individuals.

In recording line-intercept data, only those individuals that are touched by the transect line or that underlie or overlie the line should be recorded. For each individual encountered, three items of information should be recorded: the species, the length of the transect line intercepted (I), and the maximum width of the individual perpendicular to the transect line (M). These values should be recorded in table 14.2. In this table, data for different transect intervals should be placed in different columns. The length of transect segments overlying bare ground should be measured and recorded in the same manner. In the case of communities with two or more quite distinct strata, such as herb, shrub, and tree strata in a forest, each stratum should be sampled separately.

In summarizing the sampling data, the following values should first be determined for each species:

- Total number of individuals encountered (N)
- Total of intercept lengths (ΣI)
- Number of transect intervals in which species occurred
- Total of reciprocals of maximum plant widths ($\Sigma 1/M$)

With these values the standard measures of community composition can be obtained. Density and relative density are given by the equations

$$\text{density} = \left(\sum \frac{1}{M}\right)\left(\frac{\text{unit area}}{\text{total transect length}}\right)$$

$$\text{relative density} = \frac{\text{density for a species}}{\text{total density for all species}} \times 100$$

The sum of reciprocals of individual widths is necessary for the density calculation because the chance of an individual being encountered by the transect is proportional to its width perpendicular to the transect line. The unit area in this calculation is simply the size of the area, in the same units as plant width and transect length, on the basis of which density is to be expressed. For example, if plant widths and transect length are measured in centimeters, the unit-area value in the equation is the size of the area, in square centimeters, for which the density value will be stated. A complete explanation and derivation of the density calculation above is given by Strong (1966).

Dominance is given by the following equations:

$$\text{dominance or cover (as \% of ground surface)} = \frac{\text{total of intercept lengths for a species}}{\text{total transect length}} \times 100$$

$$\text{relative dominance} = \frac{\text{total of intercept lengths for a species}}{\text{total of intercept lengths for all species}} \times 100$$

Frequency values calculated from line-transect data can be difficult to interpret. Frequency is perhaps most simply expressed by the equation

$$\text{frequency} = \frac{\text{intervals in which species occurs}}{\text{total number of transect intervals}} \times 100$$

This value is somewhat misleading, however, since the chance of a species being recorded in a given transect interval is strongly related to the size of the individuals as well as to the abundance and distribution of the species. In calculating relative frequency these factors may be taken into account. A weighting factor, F, can be derived

Table 14.2. Data sheet for recording species, intercepts, and maximum width values for individuals encountered in line-intercept sampling

Date _____ Locality _____ Stand (Type or Number) _____

Observer Name _____ Interval Length _____

Species	Interval ___ # ___ I — M	Interval ___ # ___ I — M	Interval ___ # ___ I — M	Interval ___ # ___ I — M	Interval ___ # ___ I — M	Interval ___ # ___ I — M	Interval ___ # ___ I — M	Interval ___ # ___ I — M	Interval ___ # ___ I — M	Interval ___ # ___ I — M	Interval ___ # ___ I — M

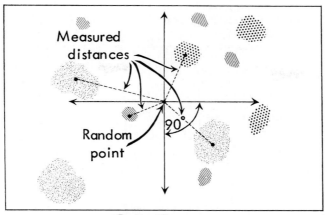

Point Quarter

Figure 14.1.
Point-quarter sampling procedure

and used to calculate a weighted frequency, which, in turn, is used to calculate relative frequency.

$$F = \frac{(\Sigma 1/M)}{n} \qquad \text{Where: } n = \text{Number of values of M}$$

$$\text{weighted frequency} = (F) \begin{pmatrix} \text{number of transect intervals} \\ \text{in which species occurs} \end{pmatrix}$$

$$\text{relative frequency} = \frac{\text{weighted frequency for a species}}{\text{total of weighted frequencies for all species}} \times 100$$

The importance value can then be calculated by summing or averaging relative density, relative dominance, and relative frequency.

An estimate of the total percentage of the ground surface covered by individuals can be obtained by totaling cover percentages if measurements of intercept distances were taken in a nonoverlapping manner. If overlap of intercept measurements did occur, owing to the sampling of individuals belonging to different strata, total coverage must be obtained by the equation

$$\text{total coverage} = \frac{\text{total transect length} - \text{total bare ground}}{\text{total transect length}} \times 100$$

POINT-QUARTER TECHNIQUE

This technique also gives estimates of density, cover, and frequency comparable to those obtained by quadrats or line-intercept measurements. In this technique (figure 14.1), the area around each sampling point is divided into four equal sectors, or quadrants. This division can be done with a compass, or, if a transect line is used, the quadrants may be formed by the line itself and a second line perpendicular to it at the sampling point. The individual nearest the point in each quadrant is then located, and the species, basal area or cover, and point-to-individual distance recorded in table 14.3. Basal area or cover is calculated from diameter or circumference measurements. Point-to-individual distances should be measured to the center of the organism (e.g., the center of the crown or rooted base) rather than to the edge of the organism.

To summarize sampling data, point-to-individual distances should first be totaled for all species and all points, and averaged to give the mean point-to-individual distance. The square of this value is the mean area per individual, or the average area of surface on which one individual occurs. Thus, the total density of individuals of all species is obtained by dividing the mean area per individual into the unit area. As an equation, this may be written

Table 14.3. Data sheet for recording species, dominance values, and point-to-individual distances in point-quarter sampling

Date

Locality

Stand (Number or Type)

Observer Name

Species Name and Abbreviation

1. _____
2. _____
3. _____
4. _____
5. _____
6. _____
7. _____
8. _____
9. _____
10. _____
11. _____
12. _____
13. _____
14. _____
15. _____
16. _____

	Quadrant 1	Quadrant 2	Quadrant 3	Quadrant 4
1				
2				
3				
4				
5				
6				
7				
8				
9				
10				
11				
12				
13				
14				
15				
16				
17				
18				
19				
20				
21				
22				
23				
24				
25				
26				
27				
28				
29				
30				

$$\text{total density of all species} = \frac{\text{unit area}}{(\text{mean point-to-individual distance})^2}$$

In this equation, the term *unit area* refers to the size of the area, in the same units as those for the mean area per plant, on the basis of which density is to be expressed. For example, if density is to be expressed per hectare, but the mean area per plant is in units of meters squared, the unit-area value in the calculation would be 10,000 (the number of square meters in a hectare); if density is to be expressed for a 100 m^2 area, the unit-area value equals 100.

The number of individuals of each species should next be determined. The basal area or areal cover values for individuals of each species should then be summed and divided by the number of individuals of the species, to give average dominance values for the various species. These average dominance values simply represent the average basal area or areal cover of one individual of each species. From these data, absolute and relative values for density, dominance, and frequency, together with the importance value for each species, can be determined by the following equations:

$$\text{relative density} = \frac{\text{individuals of a species}}{\text{total individuals of all species}} \times 100$$

$$\text{density} = \frac{\text{relative density of a species}}{100} \times \frac{\text{total density of}}{\text{all species}}$$

$$\text{dominance} = \text{density of species} \times \text{average dominance value for species}$$

$$\text{relative dominance} = \frac{\text{dominance for a species}}{\text{total dominance for all species}} \times 100$$

$$\text{frequency} = \frac{\text{number of points at which species occurs}}{\text{total number of points sampled}}$$

$$\text{relative frequency} = \frac{\text{frequency value for a species}}{\text{total of frequency values for all species}} \times 100$$

The importance value can then be calculated as the sum or average of the available relative values.

SUGGESTED ACTIVITIES

1. Sample a plant community in a field location selected by your instructor. Mark a baseline along one side of the stand to be sampled, and determine the distance at right angles to this baseline that defines the limits of the stand. Obtain random coordinates or points along the baseline to define the locations of quadrats, transects, or point-quarter sampling points. Carry out sampling and analysis, recording the field data in tables 14.1, 14.2, or 14.3, and summarizing vegetational composition in table 14.4.

    Comparisons can be made of the community analysis obtained with different procedures; some possible comparisons include

    - Equal numbers of quadrats (transects, point-quarter points) located systematically, randomly, or by a stratified random procedure
    - Equal numbers of quadrats of the same size but different shape
    - Quadrats of different size, but adjusted in number to sample the same area
    - Quadrat analyses compared to analyses of the same stand or population by line-intercept and point-quarter techniques

2. Sample the map of a desert plant community—a rich alluvial fan environment in Borrego Valley, San Diego County, California—included at the back of this manual. Locate quadrats by pairs of random coordinates (between 0 and 80 on the x-axis and 0 and 60 on the y-axis) using the meter scale along the sides of the mapped plot. Clear plastic discs, squares, or rectangles can be used to represent quadrats.

Table 14.4. Summary of community composition measurements

Date _____ Location _____ Stand (Number or Type) _____
Observer Name _____ Method and Sampling Details _____

Species	Density ( )	Relative Density	Dominance ( )	Relative Dominance	Frequency	Relative Frequency	Importance Value
Total							

The sizes of quadrats should be determined by reference to the map scale. Starting points for transects or locations for point-quarter sampling can be located in a similar manner. Use a ruler to measure point-to-plant distances and intercept lengths (calibration of these measurements with the map scale may be necessary).

Make comparisons as suggested in #1 above. In addition, compare the wash and nonwash subareas of the desert community.

3. Sample an intertidal community of marine algae and invertebrates on a rocky substrate, using quadrat, transect, or point-quarter methodology. Make comparisons as suggested in #1 above.

✪ QUESTIONS FOR DISCUSSION

1. What are the sources of error for the different methods? In what situations would each method probably be most accurate? How would you design tests of the accuracy or efficiency of various sampling methods?

2. Describe conditions under which species importance rankings based on density, dominance, and frequency values may give very different results.

3. What is the value of converting density, cover, and frequency to relative values? What is the reason for combining the three relative values into a single importance value?

✪ SELECTED BIBLIOGRAPHY

Awbrey, F. T. 1977. Locating random points in the field. Journal of Range Management 30:157–158.

Barbour, M. G., J. H. Burk, and W. D. Pitts. 1987. Methods of sampling the plant community. Chapter 9 *in* Terrestrial plant ecology. Benjamin/Cummings Publishing Company, Menlo Park, California, USA.

Bonham, C. D. 1989. Measurements for terrestrial vegetation. John Wiley & Sons, New York, New York, USA.

Butler, S. A., and L. L. McDonald. 1983. Unbiased systematic sampling plans for the line intercept method. Journal of Range Management 36:463–468.

Floyd, D. A., and J. E. Anderson. 1987. A comparison of three methods for estimating plant cover. Journal of Ecology 75:221–228.

Goldsmith, F. B., C. M. Harrison, and A. J. Morton. 1986. Description and analysis of vegetation. Pages 437–524 *in* P. D. Moore and S. B. Chapman editors. Methods in plant ecology. Blackwell Scientific Publications, Oxford, England.

Grieg-Smith, P. 1983. Quantitative plant ecology. 3rd ed. Blackwell Scientific Publications, Oxford, England.

Kershaw, K. A., and J. H. Looney. 1985. Quantitative and dynamic ecology. 3rd ed. Edward Arnold, London, England.

Kuchler, A. W., and I. S. Zonneveld. 1988. Vegetation mapping. Kluwer Academic Publishers, Hingham, Massachusetts, USA.

Strong, C. W. 1966. An improved method of obtaining density from line-transect data. Ecology 47:311–313.

EXERCISE **15**

Acclimation to Temperature

INTRODUCTION

Acclimation is the modification of physiological mechanisms of adaptation or response by an individual as a result of prolonged exposure to specific environmental conditions. Acclimation can be expressed as measurable changes in characteristics such as limits of tolerance, metabolic rate, and preferenda in a thermal gradient. These changes can usually be interpreted as compensatory responses that allow normal function under conditions away from those optimal (Bullock 1955). The capacity for acclimation is an important determinant of the ecological distribution and seasonal occurrence of many animals (Newell and Bayne 1973, Weiser, 1973, Snyder and Weathers 1975).

TEMPERATURE ACCLIMATION

Acclimation to temperature is shown by many terrestrial and aquatic invertebrates and vertebrates, both ectotherms and endotherms. Temperature acclimation is likewise a major mechanism of adjustment by plants to seasonal changes.

In many small vertebrates, the capacity for acclimation can be measured without injury to individuals by determining the *critical thermal minimum (CTMin)*, or low temperature at which normal body movement becomes impossible. This exercise describes the procedure for studying temperature acclimation in small terrestrial or amphibious vertebrates. Modifications of the procedure to accommodate aquatic organisms make possible the use of this approach for aquatic organisms such as crayfish (Claussen 1980).

For controlled laboratory studies, animals should be obtained about one month ahead of the time of testing and kept initially at an intermediate temperature for at least two weeks. Before animals are taken from the wild, their abundance and legal protection should be considered carefully, and populations that are protected or that might be adversely affected by collecting should not be used. Suitable small amphibians that have been bred in captivity are available from a number of biological supply companies. Most diurnally active lizards can be housed in an uncovered terrarium with an incandescent lamp positioned to create a temperature gradient and objects such as pieces of broken flowerpot placed to provide shelter. Terrestrial salamanders can be kept in a covered container on a substrate of moist sand with pieces of bark or broken flowerpot for shelter. Aquatic salamanders can be kept in aquaria with shallow water and a substrate of sand and stones for shelter. The water in such aquaria should be changed often to prevent bacterial build-up. More detailed recommendations for captive housing and feeding of reptiles and amphibians are given by Pough (1991).

Begon et al. (1990): 2:53–54
Brewer (1994): 2:16–18
Ehrlich and Roughgarden (1987): 2:9–12, 17–23

Krebs (1994): 3:40
Pianka (1988): 65–69, 112–113
Ricklefs (1990): 7:93–108

Ricklefs (1993): 10:189–191
Smith (1990): 6:104–109
Smith (1992): 5:79–94

Animals should be taken from these cultures and placed at one or more acclimation temperatures. The intermediate-temperature animals can serve as controls for these acclimation experiments. Complete acclimation requires about 14 days in most cases. If enough animals are available, the time course of acclimation can be followed by placing separate groups of animals at the acclimation temperatures at times ranging from two weeks to a few days before the time of testing. Alternatively, the entire group of animals can be placed at the acclimation temperature on a single occasion and individuals removed and tested at a number of subsequent times. Appropriate acclimation temperatures will vary with the animal. For lizards, 10° C, 20° C, 30° C, and 35° C might be appropriate, but for amphibians a lower range of temperatures should be selected.

Measurements of CTMin are usually done by determining the temperature at which an animal loses its righting response, that is, is unable to right itself within a specified time (e.g., 30 sec) when placed on the substrate on its back. To determine the CTMin of salamanders, individuals are placed in a half-liter pyrex dish containing 100 ml of distilled water. If the dish and its contents are cooled slowly (<1° C per min) on a cold plate or in a cold cabinet, the body temperature of a small salamander or anuran will be essentially the same as that of the water. The water temperature can be monitored with a thermocouple or a digital thermometer. For small lizards, it is better to use a dry container and to take cloacal body temperature measurements directly with a Shultheis thermometer.

Comparisons can be made with individuals from different habitats or from localities with different elevations or temperature regimes.

✷ Suggested Activities

1. Examine the time course of temperature acclimation of individuals of different body size of a common species of frog, toad, salamander, or lizard by measurement of CTMin as outlined above.
2. Using the CTMin test, compare temperature acclimation abilities of species of lizards or anurans with different seasonal activity patterns or habitats. For example, compare species of lizards with activity patterns restricted to warm summer periods, such as species of *Cnemidophorus,* with those of species active at cooler seasons, such as species of *Uta* or *Sceloporus.* Or, contrast acclimation capabilities of amphibians breeding in early spring with those of species breeding later in the season. Or, contrast acclimation abilities of closely related species or members of populations resident at different elevations or in desert versus nondesert ranges.
3. Determine the CTMin for individuals of a species of small amphibian or lizard obtained in the wild at different times of year. Determine the prevailing temperatures for the 2- or 3-week period preceding capture of individuals, as measured at a nearby U.S. Weather Bureau station, and estimate the temperatures that prevailed in the occupied habitat of the species. Compare the CTMin values of wild-caught animals with those of individuals experimentally acclimated to a range of temperatures.

Equipment for CTMin Measurement

Several means of cooling dishes or chambers with experimental animals are available. The freezer compartment of a refrigerator may be used if arrangements are made to preclude rapid temperature drop or freezing of the experimental animals. A dry-ice and water bath can also be used, with the chamber containing animals being floated on the water surface. More satisfactory is a cold plate (see appendix B), a refrigeration device that can produce a controlled pattern of cooling.

Digital thermometers with remote probes, as well as Shultheis thermometers for taking cloacal temperatures, are available from most major scientific supply houses.

🕙 QUESTIONS FOR DISCUSSION

1. In what specific habitats would temperature acclimation ability be of greatest value to an organism?
2. To what environmental factors other than temperature would you expect acclimation to be shown? In what specific ways should temperature acclimation be shown by other physiological and behavioral processes in animals such as reptiles and amphibians?
3. What possible evolutionary significance does acclimation ability have? Do you think acclimation could be the first stage of adaptation of an organism to a new range of environmental conditions?

🕙 SELECTED BIBLIOGRAPHY

Cheper, N. J. 1980. Thermal tolerance of the isopod *Lirceus brachyurus* (Crustacea: Isopoda). American Midland Naturalist 104:312–318.

Claussen, D. L. 1977. Thermal acclimation in amblystomatid salamanders. Comparative Biochemistry and Physiology 58A:333–340.

——— . 1980. Thermal acclimation in the crayfish, *Orconectes rusticus* and *O. virilis*. Comparative Biochemistry and Physiology 66A:377–384.

Edney, E. B. 1964. Acclimation to temperature in terrestrial isopods. I. Lethal temperatures. Physiological Zoology 37:364–377.

Feder, M. E. 1978. Environmental variability and thermal acclimation in neotropical and temperate zone salamanders. Physiological Zoology 51:7–16.

——— , A. G. Gibbs, G. A. Griffith, and J. Tsuji. 1984. Thermal acclimation of metabolism in salamanders: fact or artifact. Journal of Thermal Biology 9:250–260.

Fitzpatrick, L. C., and A. V. Brown. 1975. Metabolic compensation to temperature in the salamander *Desmognathus ochrophaeus* from a high elevation population. Comparative Biochemistry and Physiology 50A:733–737.

Gatten, R. E., Jr., A. C. Echternacht, and M. A. Wilson. 1988. Acclimatization versus acclimation of activity metabolism in a lizard. Physiological Zoology 61:322–329.

Layne, J. R., Jr., and D. L. Claussen. 1982. Seasonal variation in the thermal acclimation of critical thermal maxima (CTMax) and minima (CTMin) in the salamander *Eurycea bislineata*. Journal of Thermal Biology 7:29–33.

——— . 1982. The time courses of CTMax and CTMin acclimation in the salamander *Desmognathus fuscus*. Journal of Thermal Biology 7:139–141.

——— . 1987. Time courses of thermal acclimation for critical thermal minima in the salamanders *Desmognathus quadrimaculatus*, *Desmognathus monticola*, *Desmognathus ochrophaeus* and *Plethodon jordani*. Comparative Biochemistry and Physiology 87A:895–898.

Layne, J. R., Jr., M. L. Manis, and D. S. Claussen. 1985. Seasonal variation in the time course of thermal acclimation in the crayfish *Orconectes rusticus*. Freshwater Invertebrate Biology 4:98–104.

Newell, R. C., and B. L. Bayne. 1973. A review on temperature and metabolic acclimation in intertidal marine invertebrates. Netherlands Journal of Sea Research 7:421–433.

Pough, F. H. 1991. Recommendations for the care of amphibians and reptiles in academic institutions. Institute for Laboratory Animal Resources News 33:S1–S21.

Snyder, G. K., and W. W. Weathers. 1975. Temperature acclimation in amphibians. American Naturalist 109:93–101.

Tsuji, J. S. 1988. Thermal acclimation of metabolism in *Sceloporus* lizards from different latitudes. Physiological Zoology 61:241–253.

Weiser, W., editor. 1973. Effects of temperature on ectothermic organisms. Springer Verlag, Heidelberg, Germany.

EXERCISE 16
Plant Water Potential

INTRODUCTION

The movement of water into and out of plants is governed by the same principles that determine the movement of water through the physical landscape. Water that is confined behind a dam, for example, possesses a certain potential energy, and if the dam is opened, the water will flow downhill to a location at which its potential energy is less. Water in plants and the soil also has measurable potential energy, and will move through soil-plant-air systems from points of high potential energy to points of lower potential energy. This exercise examines one of the basic techniques for measuring potential energy relationships in plants.

BASIC CONCEPTS OF PLANT WATER STATUS

The potential energy status of water in soil and plants is termed *water potential,* and is measured in units of pressure and designated by the Greek letter psi, ψ. The presently accepted international unit of water potential is the *megapascal (MPa).* Past studies usually expressed water potentials in *bars (1 MPa = 10 bars).*

For a particular location, the energy potential of pure water is 0. Since water in the soil and in plants is constrained in its ability to do work, relative to an equal amount of free water at the same location, the water potential in soil and plant systems is negative. In the soil, these constraints relate to the forces with which water is held in the pore space (*matric potential*), by the solutes that the water contains (*solute potential*), and by other factors. The total potential of the soil is the sum of these additive effects. In plants, the major components of total water potential are hydrostatic potential (turgor pressure), osmotic potential, and matric potential. That these potentials are negative is evidenced by the fact that water in the xylem is almost always under tension. Thus, in soil-plant systems, water moves from a point of less negative (= higher potential energy) water potential to one of more negative (= lower potential energy) status.

In nature, soil water potential may vary from high values of about -0.03 MPa (-0.3 bars), when the soil is at field capacity (containing all the water it can hold against gravitational force), to much more negative values as the soil dries. At a soil water potential of about -1.5 MPa (-15 bars), known commonly as the permanent wilting point, many herbaceous plants (specifically the cultivated sunflower) are unable to extract water from the soil, and thus wilt—ultimately dying if water is not supplied. In general, at plant water potentials from -1 to -2 MPa physiological processes of most plants suffer stress, and below -2 MPa most plants are either wilted or physiologically dormant. Plants adapted to deserts (xerophytes) or those from saline environments (halophytes) may withstand water potentials exceeding -7 MPa.

The water potential of a plant is influenced by quite a few factors. Principal among these are the water potential of the soil and the rates of evaporation from leaf surfaces and transpiration through open stomates.

Begon et al. (1990): 3:90–95
Brewer (1994): 3:48–54
Ehrlich and Roughgarden (1987): 140–146

Krebs (1994): 7:96–114
Ricklefs (1990): 6:76–77
Ricklefs (1993): 2:26–30

Smith (1990): 10:192–196
Smith (1992): 6:100–102

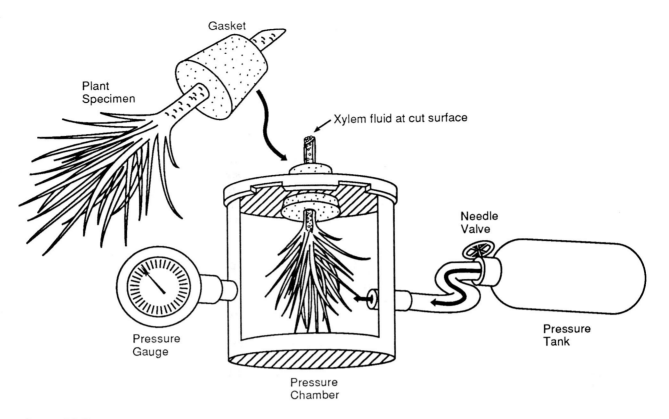

Figure 16.1.
Diagram of the arrangement of apparatus and plant shoot for measurement of plant water potential with the pressure bomb

In the daytime the water potential gradient between the plant and the atmosphere is large, producing high evaporation. This increases the tension on the columns of water in the xylem, lowering the plant water potential. At night, when water loss declines, water moves from the soil into the plant and plant water potential rises, eventually reaching an equilibrium, usually at a level close to that of the soil.

Plants typically close their stomata during times of drought, and usually possess surface coverings (e.g., cuticular waxes, dense pubescence) that also reduce evaporation. Plants from arid regions are also able to develop very low water potentials. These low potentials enable them to extract water from soils that are quite dry.

Measurements of plant water potential can thus reveal differences in physiological condition among plants of one species in different situations, providing information on stress tolerance, as well as among plants of different species in the same situation, providing information on adaptive strategies by plants to common environmental conditions. Analysis of such differences can aid in understanding overall patterns of plant adaptation to water stress.

MEASURING PLANT WATER POTENTIAL

Plant water potential is most commonly measured with a pressure chamber, or pressure "bomb" (figure 16.1). The principle of this device is simple. If a twig or shoot is cut from a plant, the water in the xylem vessels pulls back from the cut surface because the water columns that were under tension (negative ψ) have been severed. If this shoot is placed in a chamber, with its cut end extending outside the chamber, and pressure is increased inside the chamber, the xylem columns can ultimately be forced back to the cut surface. The positive pressure needed to do this equals the negative water potential of the vessels. The logic of expressing water potential in pressure units is made clear by this procedure.

Preparing the Plant Sample

The pressure bomb can be used to measure water potentials in many different kinds of plants, as long as they possess stems that can be accomodated by the gasket in the chamber cap (figure 16.1).

1. Select a twig or shoot with a diameter of about 1–3 mm and a length of bare, undamaged stem slightly longer that the depth of the rubber gasket in which it will be placed (do not plan to prune leaves after cutting to achieve this). Remember that the plant specimen must be small enough to fit inside the bomb chamber without being distorted so that tissues are ruptured or broken.
2. Enclose the shoot in a plastic sheath (to reduce water loss during measurement) and immediately cut the stem (ideally at an angle).
3. Insert the specimen into a rubber gasket (e.g., a rubber stopper with a center hole cut to a slightly smaller diameter with a cork-borer) so that the leafy portion of the shoot extends from the wide face of the gasket, and the cut end protrudes as little as possible from the narrow face of the gasket. Do not recut or trim the stem. The twig should fit snugly in the gasket hole. A slender metal tube the same diameter as the specimen may be needed to insert the shoot. This tube is first pushed through the gasket, the specimen inserted in it, and the tube pulled back through the gasket. Gaskets with different types of specimen holes may be needed to allow measurements of plant shoots ranging from cylindrical stems of herbaceous and woody dicots to the wide, flat blades of grasses. A quick-setting epoxy may be needed to anchor soft, herbaceous stems in the gasket.
4. Take the water potential measurement promptly. If substantial evaporation occurs from the specimen after it has been cut, the value of the water potential measurement will obviously be affected.

Preparing the Pressure Chamber

The pressure chamber is an instrument that needs regular maintenance and checking to assure safety of operation. The safety considerations relate to the fact that high pressure is created inside the chamber, which can lead to accidents if the chamber is improperly closed or if the specimen or gasket material is accidentally ejected by the internal chamber pressure. For these reasons, the chamber should be checked by an experienced operator prior to its use. In addition, the safety procedures noted below should be followed carefully.

Making the Measurements

To obtain the water potential readings, follow these steps:

1. Place the gasket, with the specimen inserted, in the chamber lid so that the wide end of the gasket is inside the chamber and the cut end of the specimen is visible outside the lid.
2. Place the cover on the chamber and lock it firmly in position.
3. Turn the control valve of the instrument to permit air flow into the bomb chamber.
4. Turn the needle valve of the instrument to allow pressure to increase slowly in the chamber. This slow increase of pressure is necessary to prevent the development of water potential disequilibria between different tissues within the specimen.
5. Watch the cut tip of the specimen from the side with a magnifying glass or a dissecting microscope as the pressure increases. Do not place any part of your body directly over the specimen. If the specimen or a part of the gasket is expelled, you could be injured. Wearing safety goggles is advised.

6. When moisture appears at the cut surface (evident as a change of color or sheen), note the reading on the scale and immediately turn the control valve to "off." Note that the moisture must come from the xylem (the central part of the stem). Small amounts of fluid appearing near the outer edge of the stem are probably phloem fluids. If they appear, these phloem fluids can be removed by blotting. Continue increasing the pressure slowly until xylem fluid is expelled from the interior of the cut surface. Record the reading.
7. Turn the control valve to the exhaust position, and allow the pressure to escape from the chamber.
8. Open the chamber and remove the sample.

INSTRUMENTATION FOR WATER POTENTIAL STUDY

Pressure bomb systems for measurement of plant water potential are manufactured in North America by the PMS Instrument Company in Corvallis, Oregon, and the Soil Moisture Equipment Corporation in Santa Barbara, California (see appendix B for more detailed information).

✪ SUGGESTED ACTIVITIES

1. Determine the diurnal course of water potential for one or more plants by making measurements at several times of day, beginning with a set of measurements just before dawn, and ending after nightfall. If possible, compare shallow-rooted species with deep-rooted species that may have root systems reaching a water table.
2. Compare water potentials for plants of a single species along an environmental gradient of salinity or aridity. At each location, try to obtain measurements for individuals of different ages or sizes.
3. Using plants maintained in a greenhouse, compare water potentials for drought-tolerant and drought-sensitive species that have been subjected to different watering regimes.

✪ QUESTIONS FOR DISCUSSION

1. What relationship tends to exist between predawn plant water potential and soil water potential?
2. By what sorts of physiological mechanisms are plants of arid and saline environments adapted to withstand extremely low water potentials?
3. What differences in diurnal water potential curves would you expect for C3, C4, and CAM plants occurring in the same environment?

✪ SELECTED BIBLIOGRAPHY

Bowman, W. D., and S. W. Roberts. 1985. Seasonal and diurnal water relations adjustments in three evergreen chaparral shrubs. Ecology 66:738–742.

Drivas, E. P., and R. L. Everitt. 1988. Water relations characteristics of competing single-leaf pinyon seedlings and sagebrush nurse plants. Forest Ecology and Management 23:27–38.

Fitter, A. H., and R. K. M. Hay. 1987. Environmental physiology of plants. 2nd ed. Academic Press, London, England.

Koide, R. T., R. H. Robichaux, S. R. Morse, and C. M. Smith. 1989. Plant water status, hydraulic resistance, and capacitance. Pages 161–183 *in* R. W. Pearcy, J. Ehleringer, H. A. Mooney, and P. W. Rundel, editors. Plant physiological ecology. Chapman & Hall, London, England.

Miller, R. F. 1988. Comparison of water use by *Artemisia tridentata* ssp. *wyomingensis* and *Chrysothamnus viscidiflorus* ssp. *viscidiflorus.* Journal of Range Management 41:58–62.

Nilsen, E. T., M. R. Sharifi, P. W. Rundel, and R. A. Virginia. 1986. Influences of microclimatic conditions and water relations on seasonal leaf dimorphism of *Prosopis glandulosa* var. *torreyana* in the Sonoran Desert, California. Oecologia 69:95–100.

Nobel, P. S. 1991. Physicochemical and environmental plant physiology. Academic Press, San Diego, California, USA.

Ritchie, G. A., and T. M. Hinckley. 1975. The pressure chamber as an instrument for ecological research. Advances in Ecological Research 9:165–254.

Turner, N. C. 1987. The use of the pressure chamber in studies of plant water status. Proceedings of the International Conference on Measurement of Soil and Plant Water Status 2:13–24.

<small>EXERCISE</small> 17

Field Measurement of Photosynthesis

INTRODUCTION

Technological innovations have now made possible the measurement of photosynthesis in individual plant leaves under field conditions. Self-contained, portable photosynthesis systems weighing as little as 9 kg are now available. Field photosynthesis systems use a portable *infrared gas analyzer (IRGA)* to measure the change in CO_2 elicited by the leaf or leafy branch placed in the chamber. They contain sensors that measure leaf temperature, radiation intensity, and the temperature, humidity, CO_2 concentration, and other conditions of the air in or moving through a chamber containing a leaf or leafy branch. Some systems have an open airflow (air enters the chamber from outside and is discharged to the outside) and some have a closed flow (air is recirculated between the chamber and the *IRGA*).

Although not inexpensive, portable photosynthesis systems cost no more than many items that are standard equipment in university physiology laboratories. State-of-the-art research in plant physiological ecology, dealing with questions such as how global change may influence ecosystem function, now requires the use of such equipment. Acquisition of such instrumentation should receive high priority for a university program in ecology.

BASIC THEORY OF MEASUREMENT

Several types of portable photosynthesis systems are in use by researchers (Field et al. 1989). The most popular and suitable of these systems are the LI-6200 series systems, lightweight systems with a closed-flow design manufactured by the LI-COR Corporation (appendix B). This exercise assumes that an instrument of this type is available. The following outline covers the basics of LI-6200 use, but one should consult the LI-6200 manual if difficulties arise, and make sure that someone familiar with use of the system is available.

LI-6200 systems consist of three major components. The first is a chamber or *cuvette* in which the leaf or leafy branch is placed. Attached to this chamber are sensors that measure the temperature of the leaf and the air, the intensity of photosynthetically active radiation, and relative humidity; other variables are calculated by software. Chambers with volumes of 0.25 l, 1.0 l, and 4.0 l are available, the most commonly used chamber being that with the volume of 1.0 l. One can also make a custom cuvette.

The second component of the system is the *IRGA*, which measures CO_2 concentration by the absorption of infrared wavelengths by this gas. Attached to the IRGA are tubes of soda lime (a CO_2 scrubber required for standardization of the instrument) and magnesium perchlorate (the desiccant used to control humidity in the system).

Colinvaux (1993): 3:32–56	Ricklefs (1990): 11:189–196, 15:280–291	Smith (1992) 7:110–121
Krebs (1994): 26:604–632	Ricklefs (1993): 6:106–110	Stiling (1992): 19:372–380
Odum (1993) 4:80–82	Smith (1990): 3:31–36	

The third component of the system is the *console,* essentially a microcomputer that programs parameters of the analysis, computes the values of photosynthesis and transpiration from the sensor measurements, and stores the data and results for future reference. Values for photosynthesis and transpiration are actually determined for a series of short periods (about 2 seconds) during which CO_2 is being drawn down and humidity may be increasing or decreasing. The raw data and computed estimates of physiological interest are stored in a tabular format.

Measurements with LI-6200 systems yield estimates of several physiological variables:

- Net photosynthesis, *PHOTO* (μmol CO_2 m^{-2} s^{-1})
- Stomatal conductance, *CMOL* (mol CO_2 s^{-1})
- Intercellular CO_2 concentration, *CINT* (ppm)
- Stomatal resistance, *RS* (s cm^{-1})
- Stomatal conductance, *CS* (cm s^{-1})

PREPARING THE APPARATUS

Using the Console

There are four main system modes: **PAUSE, STATUS, MONITOR,** and **LOG.** These can be activated by pressing their named keys; they function as follows:

- **PAUSE.** This is the waiting mode that the instrument enters when turned on, and from which other modes are accessed.
- **STATUS.** When activated, this mode shows (in the console window) the time, battery voltage, percentage of memory free for data storage, and number of the last data set, or "page," in memory.
- **MONITOR.** In this mode, one can observe (in the console window) current readings for each of the sensors in the cuvette and IRGA system. Alternative ways exist for using the monitoring capability (see manual). The simplest, perhaps, is to activate first **MONITOR,** and then the desired sensor key on the top row of the console keyboard. The sensors available are
    - **TIME.** Number of seconds since the start of a measurement
    - **PAR.** Photosynthetically active radiation (μmol m^{-2} s^{-1})
    - **TLEAF.** Leaf temperature ($^{\circ}$C)
    - **TCHAM.** Cuvette air temperature ($^{\circ}$C)
    - **TIRGA.** IRGA air temperature ($^{\circ}$C)
    - **CO2.** CO_2 concentration (ppm)
    - **FLOW.** Flow rate (μmol s^{-1})
    - **RH.** Cuvette relative humidity (%)
    - **LOG.** This is the operating mode during actual measurements, and is described below.

Preliminary Tests

Toggle the console power to **ON.** Check the voltage capability of the batteries by pressing the **Status** key. The voltage should be greater than 12 V for instrument use. If the voltage reading is less, the batteries should be charged.

Console Setup

Press the **Setup** key to access this menu. Press the ↑ key so that the view window shows **41 set pg param** as its top line, and then press **RTRN.** Then, simultaneously press the **SHIFT** and **EDIT** keys. This will enable you to edit or accept specific conditions of nine parameter sets. Values may be changed by pressing the **EDIT** key, typing in new values, and pressing **RTRN,** or accepted as is by pressing the **RTRN** key. The parameters are

1. *LAB.* This is an identification label (perhaps your name) with up to 12 characters.
2. *Vt(cc).* Enter the exact volume of the cuvette and *IRGA* in cm³ (380, 1140, or 4110 cm³ for the three *LI-COR* cuvettes).
3. *Vg(cc).* Enter the volume of the system, not including the cuvette in cm³ (120 cm³ for the *LI-6250*).
4. *P(mb).* Enter the normal atmospheric pressure of your location, based on its elevation, in mb (about 1013 mb minus 1 mb for each 10 m of elevation above sea level).
5. *BC(mol).* This value is the boundary layer resistance, or the resistance to gas exchange between air and leaf due to the surface air zone of the leaf. It is related to leaf size, chamber volume, and air movement within the chamber. It can be measured by use of filter-paper cutouts of leaves (see manual), but approximate values of 1.6, 1.3, or 0.8 mol m^{-2} s^{-1} can be used for 0.25 l, 1.0 l, and 4.0 l cuvettes, respectively.
6. *STOM RAT.* This is the ratio of stomatal numbers on the two sides of the leaf. Enter **0** if the leaf has stomates only on one surface, **1.0** if stomates are equally abundant on both sides, or **0.5** if their abundance is unequal.
7. *Fx(μmol).* This is the maximum airflow rate that can occur when the flow passes entirely through the desiccant tube. Determine this rate by monitoring the flow rate under this condition.
8. *Kabs.* This parameter, related to changing vapor-pressure conditions, can be measured (see manual), but an approximation of "1.1" is satisfactory for general work.
9. *pr1–pr6.* These are additional labels you might wish to use for identifying data sets. You might wish, for example, to create **LOCATION, SPECIES, LEAF #,** or other identifying labels for data sets you plan to obtain.

When these parameters have been specified, press the **RTRN** key to return to the **SETUP** menu and **PAUSE** to return to the ready mode. Other sets of parameters exist (42–49), but these for the most part involve basic format and calibration features that are assigned when the LI-6200 instrument is initially used.

PROCEDURE FOR FIELD MEASUREMENTS

Prior to taking the measurements, keep the cuvette open, with the fan on, and in a shaded location so that it does not become heated to above air temperature. It is best to allow the instrument to equilibrate to field temperatures for half an hour or so before measurements are made.

Making the Measurements

Select a leaf for measurement and enclose it in the chamber, maintaining as much as possible its original orientation. Close and latch the chamber, making sure that its seal is tight. Make sure that the leaf is not shaded by people or parts of the apparatus itself. Some systems now have a toggle switch that allows the system to be run in an open mode, allowing the leaf to equilibrate, and then switched to the closed mode for measurement.

Once the system is sealed and operating in closed mode, monitor the CO_2 concentration. Once the concentration begins to drop, press the **LOG** key. This will cause a series of readings to be taken until the CO_2 concentration has dropped a specified amount (e.g., 5 ppm). When the observations are finished, the instrument will "beep" twice.

When logging is finished, the leaf can be removed from the chamber, and its area measured. By pressing **AREA,** and then **EDIT,** the real area value can be recorded. By pressing **VIEW,** the data are recomputed for this leaf area. At this point, the additional data labels (*pr1–pr6*) can be inserted by pressing **AUX** and then **EDIT,** which permits you to type in specific versions of the data label categories.

Finally, you can press **VIEW** and then **STORE** to save the data. You are now ready to return to the **MONITOR** mode and carry out another set of measurements.

OBTAINING DATA OUTPUT

Stored data and calculated physiological values can be printed out when the LI-6200 is linked to a computer or printer.

Printing the Data Page

The LI-6200 can be interfaced to IBM PC microcomputer systems or directly to certain printers (appendix B). When this interface has been created, the data pages obtained in a series of measurements can be transferred or printed by using an **FCT** command. Pressing this key and typing **53** opens a **DUMP PAGES** window in which the number of the first data page is typed, **RTRN** pressed, and the number of the last data page entered.

Interpreting the Data Page Printouts

Each data page lists the parameters of the run, including any special labels inserted, across the top. This is followed by a set of columns giving values from the principal sensors for observations (**OB**) at different times (**TIME**) during the run. These include the sensors described earlier (**MONITOR** mode). In the printout, **PAR** may be labeled **QNTM**, and **EAIR,** the vapor pressure of the air in the cuvette (mb), is given. (The last two columns of raw data are repetition of **EAIR** and **CO2** values.)

Following the raw data columns, means (**1M, 2M, 3M**) are given for sensor values split into three groups. In the case of **CO2** and **EAIR,** the last two columns, respectively, give the means of the change in values from one measurement to the next. Next, the ranges (**1R, 2R, 3R**) of these three sets of values are given.

Finally, at the page bottom, means of the five physiological variables (**PHOTO, CMOL, CINT, RS, CS**) computed from the data are presented.

☘ SUGGESTED ACTIVITIES

1. Obtain measurements of leaf photosynthesis for plants of a given species along a gradient of general temperature or soil moisture conditions. Select a situation in which as many other complicating environmental factors as possible—soil type, slope aspect, slope steepness—are held constant.
2. Compare leaf photosynthesis rates of evergreen and deciduous species occupying the same habitat. If possible, measure photosynthesis for watered and drought-stressed plants of the same species.

3. Compare leaf photosynthesis rates for greenhouse plants of a particular species under different light intensities. Framed pieces of nylon window screening can be used to reduce the intensity of light from a source, different numbers of layers being used to provide a graded series of intensities.

QUESTIONS FOR DISCUSSION

1. What environmental factors in addition to temperature, photosynthetically active radiation intensity, and soil moisture availability might influence photosynthesis?
2. What factors about the geometry of the photosynthetic tissues of a whole plant influence its total photosynthesis? How effectively do you think measurements on individual leaves or leafy branches reflect overall plant performance?
3. What is meant by net photosynthesis, the estimate that the IRGA systems give? Could the system be used to measure leaf respiration? Under what circumstances?

SELECTED REFERENCES

Field, C. B., J. T. Bell, and J. A. Berry. 1989. Photosynthesis: principles and field techniques. Pages 209–253 *in* R. W. Pearcy, J. Ehleringer, H. A. Mooney, and P. W. Rundel, editors. Plant physiological ecology. Chapman & Hall, New York, New York, USA.

Hall, D. O., J. M. O. Scurlock, H. R. Bolhar-Nordenkampf, R. C. Leegood, and S. P. Long, editors. 1993. Photosynthesis and production in a changing environment: a field and laboratory manual. Chapman & Hall, New York, New York, USA.

Nobel, P. S. 1991. Physicochemical and environmental plant physiology. Academic Press, San Diego, California, USA.

EXERCISE 18
Ecotypic Variation in Plants

INTRODUCTION

Ecotypes are local populations of a plant that differ genetically in adaptations or responses to environmental conditions. The traits showing ecotypic variation can be either morphological or physiological. Turesson (1922), who first recognized plant ecotypes, found that the prostrate growth forms of several herbaceous perennials on cliffs along the Baltic seacoast of Sweden—strikingly different from their erect growth inland—were genetically based. His methods were simple; he compared the growth of plants from the two areas side by side in an experimental garden.

Not all differences between populations are ecotypes, of course. Some differences are simply the result of the response of a given genotype to different environmental conditions; such differences are often termed *ecophenes*. One major objective in the study of plant adaptation is to determine the relative importance of ecotypic and ecophenic responses by comparing plants under standardized conditions in growth chambers, greenhouses, or experimental gardens. Both genotypic and phenotypic plasticity are important to plants of habitats that vary spatially and temporally.

PATTERNS OF ECOGENETIC VARIATION

Ecotypic variation occurs in all types of higher plants, from annuals to trees. Ecotypes exist in plants which reproduce by apomixis or self-fertilization, as well as in outcrossing forms. Morphological differences may involve almost any feature of form and structure, vegetative or reproductive. Physiological differences, too, are varied, and may involve germination, photosynthesis, respiration, growth rate, or allocation of photosynthate to various plant components and structures.

Early physiological ecologists concluded that genetic variation within the species was discontinuous in pattern, and that adaptation to conditions along a major environmental gradient produced a series of distinct, genotypically homogeneous populations, each adapted to a segment of the gradient. This idea, in fact, led to the term ecotype. However, the general pattern has more often been found to be clinal—gradual change in genotypic characteristics of populations as the environment changes—although the pattern seen depends somewhat on the scale of analysis. Somewhat discontinuous variation is sometimes found in forms that reproduce vegetatively or by inbreeding, as well as in areas of secondary contact of populations that have been isolated geographically for a significant evolutionary period.

Ecogenetic variation is not restricted to plants. Comparable ecological races occur in animals, and for animals such variation can also involve behavior. Originally, when the phenomena of ecotypes and ecological races were first recognized, these patterns were believed to be somewhat unusual mechanisms of adjustment of species to extreme environments. Now, however, it is evident that most widespread species show significant ecogenetic variation, and that this variation can also occur over very short distances, where strong gradients of conditions affecting individual survival, growth, and reproduction exist.

Brewer (1994): 2:31   Odum (1993) 5:132–133   Ricklefs (1993): 17:312–314   Smith (1992) 3:39–40
Krebs (1994): 7:106–116   Ricklefs (1990): 7:114–115   Smith (1990) 8:155–159   Stiling (1992): 2:32–35

INVESTIGATING PLANT ECOTYPES

Studies of ecotypes involve four major stages: (1) recognizing differences among field populations, (2) determining the degree to which differences have a genetic or environmental basis, (3) testing whether the differences have an adaptive significance, and (4) defining the precise genetic basis of the inherited differences.

Ecotypic variation is most easily studied in annuals that can be grown quickly from seed, or in perennial herbs and grasses that can be grown from seed or transplanted. Weedy plants, although perhaps not representative of all species exhibiting ecotypes, are especially suitable because they are hardy and easy to propagate. Weeds also flourish in disturbed sites, and when disturbance is repetitive or continual most species tend to evolve ecotypes adapted to the particular environment and disturbance regime. In this exercise, we shall concentrate on morphology, since physiological studies of ecotypic differences usually require specialized equipment. Also, we shall concentrate on the first two stages of investigation of ecotypes.

☯ SUGGESTED ACTIVITIES

1. Studies with the ribwort plantain, *Plantago lanceolata* L.

    The ribwort plaintain is well suited to studies of ecotypic and ecophenic adaptation (Van Tienderen 1992). It is a perennial rosette plant, highly variable in its growth and morphology, and virtually ubiquitous in Europe and North America. It grows in many open to lightly shaded habitats such as lawns, athletic fields, meadows, old fields, and the herbaceous understory of open woodlands and parks. Many vegetative differences have a major genetic component, and genetic differences have also been shown for reproductive characteristics such as number of inflorescences per plant, number of capsules per inflorescence, and seed weight (Primack and Antonovics 1981). A strong genetic relationship exists between time of flowering and total reproductive effort by a plant, but reproductive effort is also strongly influenced by environmental conditions (Primack and Antonovics 1981). Environmental conditions also influence leaf and rosette morphology and life history features such as growth rate, fecundity, and mortality (Antonovics and Primack 1982).

    Analysis of field populations. These studies are designed to determine the kinds and extent of differences among populations in nature. Locate plantain populations in sites that differ in environmental factors such as moisture, light intensity, soil type, frequency of mowing or trampling, or density of other herbaceous plants. Examine the plants in these populations and select characteristics for sampling and measurement. Some vegetative and reproductive characteristics to consider are listed below:

Vegetative

Length of longest leaf (cm)	Angle of leaf tip
Maximum width of longest leaf (cm)	Angle of leaf erectness
Leaf width at base (cm)	Number of leaf veins (widest point)
Leaf length/maximum width	Distance between marginal teeth at widest point (mm)
Maximum leaf width/basal width	Prominence of teeth (qualitative scale)
Leaf length/basal width	Leaf mass per unit area (mg)
Widest point to tip/total leaf length	Pubescence (qualitative scale)

Reproductive

Number of flowering stalks	Inflorescence length/thickness
Height of tallest flowering stalk (cm)	Capsules per inflorescence
Length of longest inflorescence (mm)	Seeds per capsule
Inflorescence thickness (mm)	Seed weight (mg)
Stalk height/maximum leaf length	Number of stalk ribs

Some of these characteristics are absolute (length of longest leaf), others derived (measurement ratios). The latter enable differences in the shapes of leaves and inflorescences, independent of actual size, to be evaluated.

Record selected characteristics in table 18.1. Select the plants to be measured randomly, and use objective criteria for picking the leaves and inflorescences to be measured (e.g., the longest leaf or tallest inflorescence). Record the sample size (n), mean ($\overline{X}$), and standard deviation (s) for each characteristic in the table. Use appropriate statistical tests (exercise 7) to compare mean values for plants from different populations.

Comparison of populations under standard conditions. These comparisons require, for each population, an unbiased sample of plants that have grown to maturity under the same conditions. Young plants (not just small plants!) can be transplanted to an experimental garden, greenhouse, or growth chamber where they experience the same conditions. Better still, plants can be grown from seed (which stays viable for 2–3 years), single rhizomes (with the leaves cut back at planting), or leaf cuttings (see Wu and Antonovics 1975). These plants should be started several months ahead of time, and maintained from semester to semester for use by different classes.

Repeat the measurements done for field populations on the experimental plants. Summarize these measurements in table 18.1. Compare the means of the different population samples grown under standard conditions, and of experimental plants and plants from the same field population.

2. Survey of local variation of plant form

Examine the plant species that occur along a strong gradient of environmental conditions in your area. Examples of such gradients include

- Gradient of exposure to salt spray along a seacoast
- Microclimatic gradient from south to north slope
- Soil gradient in parent material (serpentine vs. other), texture (clay to sand), moisture (well drained to waterlogged), salinity (salt marsh to inland)
- Elevational moisture-temperature gradient
- Gradient of grazing intensity by animals, or mowing intensity by humans
- Fire frequency gradient

As a class activity, examine characteristics of as many species as possible at different points along the gradient. How many of the species show apparent differences in morphology or reproductive features? Design an experimental plan to test an apparent difference for one species.

3. Studies of germination physiology

Collect seeds from populations of a local plant that grows in habitats varying in salinity. Test seed germination ability by placing samples of seeds on filter paper moistened with water of differing salinities.

🌀 QUESTIONS FOR DISCUSSION

1. What is the relative importance of ecotypic and ecophenic adjustments of *P. lanceolata* to the habitats involved? Are ecotypic differences more frequent or more pronounced for vegetative or reproductive characteristics?
2. How do you think some of the observed ecotypic or ecophenic differences translate into physiological differences? How do these differences relate to survival and reproduction in the different habitats?
3. How important do you think ecotypic differences might be in selecting material for use in habitat restoration efforts?

Table 18.1. Comparison of vegetative and reproductive characteristics of plants of the ribwort plantain, *Plantago lanceolata,* from field populations and of plants taken from the same populations and grown under standardized greenhouse or experimental garden conditions

		Population A			Population B			Population C		
Characteristic / Habitat		n	$\overline{X}$	s	n	$\overline{X}$	s	n	$\overline{X}$	s
Field Populations											
Laboratory / Experimental Garden											

🌀 SELECTED REFERENCES

Antonovics, J., and R. B. Primack. 1982. Experimental ecological genetics in *Plantago*. VI. The demography of seedling transplants of *P. lanceolata*. Journal of Ecology 70:55–75.

Primack, R. B. 1982. Experimental ecological genetics in *Plantago*. VI. Reproductive effort in populations of the ribwort plantain, *P. lanceolata*. Evolution 36:742–752.

———, and J. Antonovics. 1981. Experimental ecological genetics in *Plantago*. V. Components of seed yield in the ribwort plantain, *Plantago lanceolata* L. Evolution 35:1069–1079.

Schlichting, C. D. 1986. The evolution of phenotypic plasticity in plants. Annual Review of Ecology and Systematics 17:667–693.

Schwagerle, K. E., and F. A. Bazzaz. 1987. Differentiation among nine populations of phlox: response to environmental gradients. Ecology 68:54–64.

Teramura, A. H., and B. R. Strain. 1979. Localized populational differences in the photosynthetic response to temperature and irradiance in *Plantago lanceolata*. Canadian Journal of Botany 57:2559–2563.

Teramura, A. H., J. Antonovics, and B. R. Strain. 1981. Experimental ecological genetics in *Plantago*. IV. Effects of temperature on growth rates and reproduction in three populations of *Plantago lanceolata* L. (Plantaginaceae). American Journal of Botany 68:425–434.

Turesson, G. 1922. The genotypical response of plant species to habitat. Heriditas 3:211–350.

Van der Toorn, J., and P. H. Van Tiederen. 1992 Ecotypic differentiation in *Plantago lanceolata*. *In* P. J. C. Kuiper and M. Bos, editors. *Plantago—A multidisciplinary study*. Springer Verlag, Berlin, Germany.

Van Groenendael, J. M. 1985. Differences in life histories between two ecotypes of *Plantago lanceolata*. Pages 51–67 *in* J. White, editor. Studies on plant demography. Academic Press, London, England.

———. 1988. The contrasting dynamics of two populations of *Plantago lanceolata* classified by age and size. Journal of Ecology 76:585–599.

Van Tienderen, P. H. 1992. Variation in a population of *Plantago lanceolata* along a topographic gradient. Oikos 64:560–572.

———, and J. Van der Toorn. 1991. Genetic differentiation between populations of *Plantago lanceolata*. I. Local adaptation in three contrasting habitats. Journal of Ecology 79:27–42.

———. 1991. Genetic differentiation between populations of *Plantago lanceolata*. II. Phenotypic selection in three contrasting habitats. Journal of Ecology 79:43–59.

Wu, L., and J. Antonovics. 1975. Experimental ecological genetics in *Plantago*. I. Induction of leaf shoots and roots for large scale vegetative propagation tolerance and testing in *P. lanceolata*. New Phytologist 75:277–282.

Age, Body Size, and Growth Curves

INTRODUCTION

The ages and body sizes of individuals in a population can influence many aspects of its dynamics. For example, individuals of different body size may use vastly different quantities of resources, and those of different age may differ greatly in reproduction. If a reliable mathematical relationship between age and body size can be derived, one can use this relationship to predict future patterns of reproduction or use of resources. With such a relationship one also can determine the age structure of a population at a given time from size-frequency data. The growth curves of organisms—both plants and animals—are thus of considerable interest to ecologists.

The most useful growth curves are the S-shaped logistic and Gompertz growth curves and the convex Brody-Bertalanffy growth curve. This exercise describes how the parameters of these curves can be estimated from data on the sizes of individuals (or cohorts) in natural populations.

MATHEMATICAL MODELS OF AGE AND BODY SIZE

To determine the parameters for these growth equations, data are required on S_0, the size of individuals when they first enter the population (birth, hatching, germination, settling from plankton, etc.), and for pairs of size values of individuals at the beginning, S_1, and end, S_2, of time intervals of fixed length (e.g., 1 year). These size values may be for individuals that are marked and are actually observed at different times, t_1 and t_2, or they may be sizes of individuals that can be aged by growth rings or other features and that differ in age by one time interval. Midpoint values of modal size classes differing in age by a constant time period can also be utilized in these analyses. An effort should be made to obtain pairs of S_1 and S_2 values spanning most of the size range of individuals in the population. From these data the asymptotic maximum size, S_∞, and growth constants can be computed.

Logistic Growth Curve

Stated in terms of body growth and age, the logistic growth equation, which describes an antisymmetrical (top and bottom halves of curve identical in form, but flipped horizontally and vertically), S-shaped curve of increase in size, can be written:

$$\frac{1}{S_t} = \frac{1}{S_\infty}(1 - be^{-Kt})$$

Where: t = Age in units of the chosen time interval
K = Logistic growth constant
b = Logistic scaling constant

The critical parameters b and K of the logistic growth equation can be determined by plotting $1/S_2$ against $1/S_1$ in figure 19.1. If the body growth of the species follows a logistic curve, these values will fall in a straight line, the slope of which is m and the intercept M. These values can be estimated by fitting the points by a line

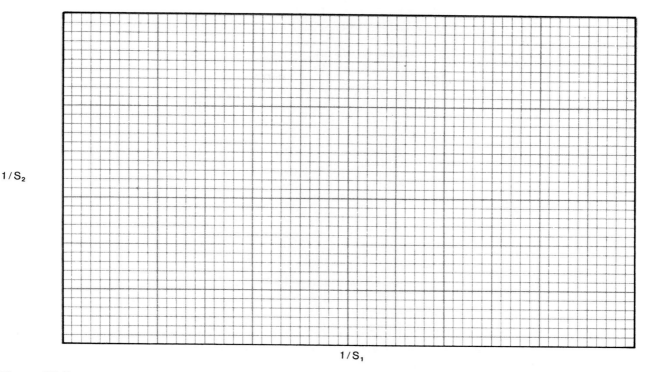

$1/S_2$

$1/S_1$

Figure 19.1.
Graphical estimation of *m* and *M* for determination of parameters of the logistic growth curve

drawn by eye and reading the intercept and calculating the slope ("rise" of the line divided by its "run"). The intercept and slope can also be estimated more objectively by calculating the linear regression relation between $1/S_2$ and $1/S_1$, as outlined in table 19.1. Using these values, the parameters of the logistic growth equation are calculated:

$$K = -\ln(m)$$
$$S_\infty = \frac{1}{M/(1 - m)}$$
$$b = (S_o - S_\infty)/S_o$$

Data on body mass or volume are often fitted best by the logistic growth curve.

Gompertz Growth Curve

The Gompertz growth curve follows the equation

$$S_t = S_\infty^{(1 - be^{-Kt})}$$

Where: K = Gompertz growth constant
b = Gompertz scaling constant

This growth curve is sigmoid in general form, but not antisymmetrical like the logistic growth curve. It describes growth patterns in which different, opposing relationships regulate growth during early and late phases of development.

The Gompertz relationship often fits body mass versus age data, as well as linear growth measures for species that change considerably in their ecology from early life to late development (e.g., species markedly changing feeding mode or habitat during development).

Table 19.1. Estimation of m, M, and other parameters of the logistic growth curve by linear regression analysis

	X $(1/S_1)$	Y $(1/S_2)$			
			$m = \dfrac{\Sigma XY - \dfrac{(\Sigma X)(\Sigma Y)}{n}}{\Sigma X^2 - \dfrac{(\Sigma X)^2}{n}}$	=	_____
			$M = \overline{Y} - m\overline{X}$	=	_____
			$K = -\ln(m)$	=	_____
			$S_\infty = \dfrac{1}{M/(1-m)}$	=	_____
			$b = (S_0 - S_\infty)/S_0$	=	_____

	_____	_____
ΣX ΣY	_____	_____
$\overline{X}$, $\overline{Y}$	_____	_____
ΣX^2, ΣY^2	_____	_____
ΣXY	_____	

Brody-Bertalanffy Growth Curve

The Brody-Bertalanffy growth curve is based on the equation

$$S_t = S_\infty (1 - be^{-Kt})$$

Where: K = Brody-Bertalanffy growth constant
b = Brody-Bertalanffy scaling constant

Rather than being S-shaped, this curve is convex in form, rising rapidly in early ages and slowing as body size approaches the maximum. The parameters K and b are estimated in a fashion similar to that for the logistic curve. The pairs of S_1 and S_2 values are graphed in figure 19.2; if growth follows the Brody-Bertalanffy relationship the plotted points will fall in a straight line. The slope, m, and intercept, M, of this line can be estimated as described earlier (from a line drawn by eye, or by linear regression, using table 19.2). With these values, equation parameters are calculated by

$$K = -\ln(m)$$
$$S_\infty = M/(1 - m)$$
$$b = (S_\infty - S_o)/S_\infty$$

The Brody-Bertalanffy growth curve is best suited to linear size measurements such as length, height, or diameter and usually does not fit mass or volume data well.

Computing Theoretical Growth Curves

The parameters of the logistic and Brody-Bertalanffy growth equations can be computed by linear regression techniques, as outlined in tables 19.1 and 19.2. The parameters of the Gompertz equation are best obtained with the software noted below. Once the parameters of the logistic, Gompertz, or Brody-Bertalanffy curves have been calculated, values along theoretical growth curves can be computed simply by selecting ages (t's) at intervals between 0 and the maximum believed likely, and calculating S_t's for each.

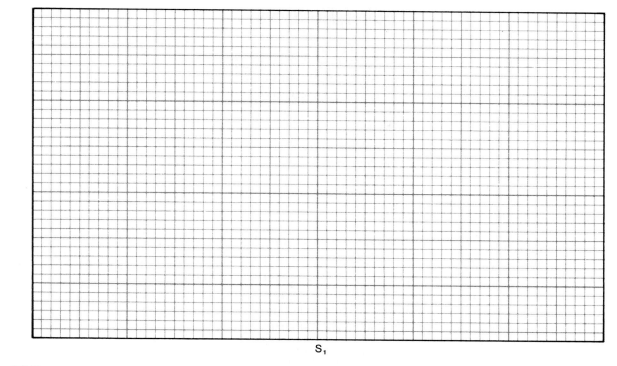

Figure 19.2.
Graphical estimation of *m* and *M* for determination of parameters of the Brody-Bertalanffy growth curve

Table 19.2. Estimation of *m*, *M*, and other parameters of the Brody-Bertalanffy growth curve by linear regression analysis

	X (S_1)	Y (S_2)	

$$m = \dfrac{\Sigma XY - \dfrac{(\Sigma X)(\Sigma Y)}{n}}{\Sigma X^2 - \dfrac{(\Sigma X)^2}{n}} = \underline{\hspace{2cm}}$$

$$M = \bar{Y} - m\bar{X} \qquad = \underline{\hspace{2cm}}$$

$$K = -\ln(m) \qquad = \underline{\hspace{2cm}}$$

$$S_\infty = M/(1-m) \qquad = \underline{\hspace{2cm}}$$

$$b = (S_\infty - S_o)/S_\infty \qquad = \underline{\hspace{2cm}}$$

$\Sigma X, \Sigma Y$ _____ _____

$\bar{X}, \bar{Y}$ _____ _____

$\Sigma X^2, \Sigma Y^2$ _____ _____

ΣXY _____

SOFTWARE FOR GROWTH CURVE ANALYSIS

Simple linear regression modules in software packages such as E-Z Stat and Ecological Analysis–PC (Vol. 1) can be used to compute values of m and M (appendix B).

A program in BASIC for Macintosh microcomputers is available from T. A. Ebert of San Diego State University (appendix B). This program computes the parameters for the logistic, Gompertz, and Brody-Bertalanffy growth curves, together with sets of values for plotting curves with these parameters. The program enables computation of parameters of growth curves from data taken at irregular, as well as regular, intervals.

Actual and theoretical t and S_t values can easily be entered or imported into several graphing programs, such as Cricket Graph, and growth curves plotted for comparison.

⊘ SUGGESTED ACTIVITIES

1. Obtain growth data that allow S_1's and S_2's to be determined. Lizards, snails, or other animals can be captured, measured, and marked. Growth in size can be determined by recapturing and remeasuring them at a later time. Annual plants can be measured at weekly intervals during the growing season, with height or diameter data utilized for growth curve analysis. Samples of shells of snails or bivalve mollusks with annual growth rings can be examined, and estimates of size and age obtained.

2. Analyze a set of data for tree age and height. Estimate age from tree-ring data or, for some conifers, number of branch whorls.

3. Analyze a published set of data on age and body mass or linear dimensions. Many sets of growth data are available in the literature, especially in journals of marine ecology, herpetology, ichthyology, and fisheries biology. Recent articles by Belk and Hales (1993), Horn (1993), Iverson et al. (1991), Marshall and Echeverria (1992) and Terceiro and Ross (1993) give sets of numerical data suitable for analysis by the above methods.

⊘ QUESTIONS FOR DISCUSSION

1. What deviations from regular growth curves are likely to result from the initiation of reproduction by organisms at a certain age?

2. What environmental factors probably influence the growth curves of different populations of a particular species? How might competition influence the values of the growth constants in these curves?

3. What cautions should be exerted in inferring age from size data by means of curves such as these?

⊘ SELECTED BIBLIOGRAPHY

Belk, M. C., and L. S. Hales, Jr. 1993. Predation-induced differences in growth and reproduction of bluegills (*Lepomis machrochirus*). Copeia 1993:1034-1044.

Chow, V. 1987. Patterns of growth and energy allocation in northern California populations of *Littorina* (Gastropoda: Prosobranchia). Journal of Experimental Marine Biology and Ecology 110:69-89.

Ebert, T. A. 1980. Estimating parameters in a flexible growth equation, the Richards function. Canadian Journal of Fisheries and Aquatic Sciences 37:687-692.

Horn, P. L. 1993. Growth, age structure, and productivity of ling, *Genypterus blacodes* (Ophidiidae), in New Zealand waters. New Zealand Journal of Marine and Freshwater Research 27:385-397.

Iverson, J. B., E. L. Barthelmess, G. R. Smith, and C. E. deRivera. 1991. Growth and reproduction in the mud turtle *Kinosternon hirtipes* in Chihuahua, Mexico. Journal of Herpetology 25:64-72.

James, C. D. 1991. Growth rates and ages at maturity of sympatric scincid lizards (*Ctenotus*) in central Australia. Journal of Herpetology 25:284-295.

Marshall, W. H., and T. W. Echeverria. 1992. Age, length, weight, reproductive cycle and fecundity of the monkeyface prickleback (*Cebidichthys violaceus*). California Fish and Game 78:57–64.

Moreau, J. 1987. Mathematical and biological expression of growth in fishes: Recent trends and further developments. Pages 81–113 *in* R. C. Summerfelt and G. E. Hall, editors. Age and growth of fish. Iowa State University Press, Ames, Iowa, USA.

Piantadosi, S. 1987. Generalizing growth functions assuming parameter heterogeneity. Growth 51:50–63.

Russell, M. P. 1987. Life history traits and resource allocation in the purple sea urchin *Strongylocentrotus purpuratus* (Stimpson). Journal of Experimental Marine Biology and Ecology 108:199–216.

Terceiro, M., and J. L. Ross. 1993. A comparison of alternative methods for the estimation of age from length data for Atlantic coast bluefish (*Pomatomus saltatrix*). Fishery Bulletin 91:534–549.

EXERCICE 20

Behavioral Preference Analysis

INTRODUCTION

Determining the preferences of organisms for habitat conditions and resources is a major objective of behavioral ecology. One of the best ways to study habitat selection is to determine the conditions chosen by individuals in gradients created in the laboratory or in the field. The selected condition is termed the *preferendum*. Simple gradients for many habitat factors can be created with materials available in most biology laboratories. Experiments with gradients of a single factor can lead to tests in which several factors are presented in combination, or in which the influence of biotic interactions on preference is tested. Experiments also can be designed to determine the types of foods, hosts, or other resources that a species prefers.

In this exercise, we limit our attention to experiments in which the investigator controls the availability of habitat conditions or resources. Techniques of evaluating preferences by animals in nature for habitats or resources as available in nature are discussed by Manly et al. (1993).

DESIGN AND ANALYSIS OF HABITAT PREFERENCE EXPERIMENTS

Habitat preference experiments are typically carried out in a chamber or enclosure in which one or more factors can be varied systematically. Preference experiments can be structured in two different ways. For some variables, such as light intensity or temperature, a continuous gradient from one extreme to another can be created. The gradient chamber can be divided into two or more sections of equal area. For other factors, such as substrate texture, it is easier to offer animals a choice of two or more distinct conditions. In this case, the areas of the chamber with each condition can be made equal. In either case, the data recorded are the numbers of individuals selecting each section of the gradient. Making the areas of the gradient sections equal permits the assumption that the same number of individuals will choose each section, if no preference exists.

A number of precautions should be observed in designing and conducting preference tests. The gradient chamber should be designed so that strong responses to complicating influences do not mask a response to the experimental factor. For example, many invertebrates have a strong negative phototaxis and a positive thigmotaxis, and become inactive or seek the edges or corners of the chamber if the experiment is carried out in bright light. This problem can be reduced or eliminated by conducting the experiment in dim light or darkness and by using a chamber with a curved bottom or rounded bottom-side junctions.

The number of individuals introduced during a single run should be determined on the basis of potential for positive or negative interactions among individuals. For testing the null hypothesis of no individual preference, such interactions must not occur. For animals with strong social behavior, this may mean that each test must be done with a single individual.

Care should also be taken when placing animals in the gradient chamber. In a continuous gradient or a gradient with several different sections, animals should be introduced in equal numbers to the various gradient

Begon et al. (1990): 9:298-304   Krebs(1994): 3:40-43, 5:61-74
Brewer (1994): 2:13-20   Stiling (1992): 8:153-174

sections. In a preference situation with only two sections, animals should be introduced near the boundary between the two sections so that they have an opportunity to encounter both conditions.

The duration of the gradient test must also be considered. This will vary with the factor being tested and with the mobility of the animals. Adequate time must be allowed, however, for the animals to recover from initial fright, shock, or disorientation associated with introduction into the gradient. For most animals, at least one-half hour should be allowed before data on orientation are taken.

A control should be conducted for each experiment, under conditions identical to those of the experimental run except that the experimental variable is held constant. The control functions to reveal any unsuspected factors, such as a slight inclination of the chamber, that might influence the animals during the experimental run.

Some interesting elaborations of simple gradient preference tests can be designed. The influence of intraspecific social behavior on habitat preference can be tested by runs in which different numbers of individuals are placed in the chamber at once. The preferendum of one species when alone can be compared with that in the presence of a second species. The second species, for example, might be a potential predator or competitor. This type of experiment was effectively used by Teal (1958) and Sutherland (1986a, 1986b) to evaluate the role of interspecific factors in the distribution of species of fiddler crabs and salamanders, respectively.

Constructing Habitat Gradients

Critical studies of gradient preferences require a carefully designed experimental chamber. A description of a chamber in which gradients or constant conditions can be simultaneously maintained for temperature, relative humidity, light intensity, and light wavelength is given by Platt et al. (1957). Much less elaborate equipment can be used for exploratory studies and class experiments, however. Suggestions for the construction of simple environmental gradients, along with selected references to more detailed descriptions of gradient apparatus, are given below.

Temperature.

For aquatic animals, either horizontal or vertical temperature gradients can be constructed. A horizontal gradient can be set up in a long, narrow trough, through which a small, constant flow is maintained. Water is introduced from an ice bath at one end and warmed by a series of separately controlled heating units as it passes to the opposite end, where it is siphoned off. Vertical temperature gradients can be created in a deep tank by passing two coiled copper tubes vertically through the tank. Hot water is passed downward through one tube and cold water upward through the other. A temperature gradient for terrestrial animals can be constructed from a long metal trough with a glass cover that has periodic openings for thermometers or thermistor leads. The ends of the trough are enclosed by watertight compartments through which hot or cold water is circulated (Getz 1959, Van der Schalie and Getz 1963). For small aquatic or terrestrial invertebrates, a satisfactory gradient can be constructed by enclosing one end of a long tube in an ice pack and the other end in a heating jacket. Temperature gradients for reptiles and amphibians can be created in a long box with a heat lamp at one end.

Humidity.

Humidity gradients can be set up in a long tube, or in a series of connected chambers, by placing a drying agent (silica jel, $CaCl_2$, KOH, NaOH, H_2SO_4) at one end and a humidifying agent (beaker of water with a series of filter-paper wicks to increase evaporating surface) at the other. Preference tests with distinct humidity levels can be carried out in a choice chamber having a false floor of a porous material such as gauze. The space beneath the false floor is divided into separate compartments in which solutions with different humidity equilibria with air are placed (Lagerspetz 1963). Several chemical agents can be used for these solutions, including H_2SO_4 and KOH (Winston and Bates 1960, Reichle 1967). Humidities in specific portions of gradients or choice chambers are measured easily with papers impregnated with cobalt thiocyanate (Solomon 1957).

Light intensity.

Gradients of light intensity can be created by placing a light at one end of a long chamber and enclosing or blackening the other end. Light intensity also can be varied with an overhead light source and a chamber having a lid transparent at one end and blackened at the other (Warburg 1964).

Body contact.

Thigmotactic responses to substrates offering different amounts of body contact can be tested in a preference chamber in which half of the floor is smooth and half covered with broken pieces of clear glass. This arrangement allows thigmotactic responses to be compared in the light and in the dark without altering the experimental chamber. Experiments can also be carried out in a chamber having a false ceiling that can be lowered over one part of the chamber so that animals can obtain contact with the dorsal body surface (Friedlander 1964).

Salinity.

Salinity preferences for amphibious animals such as gastropods or small crabs can be determined by placing water of different salinities into an aquarium that has the bottom divided by a number of low partitions. These prevent mixing of the water but allow animals to move freely between sections (Teal 1958).

Chemical factors in water.

Gradients of chemical factors, such as oxygen concentration and pH, that do not greatly influence water density can be produced in a long tank by introducing water with the same temperature but different levels of the chemical factor at opposite ends of the tank.

Substrate slope.

Geotropic responses can be studied simply by tilting the gradient container at various angles.

Substrate moisture and texture.

Preferences for substrates of different moisture level or texture are best tested in a chamber with a partitioned bottom that prevents mixing of the substrate particles or passage of moisture except at the point of contact directly above the partition (Teal 1958). With soils or sediments of different type, care should be taken that the materials do not differ significantly in factors other than particle size (e.g., presence of chemical substances in one substrate and not in the other).

Substrate color.

Preferences for substrate color or intensity of color can be tested by use of painted substrate panels. Care should be taken, however, that the surfaces do not differ in temperature, texture, or chemical factors.

Analysis of Habitat Gradient Data

The results of control and experimental tests of simple gradient preferences with individuals of a single species can be tested by chi-square goodness-of-fit procedures (see exercise 7). With experiments structured as described above, equal numbers of animals should select each part of the gradient if no differential response occurs to the experimental variable. Values expected on the basis of no differential response, therefore, can be calculated by dividing the total number of individuals counted during the run by the number of sections in the gradient. Results of separate tests of the responses of one species with and without a second species present can be tested by a chi-square contingency analysis (see exercise 7).

DESIGN AND ANALYSIS OF RESOURCE PREFERENCE EXPERIMENTS

Resources are features of the environment that are consumed by organisms (e.g., food) or are utilized in a way that makes them temporarily unavailable to other individuals (e.g., nest cavities). Some types of resources are best treated as discrete variables and quantified by counting (e.g., large seeds, nuts, prey of a carnivore) and others as continuous variables that can be quantified by measurements (e.g., volume of nectar or mass of small seeds consumed by test animals). Testing preferences for resources, such as foods, also is complicated by the fact that the availability of different resources may be changed by the activities of the test animals.

Some resource preference tests can be designed so that resource units that are taken by an animal are replaced immediately. Seed dispensers or hummingbird feeders can be designed so that the quantities of seeds or nectar removed are automatically replaced, and the amounts of different materials kept constant.

Creating Resource Preference Tests

Cafeteria-style selection experiments.

Food resources are often presented so that one or more animals can select from an array of items presented in equal amounts (numbers or masses). Different types of seeds might be placed in an array to test the food preferences of ants, desert rodents, or granivorous birds. Hummingbird feeders with simulated flower corollas of different colors might be presented in an array to test preferences for flower color.

Resource location experiments.

Some resource experiments are designed to test whether resources are utilized to the same degree in various locations. For example, nest boxes for house wrens or bluebirds might be placed at different distances from the edge of a woodlot, or at different heights, to test the importance of these variables in use of the nest-cavity resource. A similar test might involve placing artificial bird's nests containing quail eggs in different locations to test the risk of nest predation. In these cases the result might be all-or-none: the resource is either used or not used. Equal quantities of a food resource might also be made available in a similar fashion. In this case, the number or quantity of the resource removed in the different locations during replicate runs constitutes the test data.

Analysis of Resource Preference Data

In cafeteria-style preference experiments, the numbers, quantities, or proportions of different items removed constitute the basic sample data. The sample size is the number of individuals whose preferences are tested or the number of test runs, not the number of items selected. When a cafeteria-style experiment is done with immediate replacement, the most appropriate statistical test is a one-way parametric or nonparametric analysis of variance (see exercise 8).

For cafeteria-style experiments in which immediate replacement is not practical, the cumulative proportion of each item removed can be recorded at various times. With these values, a graph similar to that of figure 20.1 can be constructed. For each item, the total area under the cumulative proportion curve can be calculated by resolving the area beneath the curve into triangles [area = 0.5(base) (height)] or trapezoids [area = (base) (mean height)] and summing the set of areas. This value can be standardized so that the curve with the largest underarea is set equal to 1.00, and the other curves expressed as a decimal fraction of this maximum. This standardized index is known as *Rogers' index of preference* (Krebs 1989). Unstandardized or standardized indices of preference from replicate test runs can be analyzed by a one-way nonparametric analysis of variance (see exercise 8).

Resource location experiments with an all-or-none outcome can be analyzed by a contingency chi-square (see exercise 7). Those involving removal of numbers or quantities of material can be analyzed by the procedures outlined for cafeteria-style tests.

SOFTWARE FOR ANALYSIS OF PREFERENCE EXPERIMENTS

Krebs (1989) provides a program for calculation of Rogers' index of preference (see appendix B).

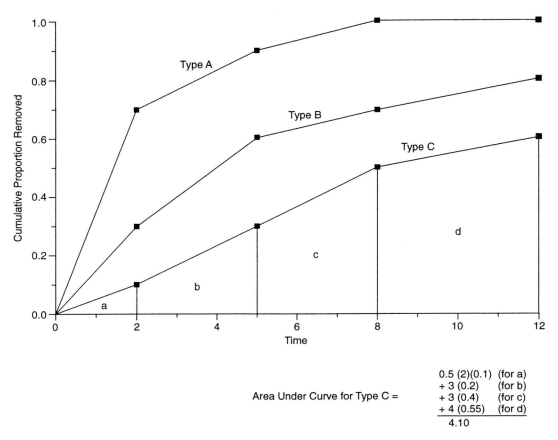

Area Under Curve for Type C =
$$
\begin{aligned}
&0.5\ (2)(0.1) &&\text{(for a)}\\
+\ &3\ (0.2) &&\text{(for b)}\\
+\ &3\ (0.4) &&\text{(for c)}\\
+\ &4\ (0.55) &&\text{(for d)}\\
\hline
&4.10
\end{aligned}
$$

Figure 20.1.
Cumulative proportion of test items of different types removed over time in a cafeteria preference test. Calculation of the total area beneath the curve for one type of item is illustrated

🐾 SUGGESTED ACTIVITIES

Laboratory or field experiments should be combined with field studies of local distribution of the species. Results bearing on questions of competition and ecological isolation may be obtained from comparative studies of the local distribution and preferences of related or ecologically similar species.

1. Plan and conduct habitat preference studies with a locally abundant aquatic or terrestrial animal. Many invertebrates and small vertebrates, both aquatic and terrestrial, are suitable for preference studies. Marine invertebrates such as brittle stars (*Ophiothrix, Ophioplocus,* etc.), gastropods (*Littorina, Tegula, Certhidea,* and many others), small crabs (*Uca, Sesarma, Pachygrapsus*), isopods (*Ligia*), and other crustacea can be used. The brittle star *Ophiothrix* responds well in gradients of light intensity, substrate slope, substrate texture, and body contact. Freshwater crayfish, immature and adult aquatic insects, and small fish (darters, small sunfish, etc.) are also satisfactory. Many ground-living terrestrial invertebrates, including isopods (*Armadillidium, Oniscus, Porcellio*), millipeds, and many insects, can be used. Most isopods respond very well in gradients of temperature, humidity, light, substrate moisture, and substrate type. Small vertebrates such as lizards, salamanders, and anurans can also be used. Lizards are interesting to study in temperature gradients, as species often sort themselves out in different regions of the gradient according to different temperature preferenda.

    Record data for simple preference experiments in table 20.1. Subdivide the data portion of this table with penciled lines into the same number of sections as the experimental gradient.

Table 20.1. Results of preference tests with various environmental factors for individuals of a single species

Gradient	Test Time	Species	Portions of Gradient with Numbers of Individuals Selecting Each	Chi-Square + d.f.	P

2. Conduct a second series of tests in which the number of individuals simultaneously introduced into the chamber is varied, or in which individuals of a second species, which may be a competitor or predator, are introduced. Record data for preferences under these conditions in table 20.2.

3. Conduct a cafeteria-style or resource preference test or resource location test. Suggested designs include

- Sets of hummingbird feeders with different "flower" colors placed at different heights or distances from a forest edge
- Cafeteria arrangements of different seeds or fruits offered in situations of differing accessibility to rodents
- Cafeteria arrangements of small seeds placed in the immediate vicinity of harvester ant nests
- Artificial bird nests placed at different heights and in different locations with respect to the border of a natural area of habitat

🔾 QUESTIONS FOR DISCUSSION

1. Design an experiment to determine the possible relationship between acclimation and preferendum for various environmental factors.
2. In what ways can differences in preferenda for certain environmental factors be important in ecological isolation of similar species?
3. What differences in gradient responses might be expected between forms inhabiting areas with little environmental fluctuation and those inhabiting areas with a great deal of environmental fluctuation?

Table 20.2. Results of preference tests with various environmental factors for individuals of a species alone and in the presence of a second species

Gradient	Test Time	Species or Species Combination	Portions of Gradient with Numbers of Individuals Selecting Each	Chi-Square + d.f.	P

SELECTED BIBLIOGRAPHY

Barry, R. E., Jr., and E. N. Francq. 1982. Illumination preference and visual orientation of wild-reared mice, *Peromyscus leucopus*. Animal Behaviour 30:339–344.

Brewer, R. H. 1978. Larval settlement behavior of the jellyfish *Aurelia aurita* (Linnaeus) (Scyphozoa: Semaeostomeae). Estuaries 1:120–122.

Friedlander, C. P. 1964. Thigmokinesis in woodlice. Animal Behaviour 12:164–174.

Getz, L. L. 1959. Notes on the ecology of slugs: *Arion circumscriptus, Deroceras reticulatum,* and *D. laeve.* American Midland Naturalist 61:485–498.

Gillis, J. E. 1982. Substrate colour-matching cues in the cryptic grasshopper *Circotettix rabula rabula* (Rehn & Hebard). Animal Behaviour 30:113–116.

Krebs, C. J. 1989. Ecological methodology. Harper & Row, New York, New York, USA.

Lagerspetz, K. 1963. Humidity reactions of three aquatic amphipods, *Gammarus duebeni, G. oceanicus,* and *Pontoporeia affinis* in the air. Journal of Experimental Biology 40:105–110.

Manly, B. F. J. 1993. Comments on design and analysis of multiple-choice feeding-preference experiments. Oecologia 93:149–152.

———, L. L. McDonald, and D. L. Thomas. 1993. Resource selection by animals. Chapman & Hall, London, England.

Marquis, R. J., and H. E. Braker. 1987. Influence of method of presentation on results of plant-host preference tests with two species of grasshoppers. Entomologia Experimentalis et Applicata 44:59–63.

Mushinsky, H. R., and E. D. Brodie, Jr. 1975. Selection of substrate pH by salamanders. American Midland Naturalist 93:440-443.

Reichle, D. E. 1967. The temperature and humidity relations of some bog Pselaphid beetles. Ecology 48:208-215.

Richkus, W. A. 1981. Laboratory studies of intraspecific behavioral interactions and factors influencing tidepool selection of the woolly sculpin, *Clinocottus analis*. California Fish and Game 67:187-195.

Roa, R. 1992. Design and analysis of multiple-choice feeding-preference experiments. Oecologia 89:509-515.

Rudolph, D. C. 1978. Aspects of the larval ecology of five plethodontid salamanders in the western Ozarks. American Midland Naturalist 100:141-159.

Solomon, M. E. 1951. Control of humidity with potassium hydroxide, sulfuric acid, and other solutions. Bulletin of Entomological Research 42:543-554.

———. 1957. Estimation of humidity with cobalt thiocyanate papers and permanent colour standards. Bulletin of Entomological Research 48:489-506.

Southerland, M. T. 1986a. Behavioral interactions among four species of the salamander genus *Desmognathus*. Ecology 67:175-181.

———. 1986b. Coexistence of three congeneric salamanders: the importance of habitat and body size. Ecology 67:721-728.

Spotila, J. R. 1972. Role of temperature and water in the ecology of lungless salamanders. Ecological Monographs 42:95-125.

Teal, J. M. 1958. Distribution of fiddler crabs in Georgia salt marshes. Ecology 39:185-193.

Van der Schalie, H., and L. L. Getz. 1963. Comparison of temperature and moisture responses of the snail genera *Pomatopsis* and *Oncomelania*. Ecology 44:73-83.

Warburg, M. R. 1964. The responses of isopods toward temperature, humidity, and light. Animal Behaviour 12:175-186.

Winston, P. W., and D. H. Bates. 1960. Saturated solutions for the control of humidity. Ecology 41:232-237.

EXERCISE **21**

Intrapopulation Dispersion Analysis

INTRODUCTION

Interactions of individuals with their environment result in a certain pattern of dispersion, or distribution in space. This pattern can vary from a random scattering of individuals through the available area or volume to a pattern tending toward either uniform spacing or aggregation. Dispersion pattern often reflects factors that influence the behavior, growth, or survival of individuals.

Pattern can be analyzed with unit-area or unit-volume sampling data by comparing observed data with values expected from the Poisson distribution, which gives the expectation for a random pattern. Pattern can also be analyzed by plotless techniques that use distances from random points to the nearest individuals, or distances between individuals and their nearest neighbors. These techniques test whether a significant deviation occurs toward either uniform spacing or aggregation.

The dispersion of individuals is often aggregated. When aggregation is detected, however, the above methods do not give a good description of its intensity. The negative binomial distribution can be fitted to many sets of sampling data from aggregated populations. One parameter of this distribution, *k,* has been suggested as a measure of the degree of aggregation.

QUADRAT TECHNIQUES OF DISPERSION ANALYSIS

Quadrat samples, combined with predictions of the Poisson distribution, can be used to assess pattern in populations with a relatively low density of individuals. Poisson techniques are invalid for situations such as dense grasslands or dense colonies of sessile invertebrates, however.

The results of quadrat analyses depend on the size of the sampling unit. Analysis of the same population with quadrats of different size may give quite different results. In general, quadrats of a certain size measure pattern as influenced by factors operating over areas of related magnitude. Quadrats of large size, containing many individuals, give a measure of pattern as influenced by general environmental factors. If one is interested in factors that influence the positions of neighboring individuals, quadrat size should be small enough that many quadrats contain no individuals. Quadrat samples should be taken randomly (see exercise 4).

To analyze such data, observed frequencies are compared to frequencies expected on the basis of random pattern (table 21.1). The expected decimal fractions of quadrats containing numbers of individuals from 0 through the highest numbers commonly observed are calculated from the expression

$$e^{-\bar{x}} \cdot \frac{\bar{X}^x}{X!}$$

Where: e = Base of natural logarithms
X = Number per sample (quadrat)
$\bar{X}$ = Mean number of individuals per sample

Begon et al. (1990): 5:159   Pianka (1994): 10:202–203   Ricklefs (1993): 14:246–250   Smith (1992): 10:158–159
Brewer (1994): 5:118–126   Ricklefs (1990): 15:280–286   Smith (1990): 14:324–336

Table 21.1. Comparison of observed frequencies of quadrats containing various numbers of individuals and frequencies expected on the basis of Poisson distribution

Number per Quadrat	Poisson Expression	Poisson Relative Frequency	Expected Number of Quadrats (E)	Observed Number of Quadrats (O)	(O-E)	$\dfrac{(O-E)^2}{E}$
0	$e^{-\bar{x}} \cdot \dfrac{\bar{x}^0}{0!}$					
1	$e^{-\bar{x}} \cdot \dfrac{\bar{x}^1}{1!}$					
2	$e^{-\bar{x}} \cdot \dfrac{\bar{x}^2}{2!}$					
3	$e^{-\bar{x}} \cdot \dfrac{\bar{x}^3}{3!}$					
4	$e^{-\bar{x}} \cdot \dfrac{\bar{x}^4}{4!}$					
5	$e^{-\bar{x}} \cdot \dfrac{\bar{x}^5}{5!}$					
6	$e^{-\bar{x}} \cdot \dfrac{\bar{x}^6}{6!}$					
7	$e^{-\bar{x}} \cdot \dfrac{\bar{x}^7}{7!}$					
8	$e^{-\bar{x}} \cdot \dfrac{\bar{x}^8}{8!}$					
9	$e^{-\bar{x}} \cdot \dfrac{\bar{x}^9}{9!}$					
>9						
Total		1.0000				

The expected decimal fraction for the last category must be determined in a different manner, since it is actually the fraction of quadrats containing X or more individuals. It is calculated by totaling the decimal fractions for categories computed by the above expression and subtracting this total from 1.0.

Once expected decimal fractions of quadrats have been obtained, they are multiplied by the observed number of quadrats (n) to obtain the expected actual numbers (table 21.1).

The direction in which observed values deviate from those expected is determined by inspection. If the deviation is toward a uniform distribution, the observed data will contain fewer high and low values than expected. If the deviation is toward a aggregated distribution, there will be more high and low values than expected.

An index of deviation, D, of the observed (O) from the expected (E) numbers of quadrats can be calculated by the equation

$$D = 1 - \sum_{k=1}^{w} \min (E/\Sigma E, O/\Sigma O)$$

where: w = Largest number of individuals observed in any quadrat

for quadrats containing one or more individuals (Hurlbert 1990). This index varies from 0 to 1, showing the maximum value when all individuals encountered are in a single quadrat.

The degree of correspondence between observed and expected sets of data can be tested by a chi-square calculation (see exercise 7). For this test, the number-per-quadrat categories at the upper end of the distribution should be grouped so that each expected category has a value of at least 5. The degrees of freedom for the calculated chi-square value are two less than the number of observed-expected comparisons.

Note that the technique of comparison of the variance and mean of a set of quadrats, a method widely used in the past, is not an acceptable approach for testing for uniformity or aggregation (Hurlbert 1990).

PLOTLESS TECHNIQUES OF DISPERSION ANALYSIS

Since quadrat methods of pattern analysis give results that are influenced by quadrat size, the detection of nonrandomness often depends on correct selection of quadrat size. To overcome this difficulty, plotless techniques for measuring dispersion pattern have been developed. These techniques use point-to-plant or plant-to-plant distance measurements.

Point-to-Plant Distance Ratio

This test (Holgate 1965) uses measurements of distances from a random point to the first (P_1) and second (P_2) nearest plants. The two measurements are squared and combined into a ratio by dividing the squared P_1 value by the squared P_2 value. The coefficient of aggregation (A) is the mean of these ratios for all sampling points.

$$A = \frac{\sum \left[\frac{(P_1)^2}{(P_2)^2} \right]}{n}$$

Where: n = Number of sampling points

This coefficient equals 0.500 for a random population. For a population tending toward uniformity, the value is less than 0.500, and for an aggregated population it is greater than 0.500.

The deviation of an observed coefficient from 0.500 is tested by a z test in which the difference between the observed value and 0.500 is compared to the standard error for a random distribution and the observed sample size. The z equation is

$$z = \frac{0.500 - A}{\left(\frac{0.2887}{\sqrt{n}} \right)}$$

Where: n = Number of sampling points
0.2887 = Standard deviation of A values for a random population

Since the Holgate statistic, A, is a ratio, and not an actual distance or density value, this test gives a measure of pattern unaffected by differences in mean density in different portions of the sampled area.

Distance to Nearest Neighbor

In this technique, devised by Clark and Evans (1954), the distance r from each of n individuals (all individuals in a plot, or a random sample of these individuals) to its nearest neighbor is measured. An observed mean distance, $\bar{r}_A$, to nearest neighbor is then calculated by the expression

$$\bar{r}_A = \frac{\Sigma r}{n}$$

To calculate an expected mean distance, an independent estimate of the density of individuals, ρ, is required. The expected mean distance, $\bar{r}_E$, for a random dispersion pattern is then given by the expression

$$\bar{r}_E = \frac{1}{2\sqrt{\rho}} .$$

The Clark-Evans dispersion index, R, is the ratio of these values:

$$R = \frac{\bar{r}_A}{\bar{r}_E} .$$

This index ranges from 0, for a maximally aggregated population, to 1.0 for a random population and 2.1491 for a perfectly hexagonal (uniform) spacing pattern. To test for significance of deviation of observed from expected mean distance to nearest neighbor, the standard error of the expected mean distance, $\bar{r}_E$, is calculated by the equation

$$\sigma_{\bar{r}_E} = \frac{0.26136}{\sqrt{n\rho}} .$$

A z test is then carried out:

$$z = \frac{\bar{r}_A - \bar{r}_E}{\sigma_{\bar{r}_E}}$$

Values for z of 1.96 and 2.58 indicate that the deviation is significant at α levels of 0.05 and 0.01, respectively.

The Clark-Evans technique assumes that the objects studied are points with no dimensions, which therefore cannot physically prevent each other from sharing an occupied space, or which cannot overlap so that they appear as one. Modification of the technique to accomodate these situations is described by Cox (1987). Clark and Evans (1979) also show how the analysis can be extended to more than two dimensions.

This technique is highly adaptable, and can be used not only for analysis of the spacing of plants, but also for the spacing pattern of burrows (e.g., fiddler crab holes), nests (e.g., ant nests), termite mounds, and many other objects.

THE NEGATIVE BINOMIAL

The negative binomial is based on two parameters: μ, the mean density of organisms, and k, the binomial exponent. The negative binomial distribution arises by expansion of the expression

$$p^k (1 - q)^{-k}$$

Where: $p = \dfrac{k}{k + \mu}$

$q = 1 - p$

The use of the negative binomial is based largely on the properties of k, which usually has a value of 0.5–3.0. As long as the mean density remains constant, the value of k decreases as aggregation increases. Thus, $1/k$ reflects the degree of aggregation for a given density. One should be cautious in the use of k as such an index, however, since different data sets are likely to differ in both mean density and aggregation. Taylor, Woiwod, and Perry (1978, 1979) also have shown that k can vary in a complicated fashion with density, and that it is sometimes highly unstable.

Sampling data like those used for Poisson pattern analysis can be fitted by the negative binomial. Any sampling data on the number of individuals or events in unit space or time can be analyzed in this way, however.

The size of the sample unit should be chosen so that the mean number of items per sample is low. Otherwise, the analysis can be exceedingly tedious.

To analyze data, the frequency of samples (f) containing various numbers of individuals (X) is determined (table 21.2). From these data the mean ($\overline{X}$) and variance (s^2) are calculated (exercise 6).

To derive expected negative binomial frequencies, one must obtain estimates, $\overline{X}$ and $\hat{k}$, of the parameters μ and k. The statistic $\overline{X}$ is simply the mean number of individuals per sample unit. Obtaining $\hat{k}$, however, requires two steps: approximation and refinement. An approximation of $\hat{k}$ is given by the equation

$$\hat{k} = \frac{\overline{X}^2}{s^2 - \overline{X}}$$

Where: $\overline{X}$ = Sample mean
s^2 = Sample variance

A refined estimate of $\hat{k}$ requires substituting $\hat{k}$ values in the equation below until the two sides are equal:

$$n \log^x \left(1 + \frac{\overline{X}}{\hat{k}}\right) = \Sigma \frac{A_x}{\hat{k} + X}$$

Where: n = Total number of samples
A_x = Total number of samples containing *more than x* individuals

The approximate value of $\hat{k}$ determined above is first substituted in the equation. If the left side of the equation is greater than the right, the $\hat{k}$ value is too large; if the left side is smaller, $\hat{k}$ is too small. The $\hat{k}$ value should be increased or decreased, as appropriate, by about 10%, and the equation reworked. This new value should be closer to, or opposite in pattern of deviation from, equality. If the latter is true, the difference between the two $\hat{k}$ values should be split, and this third value tried. By "splitting the difference" a quite exact value of $\hat{k}$ can soon be obtained.

Once $\hat{k}$ has been determined, the relative frequency, P, of samples with a specific number of individuals, X, is given by the expression

$$P(X) = \left(\frac{(X + k - 1)!}{X! \, (k - 1)!}\right) p^k q^x$$

For calculation of the expected relative frequency of samples with 0 individuals, this reduces to

$$P(0) = p^k$$

The value of this expression can be read from table 21.3, or calculated directly.

Successive relative frequencies (for $X = 1, 2, 3$, etc.) can be obtained by multiplying each previously calculated term by the expression

$$\frac{q}{X}(X + k - 1)$$

The final calculated term is actually the expected relative frequency of samples with X or more individuals, and is thus the difference between the sum of previous terms and 1.0.

Multiplying these relative frequencies by the observed number of quadrats (n) gives the expected actual number of quadrats.

COMPUTER SOFTWARE FOR DISPERSION ANALYSIS

The software package EcoStat, available for both IBM PC (and compatibles) and Macintosh microcomputers, contains modules for analysis of pattern by the point-to-plant distance and distance-to-nearest-neighbor techniques (see appendix B).

The software package Ecological Analysis–PC, Vol. 4 (appendix B) contains modules for analysis of quadrat data for pattern analysis and negative binomial analysis.

Table 21.2. Calculation of expected negative binomial distribution for an observed set of sampling data

Number per Sample (x)	(f)	fx	$f(x-\bar{x})^2$	A_x	Negative Binomial Expected Frequency P(x)	Negative Binomial Expected Number of Samples f'	$\dfrac{A_x}{\hat{k}+x}$	$\dfrac{A_x}{\hat{k}+x}$	$\dfrac{A_x}{\hat{k}+x}$	$\dfrac{A_x}{\hat{k}+x}$	$\dfrac{A_x}{\hat{k}+x}$
0											
1											
2											
3											
4											
5											
6											
7											
8											
9											
10											
Σ					1.00000		$\sum\dfrac{A_x}{\hat{k}+x}$	$\sum\dfrac{A_x}{\hat{k}+x}$	$\sum\dfrac{A_x}{\hat{k}+x}$	$\sum\dfrac{A_x}{\hat{k}+x}$	$\sum\dfrac{A_x}{\hat{k}+x}$
—	n	$\sum fx$	$\sum[f(x-\bar{x})^2]$	—	———	n					

Table 21.3. Values of p^k

k					p				
	0.1	0.2	0.3	0.4	0.5	0.6	0.7	0.8	0.9
0.5	0.316228	0.447214	0.547723	0.632456	0.707107	0.774597	0.836660	0.894420	0.948683
0.6	0.251189	0.380731	0.485593	0.577080	0.659754	0.736022	0.807040	0.874690	0.938725
0.7	0.199526	0.324131	0.430512	0.526553	0.615572	0.699368	0.779060	0.855367	0.928900
0.8	0.158489	0.275946	0.381678	0.480450	0.574349	0.664540	0.751759	0.836512	0.919160
0.9	0.125893	0.234924	0.338383	0.438383	0.535887	0.631446	0.725416	0.818050	0.909540
1.0	0.100000	0.200000	0.300000	0.400000	0.500000	0.600000	0.700000	0.800000	0.900000
1.1	0.079433	0.170268	0.265970	0.364977	0.466516	0.570120	0.675457	0.782350	0.890560
1.2	0.063096	0.144956	0.235801	0.333021	0.435275	0.541728	0.651805	0.765082	0.881240
1.3	0.050119	0.123407	0.209054	0.303863	0.406126	0.514750	0.628971	0.748200	0.871980
1.4	0.039811	0.105061	0.185340	0.277258	0.378929	0.489116	0.606928	0.731688	0.862848
1.5	0.031623	0.089443	0.164317	0.252982	0.353553	0.464758	0.585663	0.715550	0.853812
1.6	0.025119	0.076146	0.145678	0.230832	0.329877	0.441613	0.565141	0.699752	0.844850
1.7	0.019953	0.064826	0.129153	0.210621	0.307786	0.419621	0.545338	0.684300	0.836000
1.8	0.015849	0.055189	0.114503	0.192180	0.287175	0.398724	0.526231	0.669209	0.827244
1.9	0.012589	0.046985	0.101515	0.175353	0.267943	0.378868	0.507800	0.654443	0.818572
2.0	0.010000	0.040000	0.090000	0.160000	0.250000	0.360000	0.490000	0.640000	0.810000
2.1	0.007943	0.034054	0.079971	0.145991	0.233258	0.342072	0.472833	0.625883	0.801507
2.2	0.006310	0.028991	0.070740	0.133209	0.217638	0.325037	0.456263	0.612066	0.793097
2.3	0.005012	0.024681	0.062716	0.121545	0.203063	0.308850	0.490280	0.598562	0.784787
2.4	0.003981	0.021012	0.055602	0.110903	0.189465	0.293470	0.424850	0.585350	0.776560
2.5	0.003162	0.017889	0.049295	0.101193	0.176777	0.278855	0.409940	0.572431	0.768433
3.0	0.001000	0.008000	0.027000	0.064000	0.125000	0.216000	0.343000	0.512000	0.729000

☯ SUGGESTED ACTIVITIES

1. Use quadrats and plotless techniques to obtain data on individuals in a population of plants in an arid habitat, of trees in a forest or savanna, or of invertebrates on an intertidal rock surface (e.g., limpets). Compare pattern as revealed by quadrats of different size, or pattern of the same species in different habitats or stands of different age, or pattern as detected by different techniques. In the lab, analyze pattern in the mapped desert plant community in the back of this manual, using clear acetate squares or circles as quadrats, or measuring point-to-plant or interplant distances with a ruler. Compare the pattern of particular species in the nonwash and wash subhabitats with that for the area as a whole. Use quadrat techniques to analyze data on numbers of aquatic organisms in unit-volume water samples.

    When aggregation is detected or suspected, compute the binomial index, k. Compare degree of aggregation along an environmental gradient. For example, sample limpets or other intertidal molluscs at different tide levels. Examine aggregation in stands of trees as a function of age and size of the trees. Compare the aggregation index of different species in the same habitat. For animals, test the effect of habitat differences or habitat modification on aggregation.

2. Collect data on some biotic association, such as the number of insect galls per leaf, the number of thrips per flower, or the number of insect parasites per acorn (based on numbers of emergence holes). Determine whether the data differ from random, and if aggregation is found, calculate the value of the negative binomial exponent, k.

3. Examine data on numbers of parasites per host given by Morand et al. (1993) or a similar published reference. Calculate the negative binomial exponent, k, for male, female, and total parasites for different years and compare your values with those of the authors.

☯ QUESTIONS FOR DISCUSSION

1. What are the biotic and abiotic factors that may produce uniform or clumped distributions? How do random patterns originate?

2. What differences might be expected in analyses of the same population with quadrats of different size? Why is it necessary to group expected values for quadrat classes of low frequency in the chi-square test? For what types of plant and animal populations are quadrat methods and plotless methods, respectively, of dispersion analysis probably most efficient?

3. What factors produce aggregation? What are the desirable characteristics of a quantitative index of aggregation? How does k vary with changes in average density of the population due to, say, random death of individuals? Does the negative binomial distribution have the capability of fitting all types of aggregation phenomena that may occur in nature?

☯ SELECTED BIBLIOGRAPHY

Anderson, D. J. 1971. Patterns in desert perennials. Journal of Ecology 59:555–560.

Binns, M. R. 1987. Behavioral dynamics and the negative binomial distribution. Oikos 47:315–318.

Clark, P. J., and F. C. Evans. 1954. Distance to nearest neighbor as a measure of spatial relationships in populations. Ecology 35:445–453.

———, and F. C. Evans. 1979. Generalization of a nearest neighbor measure of dispersion for use in k dimensions. Ecology 60:316–317.

Cox, G. W. 1987. Nearest-neighbour relationships of overlapping circles and the dispersion pattern of desert shrubs. Journal of Ecology 75:193–199.

Ferrer, E. R., and B. M. Shepard. 1987. Sampling Malayan black bugs (Heteroptera: Pentatomidae) in rice. Environmental Entomology 16:259–263.

Gill, D. E. 1975. Spatial patterning of pines and oaks in the New Jersey Pine Barrens. Journal of Ecology 63:291-298.

Holgate, P. 1965. Some new tests of randomness. Journal of Ecology 53:261-266.

Hurlbert, S. H. 1990. Spatial distribution of the montane unicorn. Oikos, 58:257-271.

King, T. J., and S. R. J. Woodell. 1973. The causes of regular pattern in desert perennials. Journal of Ecology 61:761-765.

Morand, S., J-P. Pointier, G. Borel, and A. Theron. 1993. Pairing probability of schistosomes related to their distribution among the host population. Ecology 74:2444-2449.

Perry, J. N., and L. R. Taylor. 1986. Stability of real interacting populations in space and time: implications, alternatives and the negative binomial kc. Journal of Animal Ecology 55:1053-1068.

Phillips, D. L., and J. A. MacMahon. 1981. Competition and spacing patterns in desert shrubs. Journal of Ecology 69:97-115.

Taylor, L. R., I. P. Woiwod, and J. N. Perry. 1978. The density-dependence of spatial behaviour and the rarity of randomness. Journal of Animal Ecology 47:383-406.

———. 1979. The negative binomial as a dynamic model for aggregation, and the density dependence of k. Journal of Animal Ecology 48:289-304.

Mortality, Recruitment, and Migration Rates in Populations

INTRODUCTION

Mark and recapture techniques of estimating population size can be modified to give estimates of mortality, natality, emigration, and immigration (Southwood 1968, Seber 1973, 1986). In closed populations, where migration in or out is negligible, rates of mortality and natality can be estimated easily. In open populations, where emigration and immigration complicate matters, proper design of the mark and recapture analysis can still permit these rates to be measured. Such analyses require three or more samplings from the population, and the marking and recapture of two or more groups of animals. Techniques of this sort assume that the rates involved are constant during the study.

EXPERIMENTAL DESIGN

The following outline is one of the simplest designs permitting estimation of the above parameters. A study area is selected in which individuals can be captured, marked and released, and recaptured randomly. The study area is partitioned into a central area with a total boundary length one-half that of the larger area within which it is located. The simplest manner of arranging this is to mark out a study area in the form of a square, and locate within this area a central plot with dimensions one-half those of the large plot. For simplicity, we shall term this central plot the ''inner area,'' the portion of the larger study area outside this inner plot the ''outer area,'' and the area of the large study area the ''total area.'' When the plots are laid out in the form of nested squares, the inner area is one-fourth the size of the total area. The population must then be sampled on three dates.

Date 1. Animals are captured in both inner and outer areas, marked so that the two groups can be distinguished, and released where they were captured. Animals should be marked so that the date, as well as the part of the study area involved, is easily determined. For example, on a given date, animals from the two parts of the study area can be marked with the same color paint, but on different body parts. We shall designate these marked animals as follows:

M1a = Marked and released in inner area

M1b = Marked and released in outer area

Data on these marked animals are recorded in table 22.1.

Date 2. Animals are captured in both inner and outer areas. The total numbers captured, and the number of *M1a* and *M1b* marked recaptures, are tallied and recorded separately for the inner and outer areas (table 22.1). A second group of animals is also marked. To reduce confusion, only previously unmarked individuals should be

Begon et al. (1990): 15:141
Brewer (1994): 4:81–82, 102–114
Krebs (1994): 10:151–162

Ricklefs (1990): 15:291–295
Ricklefs (1993): 14:250–260

Smith (1990): 16:395–405
Smith (1992): 2:31–32, 13:195–196

Table 22.1. Mortality, emigration, natality, and immigration in a natural or artificial population

Date 1

Captured

Inner	Outer	Total

Marked and Released

M1a	M1b	M1a+b
a_1		a'_1

Date 2

Captured

Inner	Outer	Total
n_2		n'_2

Marked and Released

M2a	M2b	M2a+b
a_2		a'_2

Recaptured from Date 1

M1a Inner	M1b Inner	M1a Outer	M1b Outer	M1a+b Total
r_{21}				r'_{21}

Date 3

Captured

Inner	Outer	Total
n_3		n'_3

Recaptured from Date 1

M1a Inner	M1b Inner	M1a Outer	M1b Outer	M1a+b Total
r_{31}				r'_{31}

Recaptured from Date 2

M2a Inner	M2b Inner	M2a Outer	M2b Outer	M2a+b Total
r_{32}				r'_{32}

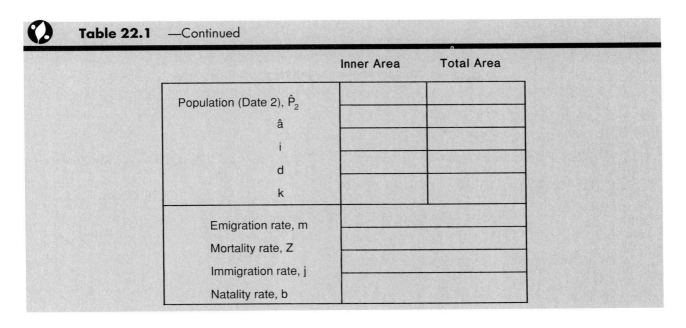

Table 22.1 —Continued

	Inner Area	Total Area
Population (Date 2), $\hat{P}_2$		
â		
i		
d		
k		
Emigration rate, m		
Mortality rate, Z		
Immigration rate, j		
Natality rate, b		

marked. The mark used should be distinct from that employed on date 1, and should distinguish individuals from inner and outer areas (e.g., a different color, applied on different body parts for inner and outer areas). We shall designate these marked animals as follows:

$$M2a = \text{Marked and released in inner area}$$
$$M2b = \text{Marked and released in outer area}$$

Data on these individuals are also recorded in table 22.1.

Date 3. A recapture sample is taken in both inner and outer areas, and the total number of individuals obtained, together with the numbers of *M1a, M1b, M2a,* and *M2b* marked recaptures, recorded separately in table 22.1.

The data in table 22.1 can then be used to estimate the population size on date 2, and also estimate rates of mortality, natality, emigration, and immigration. For this analysis we shall use the following symbols:

$$a_x = \text{Number of animals marked on date } x$$
$$n_x = \text{Total number of animals in population sample on date } x$$
$$r_{yx} = \text{Total recaptures, on date } y, \text{ of animals marked on date } x$$

To facilitate calculations, the data from table 22.1 should be summarized separately for inner and total areas as follows:

Inner Area		*Total Area*	
a_1	Marked M1b	a'_1	Marked M1a + M1b
a_2	Marked M2b	a'_2	Marked M2a + M2b
n_2	Total captured in inner area, date 2	n'_2	Total captured in inner + outer areas, date 2
n_3	Total captured in inner area, date 3	n'_3	Total captured in inner + outer areas, date 3
r_{21}	M1b recaptured in inner area on date 2	r'_{21}	M1a + M1b recaptured in total area on date 2
r_{31}	M1b recaptured in inner area on date 3	r'_{31}	M1a + M1b recaptured in total area on date 3
r_{32}	M2b recaptured in inner area on date 3	r'_{32}	M2a + M2b recaptured in total area on date 3

Using these data, estimates of population size (P_2), disappearance rate (i), and addition rate (k) are calculated separately for the inner and total areas. The disappearance rate is, of course, equal to the mortality rate if the population is closed, and to the combination of mortality and emigration rates if it is open. The addition rate is equal to natality in a closed population, and to natality plus immigration in an open population.

The population estimate for date 2 ($\hat{P}_2$), known as the *Bailey triple catch estimate,* is given by the equation

$$\hat{P}_2 = \frac{a_2\, n_2\, r_{31}}{r_{32}\, r_{21}} \tag{1}$$

The disappearance rate (i) is for the interval between dates 1 and 2, and is given by the equation

$$i = \frac{\log_e a_1 - \log_e \hat{a}_1)}{t} \tag{2}$$

$$\text{Where: } \hat{a}_1 = \text{Number of } a_1 \text{ still in the population on date 2}$$
$$t = \text{Interval in time units between dates 1 and 2}$$

For this calculation, $\hat{a}_i$ must first be obtained:

$$\hat{a}_1 = \frac{a_2\, r_{31}}{r_{32}} \tag{3}$$

In calculating the disappearance rate, the time units (days, weeks, years, etc.) should be selected as appropriate. The rate then represents an instantaneous rate, or the number of individuals disappearing per individual in the population per unit time.

Similarly, the addition rate (k) may be calculated, for the interval between dates 2 and 3, by the equation

$$k = \frac{(\log_e d)}{t} \tag{4}$$

$$\text{Where: } d = \text{Dilution rate for the interval between dates 2 and 3}$$
$$t = \text{Interval in time units between dates 2 and 3}$$

For this calculation, the dilution rate, or the rate at which the ratio of marked to total animals is diluted by the entry of unmarked individuals, is given by

$$d = \frac{n_3\, r_{21}}{n_2\, r_{31}} \tag{5}$$

Having calculated disappearance rates and addition rates for both inner and total areas of the study plot, one can partition these values to separate mortality from emigration and natality from immigration. This is done by assuming that both emigration and immigration are proportional to the length of the boundary of the area involved, which differs by a factor of 2 for inner and total areas.

Thus, for the disappearance rate, the following equations can be written:

$$i_s = Z + 2m \tag{6}$$
$$i_g = Z + m \tag{7}$$

$$\text{Where: } i_s = \text{Disappearance rate for inner area}$$
$$i_g = \text{Disappearance rate for total area}$$
$$Z = \text{Mortality rate}$$
$$m = \text{Emigration rate}$$

These two equations, by subtraction, can be combined and solved for *m* and *Z:*

$$m = i_s - i_g \tag{8}$$
$$Z = 2i_g - i_s \tag{9}$$

Likewise, for the addition rate, the following can be written:

$$k_s = b + 2j \tag{10}$$
$$k_g = b + j \tag{11}$$

Where: k_s = Addition rate for inner area

k_g = Addition rate for total area

b = Natality rate

j = Immigration rate

Solution of this pair of equations gives

$$j = k_s - k_g \tag{12}$$
$$b = 2k_g - k_s \tag{13}$$

SUGGESTED ACTIVITIES

1. Conduct a study of mortality, recruitment, and migration in a field population of a fairly abundant invertebrate or small vertebrate in an area of relatively uniform habitat. Terrestrial invertebrates such as ground beetles, sowbugs, and snails are quite suitable, as are various intertidal marine invertebrates. Such a study might be organized as a long-term project involving classes in successive school terms, or perhaps as individual special studies or course term projects.

2. Simulate such a population study with animals such as sowbugs (or for that matter, inanimate objects such as wooden beads) in an enclosure a meter or two in width. Take samples, mark the individuals, and return them to inner and outer sectors of a study plot laid out within this enclosure. Individual students can effect emigration, immigration, "mortality" (by removal), and natality (by addition) between the first and second and between the second and third capture periods. In this way, data can be obtained in a relatively short laboratory period.

3. Carry out the calculations involved in this analysis, using the following data:

	Inner Area	Total Area
a_1	155	593
a_2	98	394
n_2	145	610
r_{21}	47	169
n_3	120	539
r_{31}	18	96
r_{32}	28	112

These data should yield population estimates (P_2) of 198 and 1,215 for inner and total areas, respectively, and rate estimates of $m = 0.108$, $Z = 0.080$, $j = 0.103$, and $b = 0.045$.

QUESTIONS FOR DISCUSSION

1. What condition must be fulfilled with respect to loss of marks during this study? Why must the assumption of constancy of rates during the period be made?

2. What are the potential causes of mortality, and potential influents of emigration and immigration in the population studied? How will the population involved probably change in size in the near future, based on your estimates?

3. What sorts of interrelations exist among the various rates calculated for this population?

🕐 SELECTED BIBLIOGRAPHY

Arnason, A. N. 1972. Parameter estimates from mark-recapture experiments on two populations subject to migration and death. Researches in Population Ecology 13:97–113.

———. 1973. The estimation of population size, migration rates and survival in a stratified population. Researches in Population Ecology 15:1–8.

Bailey, N. T. J. 1952. Improvements in the interpretation of recapture data. Journal of Animal Ecology 21:120–127.

Cameron, R. A. D., and P. Williamson. 1977. Estimating migration and the effects of disturbance on mark-recapture studies in the snail *Cepaea nemoralis* L. Journal of Animal Ecology 46:173–179.

Jolly, G. M. 1965. Explicit estimates from capture-recapture data with both death and immigration—stochastic model. Biometrika 52:225–247.

Manly, B. F. J. 1985. A test of Jackson's method for separating death and migration with mark-recapture data. Researches on Population Ecology 27:99–109.

Pollock, K. H. 1981. Capture-recapture models: a review of current methods, assumptions and experimental design. Pages 426–435 *in* C. J. Ralph and J. M. Scott, editors. Estimating the numbers of terrestrial birds. Studies in Avian Biology, No. 6.

Seber, G. A. F. 1973. The estimation of animal abundance and related parameters. Griffin, London, England.

———. 1986. A review of estimating animal abundance. Biometrics 42:267–292.

Southwood, T. R. E. 1978. Ecological methods. Chapman & Hall, London, England.

Waseloh, R. M. 1985. Dispersal, survival, and population abundance of gypsy moth, *Lymantria dispar* (Lepidoptera: Lymantridae), larvae determined by releases and mark-recapture studies. Annals of the Entomological Society of America 78:728–735.

———. 1987. Emigration and spatial dispersion of the gypsy moth predator *Calosoma sycophantra*. Entomologia Experimentalis et Applicata 44:187–193.

Zeng, Z., and J. H. Brown. 1987. A method for distinguishing dispersal from death in mark-recapture studies. Journal of Mammalogy 68:656–665.

Life Tables and Survivorship Curves

INTRODUCTION

Age-specific mortality, survivorship, and reproductive data for populations of animals or plants can be summarized in a life table. A life table shows, for each age interval, the actual mortality, mortality rate, number of survivors, and future life expectancy for a certain number, or cohort, of individuals that begin life together. Age-specific fecundity data are often included, as well. Life tables thus furnish a quantitative tool for analyzing the effects of ecological factors on population dynamics.

Survivorship curves can be constructed by plotting numbers of survivors against age. When the number scale is logarithmic, straight-line sections of the curve indicate periods of constant survival rate, whereas changes in slope indicate increase or decrease in survival rate.

Life tables have been used most often with animal populations, where they have proved valuable in developing game management and pest control strategies. Recently, the rapidly growing field of plant demography has recognized that life tables are useful for population studies of plants (Leverich and Levin 1979).

LIFE-TABLE COMPUTATIONS

A life table consists of columns of age-specific information on aspects of mortality, survivorship, and reproduction for a cohort of individuals that begin life together. Since most life tables have utilized a cohort of 1,000 individuals, the calculations in this exercise are based on this number.

The first column (x) in the life table gives age in time units appropriate for the species. A survivorship (l_x) column begins with 1,000 individuals that enter the first age interval, and gives the number remaining at the start of each subsequent age interval. A mortality (d_x) column gives the number dying during each age interval out of the initial cohort of 1,000 individuals. These two columns are easily interconverted. Mortality in a given age interval is the difference between the number of individuals entering that age interval and the number surviving to enter the next interval:

$$d_x = l_x - l_{x+1}$$

In related fashion, survivorship for a given age interval is the survivorship for the preceding age interval minus mortality for the preceding age interval:

$$l_x = l_{x-1} - d_{x-1}$$

The sum of mortality values for all age groups thus is 1,000.

Begon et al. (1990): 4:131–148, 15:522–529

Brewer (1994): 4:82–86

Ehrlich and Roughgarden (1987): 5:75–90

Krebs (1994): 11:168–197

Pianka (1994): 8:144–152

Ricklefs (1990): 15:295–300, 28:560–579

Ricklefs (1993): 14:256–260

Smith (1990): 14:337–361

Smith (1992): 11:169–178

Stiling (1992): 9:184–191

The age-specific mortality rate (*1,000q_x*) column gives the mortality per 1,000 individuals entering each age interval. This value is calculated from mortality and survivorship data by the equation

$$1{,}000q_x = \frac{1{,}000 \; d_x}{l_x}$$

The future-life-expectancy (*e_x*) column gives the mean lifetime remaining to individuals entering each age interval. This value is given by the equation

$$e_x = \frac{T_x}{l_x}$$

Where: T_x = Total individuals times age unit still to be lived

l_x = Total individuals still to live them

To obtain T_x, it is assumed that individuals dying during an age interval do so uniformly through the interval so that each lives, on the average, 1/2 of the age interval. The number of individuals living in a particular age interval (L_x) is therefore the number surviving through the entire period plus 1/2 the number dying during the period:

$$L_x = l_{x+1} + \frac{d_x}{2}$$

The T_x value is thus the sum of L_x values for age intervals older than and including the one for which the life-expectancy calculation is being made. In constructing a life table, it is useful to include L_x and T_x columns next to the life-expectancy column for purposes of calculation.

The life expectancy for individuals of age 0 is equal to the mean life span. The mean life span can also be obtained from the equation

$$\text{mean life span} = \frac{\Sigma[(\text{age interval midpoint}) \; (d_x)]}{1{,}000}$$

Many life tables have a column showing age at the start of each age interval as percent deviation from mean length of life (x'). This age scale facilitates the comparison of life tables for organisms with very different actual life spans. Age as percent deviation from mean length of life is given by the equation

$$x' = \frac{\text{age at start of interval} - \text{mean life span}}{\text{mean life span}} \times 100$$

Age-specific fecundity data, m_x, the number of offspring per individual during the age interval, also can be included in a life table when these data are available. Survivorship and fecundity data from a life table can be used to project population growth (see exercise 28).

SOURCE OF LIFE-TABLE DATA

The raw data for a life table can be either mortality or survivorship data. Suitable data can be obtained in three principal ways.

1. Data on age at death. Carcasses or remains of individuals dying of natural causes yield age-specific mortality (d_x) data if they involve marked individuals of known age or if the remains can be aged. The sample must be an unbiased cross section of individuals dying at various ages. For example, recoveries of banded birds dying of natural causes furnish raw data for life tables if the individuals were marked as nestlings or juveniles. The age at death of some animals can be estimated if body structures show growth patterns correlated with age. In many vertebrates, horns, teeth, scales, or eye lenses may show annual growth increments. Life tables have been constructed, for example, from age-at-death data determined from growth increments on horns of Dall sheep skulls found in the wild. The age at death of some mollusks, such as cockles or scallops, can be determined or at least estimated from growth lines on the shells or opercula. Growth rings of trees can likewise be used to determine age at death.

2. Direct observation of survivorship. If a cohort of individuals of known age is marked and isolated in some manner, the number of survivors (l_x) can be observed subsequently until all have died. This can be done with many invertebrates in the laboratory or under seminatural conditions. A simple laboratory experiment might involve the comparison of survivorship of wild-type and mutant strains of *Drosophila*. In the field, the survival of plants germinating at the same time or of sessile animals, such as barnacles, colonizing cleared areas during a certain period can be followed.

3. Population age-structure data. In stable populations with natality and mortality rates that have remained constant over a period of time, the numbers of individuals of various ages form a survivorship (l_x) curve. If these individuals can be aged, the numbers in the various age intervals can be determined. Growth increments on hard parts of the body may furnish a basis for aging, as described above. For organisms with indeterminate growth, modal body-size groups corresponding to ages can sometimes be recognized. Many crustaceans, such as terrestrial isopods and the marine sand crab, *Emerita* spp., show modal size groups corresponding to ages. These species are abundant enough to yield size-frequency data adequate for life-table construction. The age structure of laboratory populations of certain insects can be studied if immatures can be separated according to instar.

It is often difficult to obtain an accurate estimate of the size of the cohort entering the population at time 0 (birth or hatching), or of the mortality values for the earliest age intervals. In such cases, it may be necessary to let the age of the youngest fully represented individuals equal 0, and to calculate a life table beginning with 1,000 individuals of this age.

The first step in analyzing raw data is their conversion to a base of 1,000 individuals at time 0. For mortality (d_x) data, the value for each interval must be multiplied by the expression

$$\frac{1,000}{\Sigma \, d_x}$$

For survivorship (l_x) data, the number of individuals in the first age interval should be set equal to 1,000 and the values for the remaining intervals multiplied by the expression

$$\frac{1,000}{l_{0-1}}$$

Once the raw data have been converted to the basis of an initial cohort of 1,000 individuals, the remaining life-table information can be calculated as described earlier (table 23.1).

SURVIVORSHIP CURVES

Survivorship can be shown graphically by plotting a curve of the $\log_{10}$ of the number of survivors on the ordinate against age on the abscissa (figure 23.1). Age may be plotted either in actual time units or as percent deviation from mean length of life. The $\log_{10}$ scale causes periods of constant mortality or survival rates to appear as straight-line sections of the curve. The survivorship curve for a species having the same mortality rate at all ages would, therefore, be a straight line extending from 1,000 at time 0 to 0 at the age at which the oldest individuals die. On such a plot, periods of high mortality rate appear as steep, vertical segments of the curve and periods of low mortality rate as flat, horizontal segments of the curve.

COMPUTER SOFTWARE FOR LIFE-TABLE ANALYSIS

The IBM PC–compatible software package Ecological Analysis–PC Vol. 1, distributed by Oakleaf Systems (appendix B), contains a life-table module. This module requires only the input of the l_x or d_x values, and calculates the columns of the table that relate to survivorship, mortality, and future life expectancy.

Table 23.1 Life table for _____

x	x'	d_x	l_x	$1,000q_x$	L_x	T_x	e_x
Age in _____ (Age _____ = 0)	Age at start of interval as % deviation from mean life span	Mortality within interval per 1,000 entering population at age 0	Number surviving to start of interval per 1,000 starting at age 0	Morality per 1,000 entering age interval			Mean lifetime remaining to those entering age interval

Male

Female

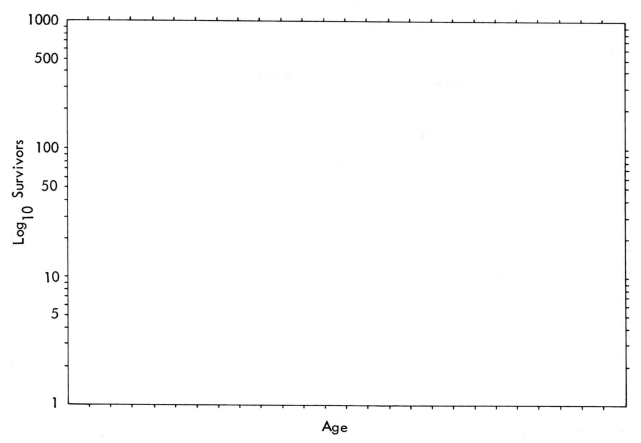

Figure 23.1
Survivors remaining at various ages out of a cohort of 1,000 individuals beginning life together

The EcoStat software package, available for both IBM PC–compatible and Macintosh microcomputers, and distributed by Trinity Software (appendix B), also contains a life-table module. This module uses raw survivorship and fecundity data as input, and gives proportional survivorship and future-life-expectancy values, together with other statistics relating to growing populations (see exercise 28).

SUGGESTED ACTIVITIES

1. Collect an unbiased sample of about 200–300 individuals from a population of a terrestrial isopod (how should you do this?). Using dissecting microscopes, sex and measure the body length of individuals (consult an invertebrate morphology text for information on sexing individuals). Construct a frequency distribution of body lengths of individuals of each sex, and estimate the relative numbers of individuals of different ages. Dissect adult females and determine the fraction of females carrying eggs, and the average size of the egg clutch. Construct a life table, using numbers of individuals in modal size groups as age groups, and an estimate of the total eggs produced by females as the size of the cohort entering the 0–1 age interval.

2. Collect an unbiased sample of 200–300 crayfish from a stream habitat. Sex and measure the carapace length of individuals (consult an invertebrate text for information on sexing individuals). Construct a frequency distribution of body lengths of individuals of each sex, and estimate the relative numbers of individuals in different age intervals. Estimate the numbers of eggs for gravid females. Construct a life table, using numbers of individuals in modal size groups as age groups, and an estimate of the total eggs produced by females as the size of the cohort entering the 0–1 age interval.

3. Follow the fate of individuals in a population of an annual plant, such as a species of *Erodium*. Count the number of plants that germinate, and census the number of survivors at regular intervals. For the plants that mature, harvest and count the number of seeds. Use these data to construct a life table, considering the cohort entering the first age interval to be the number of seeds produced.

✪ Questions for Discussion

1. What factors influence the accuracy of life tables constructed with various types of raw data? What factors can limit the application of a life table calculated for one population to other populations? What sampling conditions must be satisfied for a valid life table to be constructed?

2. For species showing markedly different survivorship curves, describe the manner in which mortality rate and life expectancy change as one proceeds from younger to older ages. Must mortality rate always increase? Must life expectancy always decrease? At what age is mortality rate greatest? When life expectancy is calculated as described above, what value will always be shown for the last age interval? Why?

3. How would you expect the form of the survivorship curve to vary with variation in the number of reproductive bodies produced by an organism during its lifetime? With the degree of parental care afforded the young? For populations of the same species in environments differing in severity of action of biotic or abiotic factors?

✪ Selected Bibliography

Baker, M. C., L. R. Mewaldt, and R. M. Stewart. 1981. Demography of white-crowned sparrows (*Zonotrichia leucophrys nuttalli*). Ecology 62:636–644.

Boer, A. H. 1988. Mortality rates of moose in New Brunswick: a life table analysis. Journal of Wildlife Management 52:21–25.

Carey, J. R. 1993. Applied demography for biologists with special emphasis on insects. Oxford University Press, New York, New York, USA.

Caughley, G. 1977. Analysis of vertebrate populations. John Wiley & Sons, New York, New York, USA.

Frazer, N. B., J. W. Gibbons, and J. L. Greene. 1991. Life history and demography of the common mud turtle *Kinosternon subrubrum* in South Carolina, USA. Ecology 72:2218–2231.

Jones, S. M., and R. E. Ballinger. 1987. Comparative life histories of *Holbrookia maculata* and *Sceloporus undulatus* in western Nebraska. Ecology 68:1828–1838.

Knight, R. R., and L. L. Eberhardt. 1985. Population dynamics of Yellowstone grizzly bears. Ecology 66:323–334.

Lander, R. H. 1981. A life table and biomass estimate for Alaskan (USA) fur seals (*Callorhinus ursinus*). Fisheries Research 1:55–70.

Leverich, W. J., and D. A. Levin. 1979. Age-specific survivorship and reproduction in *Phlox drummondii*. American Naturalist 113:881–903.

Royama, T. 1981. Evaluation of mortality factors in insect life table analysis. Ecological Monographs 51:495–505.

Tinkle, D. W., J. D. Congdon, and P. C. Rosen. 1981. Nesting frequency and success: implications for the demography of painted turtles. Ecology 62:1426–1432.

Tryon, C. A., and D. P. Snyder. 1973. Biology of the eastern chipmunk, *Tamias striatus*: life tables, age distributions, and trends in population numbers. Journal of Mammalogy 54:145–168.

Warren, M. S., E. Pollard, and T. J. Bibby. 1986. Annual and long-term changes in a population of the wood white butterfly *Lepidea sinapis*. Journal of Animal Ecology 55:707–720.

Zammuto, R. M., and P. W. Sherman. 1986. A comparison of time-specific and cohort-specific life tables for Belding's ground squirrel, *Spermophilus beldingi*. Canadian Journal of Zoology 64:602–605.

EXERCISE 24

Ecological Isolation

INTRODUCTION

Speciation is the divergence of two or more parts of a parental species population to the point of reproductive isolation. Reproductive isolation means that, in nature, individuals of the different population segments do not interbreed commonly, even when they occur together, because of certain morphological, physiological, or behavioral differences. These differences are termed reproductive isolating mechanisms.

Reproductive isolation alone, however, may not guarantee the survival of newly diverged species populations. Divergence in characteristics affecting reproduction can occur without much change in the ecological characteristics of the populations. If this is the case, recontact of descendant populations may result in intense interspecific competition for environmental resources such as food or nest sites. The result of this competition may be elimination of all but one of the newly evolved species. This exemplifies the *competitive exclusion principle,* which states that in a stable environment, two or more resource-limited species with identical patterns of resource utilization cannot coexist indefinitely. If two or more such species come together, one should prove better able to exploit the available resources and will outcompete and eliminate the others.

Thus, a second requirement for "successful" speciation is usually the development of ecological differences that prevent competitive elimination of all but one of the descendant species. In other words, to survive, descendant species must possess ecological isolation as well as reproductive isolation. Ecological isolation can be achieved in a variety of ways, termed *ecological isolating mechanisms.* The major types of ecological isolating mechanisms are outlined below:

Differences in spatial occurrence.

1. Occupation of nonoverlapping geographical ranges (allopatry). Species may be separated by a strong barrier to dispersal, which prevents their contacting and competitively challenging each other. In other cases, they may replace each other geographically so that only areas of contact or limited overlap occur. In these latter cases, each species likely possesses adaptations to its own geographical area that make it competitively superior there.
2. Occupation of separate habitats. Although the range between geographical and habitat separation is a continuum, habitat isolation generally involves occupation of different environmental situations within easy dispersal distance of individuals of all species. Cases of habitat isolation vary from situations in which species are restricted to their separate habitats even in the absence of competitors, to situations in which restriction is maintained by competitive elimination of species from all but their optimal habitats.

Begon et al. (1990): 20:717–738
Brewer (1994): 8:230–246, 10:282–284
Colinvaux (1993): 9:152–170
Ehrlich and Roughgarden (1987): 16:352–355, 18:368–385

Krebs (1994): 13:230–261
Pianka (1994): 13:268–293
Ricklefs (1993): 19:347–357

Smith (1990): 18:464–477
Smith (1992): 15:225–231
Stiling (1992): 11:220–225

Differences in time of occurrence (in same geographical area and habitat).

1. Activities concentrated at different times in the diurnal cycle (or other short-term cycles, such as tidal cycles). To the extent that such differences lead to use of different resources, ecological isolation can result. Competition can occur, however, between species with different activity patterns, if the same resources are taken.

2. Activities concentrated at different seasons. Such isolation may vary from actual seasonal replacement of one species by another, to situations in which the species differ in the season of activities with great resource demands (e.g., nesting or breeding in animals, flowering or fruiting in plants).

Differences in required resources and adaptations for their exploitation.

1. Specialization in morphology. Many animals, especially birds, have specialized feeding structures that adapt species to use different types of food resources. Plants often have flowers specialized in structure for different pollinators.

2. Specialization in physiology. Differences exist among animals, for example, in ability to digest and absorb certain food materials. Such differences can result from symbiotic relationships with intestinal microorganisms. In plants and microorganisms, differences in biochemical systems of uptake and excretion of water, salts, and specific nutrients may be important in ecological isolation.

3. Specialization in behavior. Animals may show differences in feeding microhabitats, foraging behaviors, and techniques of food capture. These may cause the species to encounter and take different foods.

Rarely, if ever, is ecological isolation achieved by any single mechanism listed above. For example, temporal isolation, by restriction of activity to different parts of the diurnal cycle (day, dusk, night), is very likely accompanied by differences in behavior and in the structure and physiology of the eyes. When different foods are taken, the differences among species likely will include a combination of behavioral, morphological, and physiological features.

Some recent workers (e.g., Wiens 1984) have questioned whether habitat and resource conditions in many ecosystems are stable enough that strong patterns of ecological isolation are essential to coexistence. This hypothesis has stimulated new, more rigorous investigations of resource use designed to test hypotheses about resource selection versus opportunistic use (e.g., Kaspari and Joern 1993). Such studies promise new insight into the role of ecological isolating mechanisms in community structure.

✪ SUGGESTED ACTIVITIES

1. Analysis of ecological isolation in Australian lizards

    Some 14 species of lizards of the genus *Ctenotus* occur in desert areas of western Australia. Pianka (1969a, 1969b, 1986) and James (1991) give data on the geographical distribution of these species and on their habitats, seasonal activity patterns, activity times and temperatures, foraging behavior, and foods taken. Copies of these publications will be made available by your instructor. From the data in these publications, you should complete the ecological isolation chart (table 24.1) so that the following types of questions can be answered:

    1. What are the principal ecological differences between a particular pair of species?
    2. How are the species that occur together in a particular geographical area or particular habitat separated in their patterns of resource use?
    3. What species are separated primarily by differences in habitat, time of activity, foods taken, or other differences?
    4. What are the most important types of ecological isolating mechanisms in this group of species?

Table 24.1 Summary of differences contributing to ecological isolation among *Ctenotus* lizards* of Western Australia

Name Code	Allopatric	Habitat Restriction	Foraging Microhabitat	Activity Air Temp. °C	Seasonal Activity	Diurnal Foraging Time	Foods	Other
dux								
bro								
lea								
col								
sch								
cal								
ari								
leo								
gra								
hel								
pan								
qua								
pia								
atl								

*Name code is the first three letters of the specific name.

You should first read the general description of Australian desert habitats (Pianka 1969a, p. 498), and then seek the data required to complete table 24.1, using the several publications. Fill in each column of the table as indicated below, marking an X through the space for species for which data are inadequate.

Allopatric

For each species, examine geographical range maps (Pianka 1969b) and record by name code the other species which have completely (++) or largely (+) nonoverlapping geographical ranges.

Habitat restriction

Determine the habitat type to which each species is restricted. Block off portions of this column to show these habitats for individual species or groups of species.

Foraging microhabitat

Block off portions of this column to show which species or groups of species forage primarily in sunny areas, shaded areas, or both.

Activity air temperature

Determine whether the species forages primarily at air temperatures that are low (<21° C), intermediate (25–26° C), or high (>27° C), and block off portions of this column to show this relationship.

Seasonal activity

Block off this column to show which species are active primarily in summer, in winter, or year-round.

Diurnal foraging time

Divide this column to show whether foraging is unimodal (midday), bimodal (activity peaks in the morning and late afternoon), or continuous throughout the day.

Foods

Record the most important foods for each species, contrasting the food-use patterns of the pairs of species with adjacent spaces separated by a dotted line. For this purpose, the greatest importance should probably be attached to data from Pianka (1969b) and James (1991) on percent composition by volume of various food types. However, consider other food information in the light of what the data might suggest about where, when, and how these lizards forage.

Other Foraging behavior

Record any peculiar patterns of foraging behavior or other differences that cannot be shown in other columns.

2. Analysis of ecological isolation in subalpine birds

Some 20 species of insectivorous passerine birds occur in subalpine forests at elevations of 750–1450 m in the White Mountains of New Hampshire. Patterns of elevational distribution, habitat selection, and foraging behavior of these species have been summarized by Sabo (1980). Copies of this publication will be made available by your instructor. Develop an ecological isolation chart similar to table 24.1. This table should be structured so that the four types of questions posed for the Australian lizards can be answered. In particular, consider the following kinds of differences in designing the structure of the table:

- Major differences in elevational range of species
- Preference for coniferous vs. broadleaf trees
- Foraging on the ground or understory vs. in the canopy
- Preference for foraging at different heights in the canopy
- Tendency to forage in foliage vs. on twigs and branches
- Use of special foraging behaviors or microhabitats

3. Other data sets for analysis of ecological isolation. Examine the data given by Toft (1980) on the feeding ecology of 13 species of anurans in Amazonian Peru. Construct a table summarizing the differences among these species, based on foraging strategies (searching vs. sit-and-wait), food types, and food sizes. Or, examine the data given by Courtney and Chew (1987) on seasonal activity, plant foods (species and plant parts), and habitats of Pierid butterflies in Morocco. Construct a table showing the ecological isolation pattern among these species.

✪ QUESTIONS FOR DISCUSSION

1. Are all of the species in your analysis clearly distinct in their ecology? Is ecological isolation achieved by a single, clear-cut difference between a species and the other members of the genus, or is it the result of a combination of differences?

2. Does the size of the geographical range of a species correlate with the strength of its ecological isolation pattern? Do species that are most similar in patterns of spatial occurrence differ most in other characteristics?

3.   Is it certain that the abundances of the species you examined are limited by food resources, or might other limiting factors exist? What predators do these animals have, and what influence might predation have on interspecific competition and ecological isolation?

4.   What experiments might you conduct to test hypotheses about competition and ecological isolation in this group of animals?

SELECTED BIBLIOGRAPHY

Arthur, W. 1987. The niche in competition and evolution. John Wiley & Sons, New York, New York, USA.

Courtney, S. P., and F. S. Chew. 1987. Coexistence and host use by a large community of Pierid butterflies: habitat is the templet. Oecologia 71:210–220.

James, C. D. 1991. Temporal variation in diets and trophic partitioning by coexisting lizards (*Ctenotus*: Scincidae) in central Australia. Oecologia 85:553–561.

Kaspari, M., and A. Joern. 1993. Prey choice by three insectivorous grassland birds: reevaluating opportunism. Oikos 68:414–430.

Pianka, E. R. 1969a. Habitat specificity, speciation, and species density in Australian desert lizards. Ecology 50:498–502.

———. 1969b. Sympatry of desert lizards (*Ctenotus*) in western Australia. Ecology 50:1012–1030.

———. 1986. Ecology and natural history of desert lizards. Princeton University Press, Princeton, New Jersey, USA.

Sabo, S. R. 1980. Niche and habitat relations in subalpine bird communities of the White Mountains of New Hampshire. Ecological Monographs 50:241–259.

Toft, C. A. 1980. Feeding ecology of thirteen syntopic species of anurans in a seasonal tropical environment. Oecologia 45:131–141.

Wiens, J. A. 1984. On understanding a non-equilibrium world: myth and reality in community patterns and processes. Pages 439–457 *in* D. R. Strong, Jr., D. Simberloff, L. G. Abele, and A. B. Thistle, editors. Ecological communities: conceptual issues and the evidence. Princeton University Press, Princeton, New Jersey, USA.

Niche Breadth, Niche Overlap, and the Community Matrix

INTRODUCTION

Studies of competition and ecological isolation in small sets of species lead automatically to questions about such relationships in larger assemblages. For example, one might wish to compare the average range or degree of overlap of food resources used by members of one community with those used by species in another community. Or, one might wish to describe the range or degree of overlap in habitats occupied by various species along some environmental gradient.

Much of the interest in these topics can be traced to Hutchinson's (1957) definition of the ecological niche as the total set of conditions within the limits of survival and reproduction for all ecological factors pertinent to the organism. Since a given factor can be visualized as an axis in space with upper and lower limits to survival and reproduction indicated, and since other factors can be visualized as axes in other dimensions with limits similarly indicated, the Hutchinsonian niche can be described as an *n*-dimensional hypervolume defined by upper and lower limits on axes of all of the *n* ecological factors relevant to the species. Using this concept, Hutchinson distinguishes the *fundamental niche* as that set of conditions under which the species could potentially exist in the absence of interspecific competition. The *realized niche,* in contrast, constitutes that portion of the fundamental niche actually occupied in a "real world" situation involving competition with other species.

The niche formulation of Hutchinson can be viewed as defining the way in which resources, including space, food, and others, are allocated to species. For food resources, quantitative measures that may correspond to niche axes include size, texture, palatability, the nature of defense or escape capacities (for prey species sought by a predator), and many others. For the space resource, the descriptive characteristics of niche axes might be temperature, humidity, salinity, vegetational geometry, or any other factor of the physical or biotic environment to which an organism must adjust.

A number of other formulations of the niche concept exist. Most students of niche theory have employed the concept of the Hutchinsonian niche in examining the breadth and overlap of the ecological characteristics of species. Many of these studies have dealt with *niche breadth,* the amplitude of ecological conditions utilized by individual species, and *niche overlap,* the extent to which different species share the same conditions. Often, these studies have been carried out in highly abstract terms. One can, however, apply the techniques of analysis of breadth and overlap to specific aspects of the ecology of species, such as food resources utilized, habitats occupied, or hosts parasitized. In such cases, an appropriate modification of terminology is desirable, e.g., habitat overlap or resource overlap. This approach is equivalent to the examination of one or a few of the total axes of the Hutchinsonian niche.

Begon et al. (1990): 7:269–272
Brewer (1994): 10:281–282
Colinvaux (1993): 16:314–315
Krebs (1994): 23:529–532

Pianka (1994): 13:268–293, 16:343–347
Ricklefs (1990): 35:728–737
Ricklefs (1993): 24: 455–458

Smith (1990): 18:473–481
Smith (1992): 15:226–233
Stiling (1992): 11:220–225

Niche Breadth and Overlap

The basic data for analysis of niche breadth and overlap are quantitative data on habitat occupation or resource use for a set of species. These data should be arranged in a table, with the various species listed along the left side of the table, and habitat or resource categories shown as column headings (including a heading for "total"). For habitats, the data will typically consist of some measure of the numbers or biomass of a species per unit of habitat (or sometimes for the total extent of a habitat category). For resources, the data will consist of estimates of the quantity of each resource utilized. Some indices of niche breadth and overlap also require estimates of the total carrying capacity of the habitat for the general type of organism under consideration, or of the total quantity of each resource type available to the species. If such information is available, these categories should also be listed along the side of the table.

Niche Breadth

Several indices of niche breadth have been proposed. Levins (1968) formulated two of the most widely used indices, H and B:

$$H = -\sum_{i=1}^{R} p_i \ln p_i$$

and

$$B = 1/\sum_{i=1}^{R} p_i^2$$

Where: p_i = Decimal fraction of total abundance or resource use (across all categories) in category i

R = Number of categories of habitats or resources

These indices increase from 0 (for H) or 1.0 (for B), when the niche includes only one category, to larger numbers when several or many categories are represented. For a given number of categories, both indices give larger values the more even the values are for the various categories.

Culver (1972) suggested a modified form of the first index:

$$H_1 = \frac{-\sum_{i=1}^{R} p_i \ln p_i}{\ln R}$$

This index varies from a minimum value of 0, when the niche includes only one category, to a maximum of 1.0 when habitat or resource use is equal for all niche categories.

A weakness of all three of the above indices is their failure to consider actual habitat or resource use relative to potential use (Hurlbert 1978). If information is available on the potential habitat carrying capacity for the type of organism in question, or on the quantity of resources of various types actually available to a species, two more refined indices can be calculated. The first, devised by Hurlbert (1978), is

$$B' = 1/\sum_{i=1}^{R} (p_i^2/q_i)$$

Where: q_i = Decimal fraction of habitat carrying capacity or available resource in category i

A second such index (Smith 1982) is

$$FT = \sum_{i=1}^{R} (p_i q_i)^{0.5}$$

Both of these indices also have the convenient property of ranging from a minimum near 0 to a maximum of 1.0.

Measures of niche breadth all have particular quirks of behavior when used with unusual data sets. Many of these quirks are discussed by Hurlbert (1978) and Smith (1982). Caution should be used in interpreting indices when particular niche categories have very large or very small values relative to most of the categories utilized.

Niche Overlap

There are likewise several measures of overlap of species in the use of niche categories (habitat types or resource categories). The simplest of these is an index of proportional overlap:

$$C_{xy} = 1 - 0.5 \sum_{i=1}^{R} | p_{xi} - p_{yi}|$$

Where: p_{xi} = Value for species x in category i
p_{yi} = Value for species y in category i

Another widely used index is

$$O_{xy} = \frac{\Sigma\, p_{xi}\, p_{yi}}{\Sigma\, p_{xi}^2 \Sigma\, p_{yi}^2}$$

Both of these indices vary from 0, when the two species share no niche categories, to 1.0 when they share identical values across niche categories.

Neither of these overlap indices considers the potential carrying capacity of the habitat categories or the available resources of various categories. To allow an overlap comparison that also takes these relations into account, Hurlbert (1978) formulated still another index of overlap:

$$L = \left(\sum_{i=1}^{R} q_i \middle/ \sum_{i=1}^{R} p_{xi} \sum_{i=1}^{R} p_{yi} \right) \sum_{i=1}^{R} (p_{xi}p_{yi}/q_i)$$

This index varies from 0, when the species share no niche categories, to 1.0 when both utilize categories in direct proportion to their potential value, and to values greater than 1.0 if they coincide in using certain categories to a greater than proportional degree.

Like niche breadth indices, indices of niche overlap possess their own peculiarities of behavior. Most perform in a relatively satisfactory manner unless actual overlaps become very small or very high (Linton et al. 1981).

THE COMMUNITY MATRIX

Levins (1968) suggested that in a system of k potentially competing species, equilibrium conditions for the ith species are defined by the equation

$$K_i = X_i + \sum_k \alpha_{ik}X_k \qquad k \neq i$$

Where: K_i = Carrying capacity of the environment
for species i
X_i = Population of species i
X_k = Population of species other than i
α_{ik} = Effect of species k on species
i (competition coefficient)

The reciprocal interspecific competitive effects, or competition coefficients, for a community of k species, therefore, can be summarized in a matrix of the sort outlined below:

1	α_{21}	α_{31} .... α_{k1}
α_{12}	1	α_{32} .... $\alpha_{k/k2}$
α_{13}	α_{23}	1 .... α_{k3}
α_{1k}	α_{2k}	α_{3k} 1

Where: α_{ab} = Competition coefficient showing effect of species b on species a

Table 25.1. Community matrix and species packing index

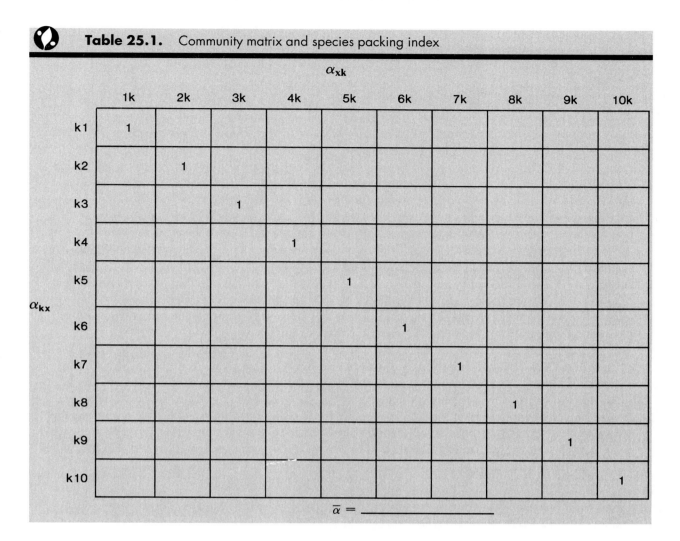

	α_{xk}									
	1k	2k	3k	4k	5k	6k	7k	8k	9k	10k
k1	1									
k2		1								
k3			1							
k4				1						
k5					1					
k6						1				
k7							1			
k8								1		
k9									1	
k10										1

α_{kx} (row label)

$$\overline{\alpha} = \underline{\hspace{4cm}}$$

Levins (1968) also proposed that these coefficients can be approximated by the relationship

$$\alpha_{ik} = \sum_j p_{ij}p_{kj} \Big/ \sum_j p_{ij}^2$$

Where: p_{ij} = Proportion of the total of values for species i that occur in category j

p_{kj} = Proportion of the total of values for species k that occur in category j

Furthermore, the mean value of α, in the case of data for abundance of species in various habitats, is a measure of what may be termed species packing, the degree of packing increasing with increase in mean value of this index. Species packing, in descriptive terms, is the degree of coadjustment of species so that they can occupy the same habitat rather than being segregated in different habitats.

Table 25.1 is designed for the calculation of a community matrix based on the data analyzed in this exercise.

Matrix analyses of this type can be conducted in various ways, leading to different interpretations of considerable ecological interest. For example, if data are available on the distribution of species among niche categories when each species is alone, versus when the various species co-occur in mixed natural populations, two matrices can be calculated. The difference between the two represents the modifying influence of competition, or, in a sense, the difference between the fundamental and realized niches (Culver 1970).

🔊 SUGGESTED ACTIVITIES

1. Examine the extensive data given by Scott et al. (1986) on the distribution of forest birds, both native and introduced, in the Hawaiian Islands. These data include occurrence in 200 m altitudinal zones and in eight vegetation types. Formulate and test hypotheses about habitat breadth and overlap of native and introduced species. For example, test whether native and introduced bird species along an elevational transect on a particular island differ in breadth of elevations or vegetation types occupied.

2. Examine the extensive data given by Pianka (1986) for habitat utilization, activity times, and food habits of congeneric lizard species in desert regions of several continents. Formulate and test various hypotheses about niche breadth and overlap. For example, compare overlap of food resource use for allopatric and sympatric species, for species using the same as opposed to different microhabitats, and for species active at the same times of day or seasons compared to species active at different times. Compare breadth of food resource use for these same situations, as well as for species active at high versus low temperatures, or species foraging in the open versus in shaded sites. Contrast the performance of different indices of breadth and overlap in these comparisons.

3. Examine the data presented by Whittaker and Fairbanks (1958) on distribution of copepod species in lakes with different solute contents in southeastern Washington. Using data from their tables I and II, calculate measures of habitat breadth and overlap along one or more gradients of water chemistry (e.g., total salts, Na, Ca). This will require recalculating the percentage occurrences for species in table I as relative percentages of the total for each species across all lakes. Test, for example, the hypothesis that niche breadth increases as the ability to occupy lakes of high salt content increases. Using recalculated data of this sort for all lakes except the Park Lake group and Lily Pad Pond, calculate a community matrix and determine the index, α, of species packing for the 10 species of widest distribution.

🔊 QUESTIONS FOR DISCUSSION

1. When numbers of individuals of various species constitute the data for analyses of this type, what assumptions are made about such characteristics as trophic level and individual resource use for individuals of these species? What is assumed about resource limitation for the assemblage of species as a whole?

2. What problems exist in comparing niche breadth or overlap for different groups of species in different environments?

3. What assumptions are made in using α as an index of niche overlap or competitive stress? Under what conditions would α be a good index of such relationships? A poor one?

🔊 SELECTED BIBLIOGRAPHY

Bouchon-Navarro, Y. 1986. Partitioning of food and space resources by chaetodontid fishes on coral reefs. Journal of Experimental Marine Biology and Ecology 103:21–40.

Colwell, R. K., and D. J. Futuyuma. 1971. On the measurement of niche breadth and overlap. Ecology 52:567–576.

Culver, D. C. 1972. A niche analysis of Colorado ants. Ecology 53:126–131.

Feinsinger, P., E. E. Spears, and R. W. Poole. 1981. A simple measure of niche breadth. Ecology 62:27–32.

Holt, R. D. 1987. On the relation between niche overlap and competition: the effect of incommensurable niche dimensions. Oikos 48:110–114.

Hurlbert, S. H. 1978. The measurement of niche overlap and some relatives. Ecology 59:67–77.

Hutchinson, G. E. 1957. Concluding remarks. Cold Spring Harbor Symposia in Quantitative Biology 22:415–427.

Krebs, C. J. 1989. Ecological methodology. Harper & Row, New York, New York, USA.

Lawlor, L. R. 1980. Overlap, similarity, and competition coefficients. Ecology 61:245–251.

Levins, R. 1968. Evolution in changing environments. Princeton University Press, Princeton, New Jersey, USA.

Linton, L. R., R. W. Davies, and F. J. Wrona. 1981. Resource utilization indices: an assessment. Journal of Animal Ecology 50:283–292.

MacNally, R. C., and J. M. Doolan. 1986. An empirical approach to guild structure: habitat relationships of nine species of eastern Australian cicadas. Oikos 47:33–46.

Maurer, B. A. 1982. Statistical inference for MacArthur-Levins niche overlap. Ecology 63:1712–1719.

Petraitis, P. S. 1981. Algebraic and graphical relationships among niche breadth measures. Ecology 62:545–548.

Pianka, E. R. 1986. Ecology and natural history of desert lizards. Princeton University Press, Princeton, New Jersey, USA.

Scott, J. M., S. Mountainspring, F. L. Ramsey, and C. B. Kepler. 1986. Forest bird communities of the Hawaiian Islands: their dynamics, ecology, and conservation. Studies in Avian Biology No. 9, xii +431 pages.

Slobodchikoff, C. N., and W. C. Schultz. 1980. Measures of niche overlap. Ecology 61:1051–1055.

Smith, E. P. 1982. Niche breadth, resource availability, and inference. Ecology 63:1675–1681.

Whittaker, R. H., and C. W. Fairbanks. 1958. A study of plankton copepod communities in the Columbia Basin, southeastern Washington. Ecology 39:46–65.

Wissinger, S. A. 1992. Niche overlap and the potential for competition and intraguild predation between size-structured populations. Ecology 73:1431–1444.

EXERCISE 26

Experimental Studies of Plant Competition

INTRODUCTION

Many questions about intraspecific and interspecific competition can be investigated by experiments with single- and mixed-species populations of annual or perennial plants. Density-dependent effects of competition for space, moisture, or soil nutrients may be shown by plants growing in different total single-species or mixed-species densities. Mixed-species experiments can also be designed to investigate how conditions such as moisture and nutrient availability determine which species is the stronger competitor under given conditions. This exercise furnishes an outline for studying the effects of competition on total yield, survival, and individual growth of different species at various densities and at different soil moisture or nutrient levels.

INDICES OF COMPETITIVE EFFECTS

Several indices have been used in the evaluation of competitive effects in plant experiments. These indices compare the complementarity of species in use of growth resources, their competitive ability, and the intensity of competitive effects on individuals.

The *relative yield total (RYT)* is a measure of the degree to which the performances of the species in a mixture are complementary. *RYT* is calculated as the sum of ratios of biomass yield, *Y*, of each species in mixed culture to biomass yield in the respective pure culture:

$$\text{RYT} = Y_{Am}/Y_{Ap} + Y_{Bm}/Y_{Bp}$$

Where *m* and *p* indicate the biomasses of species *A* and *B* in mixed and pure cultures, respectively

An *RYT* of 2.0 indicates complete complementarity (a lack of any interspecific competition), whereas a value of 1.0 indicates that interspecific competition is as strong as competition within each species. Values greater than 2.0 would indicate stimulation of biomass production by one or both species.

The *relative crowding coefficient (RCC),* a measure of competitive ability, is the ratio, rather than the sum, of the individual biomass ratios of each species alone and in mixed culture. This coefficient can only be calculated for experiments involving 1:1 mixtures. For example, the *RCC*s for species *A* and *B* would be

$$\text{RCC}_A = (Y_{Am}/Y_{Ap})/(Y_{Bm}/Y_{Bp})$$
$$\text{RCC}_B = (Y_{Bm}/Y_{Bp})/(Y_{Am}/Y_{Ap})$$

The severity of intraspecific or interspecific competition on individual plants can also be expressed by coefficients based on individual plant biomass, *W*. This involves comparing individual biomass under a particular

Begon et al. (1990): 7:261–264
Brewer (1994): 8:229–249
Colinvaux (1993): 8:136–149
Krebs (1994): 13:230–261

Pianka (1994): 12:241–267
Ricklefs (1990): 22:439–478
Ricklefs (1993): 19:341–360

Smith (1990): 18:450–473
Smith (1992): 15:217–225
Stiling (1992): 11:209–220

density d with that under a situation of no competition. For intraspecific competition, an *absolute severity of competition (ASC)* index is calculated:

$$ASC = \log_{10}(W_0/W_d)$$

Where the subscript 0 indicates the individual biomass under no competition.

ASC indices also can be calculated for species in a mixture, by comparing their biomasses to respective values under no competition.

DESIGN OF COMPETITION STUDIES

Three basic approaches have been taken to the design of experimental plant-competition studies (figure 26.1). The relative advantages of these approaches are now a topic of debate (e.g., Snaydon 1991, Cousens and O'Neill 1993).

Replacement Design

Most studies of competition between pairs of species have followed Harper's (1977) recommendation and used a *replacement design,* in which the total number of individual plants per experimental unit is kept constant, but the ratio of numbers of plants of the two species is varied. A replacement series for studies with species A and B, for example, might include the following combinations of individuals: (1) 20 A, (2) 15 A + 5 B, (3) 10 A + 10 B, (4) 5 A + 15 B, and (5) 20 B.

This design has been criticized by several plant ecologists because the density of each species in mixtures varies inversely with the density of its competitor. Thus, it is difficult to separate effects of change in a species' own density from the effects of changing density of the second species.

In addition, the *RYT* (relative yield total) index may not give a clear indication of resource complementarity in replacement experiments, especially at low plant densities. Values close to 1.0, suggesting strong competition, may result if the total density of a mixture is low enough that no competition, intraspecific or interspecific, occurs.

Additive Design

Many other studies have used an *additive design,* in which the ratio of numbers of the two species in mixed combinations is held constant (usually 1:1) and the total density varied. An additive series for studies with species A and B might include the following combinations of individuals: (1) 20 A, (2) 20 B, (3) 10 A + 10 B, (4) 15 A + 15 B, and (5) 20 A + 20 B.

Additive designs have been criticized because the introduction of a competitor species simultaneously changes both the quantitative (i.e., total plant density) and the qualitative (i.e., neighboring plant characteristics) nature of the competitive environment of a given species.

Bivariate Factorial Design

Another alternative is a *bivariate factorial design,* in which densities and ratios are created in a balanced matrix of replacement and additive combinations. A bivariate factorial series for studies with species A and B might include the following combinations: (1) 10 A, (2) 10 B, (3) 20 A, (4) 20 B, (5) 10 A + 10 B, (6) 20 A + 10 B, (7) 10 A + 20 B, and (8) 20 A + 20 B.

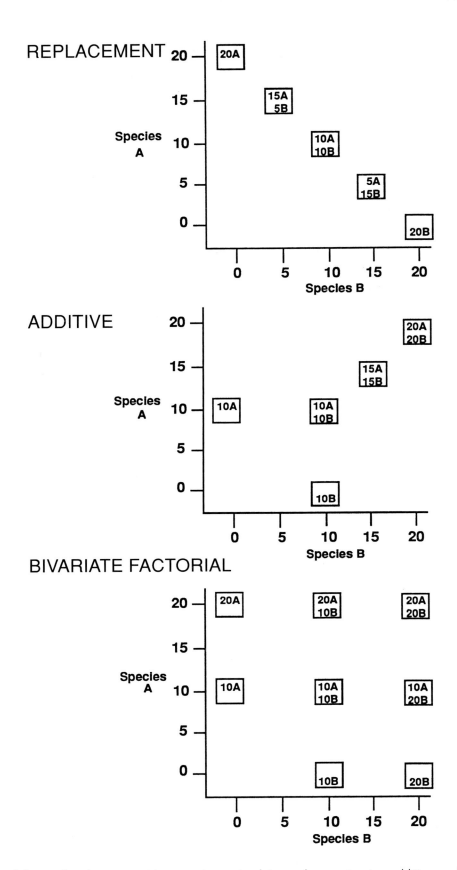

Figure 26.1.
Simple experimental designs for plant competition experiments involving replacement series, additive combinations, and bivariate factorial combinations of two species

This design requires a greater number of experimental combinations, but effectively incorporates variation in both total density and ratios of two species.

STATISTICAL ANALYSIS OF EXPERIMENTAL RESULTS

Experiments carried out with replacement series or additive series designs are usually amenable to evaluation by analysis of variance (ANOVA) (see exercise 8). Individual treatments (specific combinations of densities of one or more species, together with application levels of water, nutrients, or other factors) should be replicated (see exercise 3). More complex experiments, such as those using the bivariate factorial design, are more easily analyzed using multiple regression analysis (see exercise 8). Some specific experimental designs are outlined later in this exercise.

SELECTING SPECIES AND CONDUCTING EXPERIMENTS

Many wild or cultivated annuals or herbaceous perennials are suitable for competition experiments. Rapidly growing annuals are probably best for experiments during a single university term, although some rapidly growing biennials or perennials are also satisfactory. Choosing species with high germination percentages facilitates the creation of desired densities or density combinations. Experiments with closely related species of wild annuals are especially interesting and may produce results that bear on problems of competitive exclusion and the distribution of the species in nature. The pairing of a legume and a nonlegume in experiments having soil nitrogen level as a variable may furnish results relating to the basis of competitive advantage under different environmental conditions.

Widely distributed groups of wild plants suitable for such experiments include stork's bill (*Erodium*), mustard (*Brassica*), sorrel (*Oxalis*), brome grass (*Bromus*), fescue grass (*Festuca*), wild oats (*Avena*), rye grass (*Elymus* or *Lolium*), and many others. Suitable cultivated plants include oats (*Avena sativa*), barley (*Hordeum vulgare*), wheat (*Triticum aestivum*), clover (*Trifolium*), sweet clover (*Melilotus*), and alfalfa (*Medicago sativa*).

For experiments with most herbaceous plants, 6-inch flowerpots are satisfactory. Various patterns of distributing seeds over the soil surface have been used, but the most convenient method is to distribute them uniformly over the soil surface, interspersing seeds of different species so that individuals of the different species are adjacent. Extra seeds should be planted in a flat and the seedlings used to replace those not germinating in the pots.

Soil moisture and soil nutrient variables are easily incorporated into these experiments. Different soil moisture levels can be maintained by watering different series of pots with different measured amounts of water at regular intervals. Soil nitrogen and phosphate levels can be introduced as a variable by watering the pots initially with different amounts of ammonium nitrate, NH_4NO_3, or monocalcium phosphate, $Ca(H_2PO_4)_2 \cdot H_2O$, solutions being made up as described below:

0.0476 molar ammonium nitrate. Dissolve 3.8104 g of NH_4NO_3 in water and dilute to 1,000 ml. This solution contains 2 mg nitrogen per ml.

0.0182 molar monocalcium phosphate. Dissolve 4.5880 grams of $Ca(H_2PO_4)_2 \cdot H_2O$ in water and dilute to 1,000 ml. This solution contains 2 mg available P_2O_5 per ml.

When applied to 6-inch pots, these solutions are roughly equivalent to 25 pounds per acre per ml of solution. Suggested levels of application are 8, 16, 32, and 64 mg per pot.

🌐 SUGGESTED ACTIVITIES

1. Conduct a five-combination replacement-series experiment like that diagrammed in figure 26.1. Couple the five combinations with two levels of watering or fertilization, giving 10 experimental groups. Replicate each experimental group three times, creating an experiment involving 30 experimental units.

Also grow individual plants alone in pots large enough that no competition can be assumed. When the plants have achieved maximum growth, harvest the shoot tissue and measure the wet biomass for each species. Calculate Y and W values for each species and compute the indices *RYT, RCC,* and *ASC.*

Use ANOVA (see exercise 8) to test for differences among species combinations and watering/ fertilization treatments in total yield and mean biomass per plant (of each species). The ANOVA for total biomass will use data from all five combinations, but those for individual species will use data only from combinations containing that species, of course. ANOVAs for these tests will have the following structure:

| | *Degrees of Freedom* | |
	(Total Biomass)	(Per-plant Biomass)
Combination	4	3
Water/fertilizer regime	1	1
Interaction	4	3
Subtotal	9	7
Within groups	20	16
TOTAL	29	23

2. Conduct a five-combination additive-series experiment like that diagrammed in figure 26.1. Couple the five combinations with two levels of watering or fertilization, replicating each experimental group three times, so that the experiment also involves 30 experimental units. Likewise, grow individuals of each species alone in pots large enough that no competition can be assumed. Compute the various indices and carry out ANOVAs as described for the replacement-series experiment above.

3. Design a bivariate factorial experiment like that outlined in figure 26.1. Replicate each of the 15 density combinations twice, giving 30 experimental units. Measure the final biomass of each species. Calculate Y and W values for each species and compute the indices *RYT, RCC,* and *ASC.*

Use multiple regression analysis (see exercise 8) to analyze the results. In this analysis, Y (total for both species) or W (for one species) will represent the dependent variable, and the density (D) values of species A and B the independent variables.

The form of the regression equations will be

$$Y_{total} = a + b_A D_A + b_B D_B$$
$$W_A = a + b_A D_A + b_B D_B$$
$$W_B = a + b_A D_A + b_B D_B$$

For the equations for mean biomass *(W)* of individual species, the intercept *(a)* represents the biomass of an individual without any competition. This can be compared to the actual biomass of plants grown alone in large pots.

SOFTWARE FOR DATA ANALYSIS

The IBM PC–compatible software package Ecological Analysis–PC Vol. 2, distributed by Oakleaf Systems (appendix B), contains a stepwise multiple regression module. This module requires only the input of the dependent and independent variables. It computes the best two-factor regression equation and provides a test of significance of each of the regression coefficients.

⟐ QUESTIONS FOR DISCUSSION

1. How do the effects noted in these experiments illustrate density-dependent action of intraspecific or interspecific competition? By what specific mechanisms may the observed competition effects in this experiment be produced?

2. From the relative intensities of interspecific and intraspecific competition for the species used, what can be inferred about similarities of resource requirements by the species? Assuming that the necessary condition for coexistence is that each species inhibits its own further increase more than it inhibits the further increase of the other species, what do the data obtained in the present experiment suggest about coexistence of the species used?

3. In what ways may yield, as measured in the present experiments, be an inadequate reflection of the effect of interspecific competition?

⟐ SELECTED BIBLIOGRAPHY

Aarssen, L. W. 1985. Interpretation of the evolutionary consequences of competition in plants: an experimental approach. Oikos 45:99–109.

Austin, M. P., L. M. F. Fresco, A. O. Nicholls, R. H. Groves, and P. E. Kaye. 1988. Competition and relative yield: estimation and interpretation at different densities and under various nutrient concentrations using *Silybum marianum* and *Cirsium vulgare*. Journal of Ecology 76:157–171.

Benner, B. L., and F. A. Bazzaz. 1987. Effects of nutrient addition on competition within and between two annual plant species. Journal of Ecology 75:229–245.

Connolly, J. 1986. On difficulties with replacement series methodology in mixture experiments. Journal of Applied Ecology 23:125–137.

Cousins, R., and M. O'Neill. 1993. Density dependence of replacement series. Oikos 66:347–352.

Firbank, L. G., and A. R. Watkinson. 1985. On the analysis of competition within two-species mixtures of plants. Journal of Applied Ecology 22:503–517.

———. 1990. On the effects of competition: from monocultures to mixtures. Pages 165–192 *in* J. B. Grace and D. Tilman, editors. Perspectives on plant competition. Academic Press, San Diego, California, USA.

Harper, J. L. 1977. Population biology of plants. Academic Press, London, England.

Pacala, S. W., and J. A. Silander. 1986. Neighbourhood interference among velvet leaf (*Abutilon theophrasti*) and pigweed (*Amaranthus retroflexus*). Oikos 48:217–224.

Rejmanek, M., G. R. Robinson, and E. Rejmankova. 1989. Weed-crop competition: experimental designs and models for data analysis. Weed Science 37:276–284.

Silander, J. A., and S. W. Pacala. 1990. The application of plant population dynamic models to understanding plant competition. Pages 67–91 *in* J. B. Grace and D. Tilman, editors. Perspectives on plant competition. Academic Press, San Diego, California, USA.

Snaydon, R. W. 1991. Replacement or additive designs for competition studies? Journal of Applied Ecology 28:930–946.

EXERCISE 27

Field Studies of Plant Competition

INTRODUCTION

Competition for light, water, or nutrients can influence the survivorship, growth, and reproduction of plants. Mortality of individuals that begin to grow too close to an established, mature individual tends to lead toward uniform spacing of individuals in the population. Individuals growing close together are also likely to show stunted growth and reduced reproduction due to shortage of resources.

This exercise describes some observational and manipulative techniques for measuring the intensity of intraspecific and interspecific competition among the predominant species of a plant community.

OBSERVATIONAL FIELD STUDIES

Competition often leads to a positive correlation between plant size and interplant distance. For plants of the same species, a correlation between the mean size, or the combined sizes, of a pair of individuals and the distance between them is both an indication of intraspecific competition and a measure of its intensity. The greater the intensity of competition, the stronger the correlation. The same is true for plants of different species, but in this case differences in form and mean size need to be taken into account.

To evaluate competition, one should choose an area as uniform in resources for plant growth as possible. Differences in topography, moisture supply, or fertility can impose variations on the growth of plants the same distance apart. These variations make detection of competitive influences very difficult, even if such influences are important.

A basic, observational technique for evaluating competitive interactions was developed by Yeaton and Cody (1976) and Yeaton, Travis, and Gilinsky (1977). Several common species are selected for sampling. Data are then obtained on the sizes and interplant distances of individuals of all possible intraspecific and interspecific pair combinations. If four species (*A, B, C,* and *D*) are chosen, for example, the following combinations of individuals should be examined:

Intraspecific	A-A	B-B	C-C	D-D		
Interspecific	A-B	A-C	A-D	B-C	B-D	C-D

In sampling, the following protocol should be followed:

1. Select the first member of a plant pair randomly. This can be done by extending a tape-measure transect through the stand of vegetation. Random points are selected along this transect (see appendix A, table A.1). The plant of the desired species nearest each random point is designated as the first member of the plant pair.

Brewer (1994): 8:229–249
Colinvaux (1993): 8:136–149
Krebs (1994): 13:230–261
Odum (1993): 6:166–170

Pianka (1994): 12:241–267
Ricklefs (1990): 22:439–478
Ricklefs (1993): 19:341–360

Smith (1990): 18:450–473
Smith (1992): 15:217–225
Stiling (1992): 11:209–220

2. The second member of the plant pair is the individual of the desired species that is the nearest neighbor of the first member of the pair. This neighbor is only acceptable, however, if no individual of an undesired species intercepts the line between it and the first member (for trees, this applies to the overhead crown foliage, not just the trunk).

3. The distance between the center of the rooted base of the first plant and that of its nearest neighbor is measured. A maximum distance, beyond which competitive influence is biologically improbable, is determined prior to sampling. Measurements are recorded only if they are less than this distance.

4. The sizes of the plant pair members are measured in appropriate units. These units will not necessarily be the same for all species. These measurements might be of trunk diameter (for trees or single-stemmed cacti with bushy crowns), crown diameter (for shrubs, trees in open woodlands, or clump-forming cacti), or height of plant (for columnar cacti and other slender plants).

5. An effort should be made to obtain equal numbers of measurements for the various intraspecific and interspecific pair combinations.

For intraspecific pair combinations, data should be recorded in three columns, two to record individual plant size values and one to record interplant distance:

Plant #1 Size	Distance	Plant #2 Size

For interspecific pair combinations, four columns are used, the last to allow values of plant size of one species to be rescaled:

Species A		Species B	Species _____
Plant Size	Distance	Plant Size	Rescaled Size

These data can be analyzed by the parametric correlation technique (see exercise 8). For intraspecific combinations, correlations can be calculated directly between (1) the sum of size values of the two pair members and (2) the interplant distance for these individuals.

For interspecific combinations, the mean, variance, and standard deviation of the size values are first determined for each species (see exercise 6). The values for the species with the smaller variance are then rescaled by multiplying them by the ratio s_1/s_2, where s_1 is the standard deviation for the species with the higher value. The effect of this rescaling is to create sets of sample values with equal variances. Correlations should then be calculated between (1) the sum of the size value of one species and the rescaled value of the other and (2) the interplant distance for these individuals.

Values of sample size, correlation coefficient, and significance level for these analyses can be recorded in table 27.1.

MANIPULATIVE FIELD STUDIES

Intraspecific and interspecific competition can also be studied experimentally in the field by manipulating the density and composition of the neighborhood of target plants (e.g., Gurevich 1986, Bauder 1989, Reader and Best 1989, Pantastico-Caldas and Venable 1993). This approach is most practical for herbaceous plants, especially annuals, because responses can be observed over a few weeks or months. Longer-term studies can be set up with perennials such as bunch grasses or desert shrubs.

Select a target species for study, and identify one or more potential competitor species if interspecific competition is to be examined. The following general protocol can be used for such studies:

1. Determine the diameter of the area that represents the competitive neighborhood for the plant in question. This may range from 8–12 cm in diameter for small annual plants of desert or temporary pond habitats to 1 m for perennial bunch grasses. Neighborhood areas for shrubs would necessarily be even larger. Effectively, the competitive neighborhood represents the area within which the roots or shoots of a plant might be influenced by those of other plants.

Table 27.1. Results of correlation analyses relating plant size and interplant distance

Intraspecific or Interspecific Combination	Number of Measurements	Correlation Coefficient	Significance Level

2. Choose one or more appropriate measures of growth and reproduction, such as biomass, height, diameter, number of tillers, number of flowers, or number of seeds produced.

3. Select individuals of target species and mark them with a colored toothpick, a painted nail, or a colored string or loop of wire loosely enclosing the stem base. If plants are seedlings, initial measurements of size or other characteristics may be unnecessary. For perennial plants, however, initial measures of size may be necessary. Remove other individuals of the target and possible competing species to create two or more conditions of neighborhood densities of intraspecific and interspecific competitors.

4. Check the neighborhood plots at regular intervals, and remove any new plants that germinate or grow into the experimental neighborhood.

5. At plant maturity or at the end of a designated study period, record quantitative features of the target plants in each neighborhood unit. For plants on which initial measurements were taken, calculate an appropriate measure of change (e.g., change in height, ratio of final to initial value).

Data on these quantitative characteristics can be tested with parametric or nonparametric two-sample tests (see exercise 7) if only two levels of neighborhood density were used, or by parametric or nonparametric analysis of variance (see exercise 8) if more than two levels were created.

SUGGESTED ACTIVITIES

1. Using the observational approaches outlined above, sample a community such as desert scrub, in which shrubs and cacti show wide and highly variable spacing, and for which water is almost always a resource in short supply. Or, examine a woodland or forest, where variation in spacing and trunk size of trees is great. Comparison of pine or aspen stands of varying age offers the chance to examine intensity of intraspecific competition at different stages of stand development.

2. Using the observational approaches outlined above, sample the more abundant species of the mapped desert plant community included in this manual. Compare competition intensity for pairs of species in the wash and nonwash portions of the plot.

3. Using the manipulative approach outlined above, examine the roles of intraspecific and interspecific competition on annual plants in an old field, grassland, or desert habitat.

QUESTIONS FOR DISCUSSION

1. For which of the species examined in observational studies was the level of intraspecific competition most intense? Why might this be so? Which pairs of species showed the most intense interspecific competition? Are these the most closely related or most similar in their growth, resource needs, or phenologies?
2. What factors may be important in determining the location and size of plants of species for which strong correlations of size and interplant distance do not occur?
3. How do you think the intensity of competition should change during the course of plant succession?

SELECTED BIBLIOGRAPHY

Bauder, E. T. 1989. Drought stress and competition effects on the local distribution of *Pogogyne abramsii*. Ecology 70:1083–1089.

Fonteyn, P. J., and B. E. Mahall. 1981. An experimental analysis of structure in a desert plant community. Journal of Ecology 69:883–896.

Fowler, N. 1986. The role of competition in plant communities in arid and semiarid regions. Annual Review of Ecology and Systematics 17:89–110.

Grace, J. B., and R. G. Wetzel. 1981. Habitat partitioning and competitive displacement in cattails (*Typha*): experimental field studies. American Naturalist 118:463–474.

Gurevitch, J. 1986. Competition and the local distribution of the grass *Stipa neomexicana*. Ecology 67:46–57.

Howe, H. F., and S. J. Wright. 1986. Spatial pattern and mortality in the desert mallow (*Sphaeralcea ambigua*). National Geographic Research 2:491–499.

Nobel, P. S. 1981. Spacing and transpiration of various sized clumps of a desert grass, *Hilaria rigida*. Journal of Ecology 69:735–742.

Pacala, S. W., and D. Tilman. 1994. Limiting similarity in mechanistic and spatial models of plant competition in heterogeneous environments. American Naturalist 143:222–257.

Pantastico-Caldas, M., and D. L. Venable. 1993. Competition in two species of desert annuals along a topographic gradient. Ecology 74:2192–2203.

Phillips, D. L., and J. A. MacMahon. 1981. Competition and spacing patterns in desert shrubs. Journal of Ecology 69:97–115.

Reader, R. J., and B. J. Best. 1989. Variation in competition along an environmental gradient: *Hieracium floribundum* in an abandoned pasture. Journal of Ecology 77:673–684.

Schlesinger, W. H., and C. S. Jones. 1984. The comparative importance of overland runoff and mean annual rainfall to shrub communities of the Mojave Desert. Botanical Gazette 145:116–124.

Wright, S. J., and H. F. Howe. 1987. Pattern and mortality in Colorado Desert plants. Oecologia 73:543–552.

Yeaton, R. I., and M. L. Cody. 1976. Competition and spacing in plant communities: the northern Mojave Desert. Journal of Ecology 64:689–696.

———, J. Travis, and E. Gilinsky. 1977. Competition and spacing in plant communities: the Arizona upland association. Journal of Ecology 65:587–595.

<small>EXERCISE</small> **28**

Population Growth, Limitation, and Interaction

INTRODUCTION

Population ecology, from its inception, has been the most mathematical and theoretical area of ecology. Drawing initially from the field of human demography, population ecologists have developed a variety of mathematical models against which the behavior of real-world populations can be evaluated. These include models of population growth, limitation, interspecific competition, and predation.

In this exercise we shall explore several modules of the *EcoSim* (version 1.0) software package developed for instructional use in population biology and evolutionary ecology by Howard Towner at Loyola Marymont College (appendix B).

Key in **EcoSim** to open this simulation package. Pull down the **Help** menu and review the information on **Ecological Models.** Pull down the **Model** menu, and select one of the first models. Pull down the **Run** menu, highlight **Start Simulation,** and note how population growth is displayed in the **Results** and **Graph** windows.

DENSITY-INDEPENDENT POPULATION GROWTH

Density-independent population growth is that expected in an environment in which the growing population does not encounter any "limits to growth." This pattern of population can be described as *geometric* or *exponential.*

Geometric population growth is described by the simple difference equation

$$dN/dt = rN$$

This equation says that the change in number of individuals (*dN*) per unit time (*dt*) equals the per capita growth rate (*r*) times the number of individuals present at the start of the time period (*N*). For a closed population, *r* is the difference between per capita birth and death rates. Geometric growth would proceed as a sequence of steps in which a population experiences growth in discrete steps.

Pull down the **Model** menu, and select **Geometric.** Pull down the **Options** menu, and note that a **Set Scale** option is available. Select this option, and observe how you can relabel the *x*- and *y*-axes of the graph window. Examine the **Run** menu again, and note the options to **Reset Initial Population Size** and **Reset Parameter Values.** Using these menus, scale the graph so that the *x*-axis has 10 units and the *y*-axis 200 units,

Begon et al. (1990): 4:148–152, 6:215–226, 7:246–250, 10:337–343
Brewer (1994): 4:81–117, 6:187–198, 8:229–245
Colinvaux (1993): 8:136–149, 10:179–202
Ehrlich and Roughgarden (1987): 5:65–72, 102–107, 12:247–258, 13:261–265, 16:346–352
Krebs (1994): 12:198–229, 13:230–261, 14:262–287
Odum (1993): 6:151–172

Pianka (1994): 8:157–160, 9:182–187, 12:242–249, 15:306–314
Ricklefs (1990): 16–19:302–382, 21–24:402–501
Ricklefs (1993): 15:262–280, 19:341–360, 20:361–384
Smith (1990): 15:362–389, 18:450–464, 19:484–497
Smith (1992): 12:180–188, 13:189–203, 15:217–233, 16:236–252
Stiling (1992): 11:209–220, 12:234–247

and set the birth rate to 0.4 and the death rate to 0.05, so that $r = 0.35$. Pull down the **Run** menu, highlight **Start Simulation,** and note how population growth is displayed in the **Results** and **Graph** windows. Does the population grow continuously or in a series of steps? What is the population size after 10 time periods? What is the doubling time for this population?

Some populations do not grow in discrete steps, but continuously or exponentially. The integrated form of the above equation is

$$N_t = N_0 e^{rt}$$

This equation says that N at any number of time units (t) after time zero (N_0) equals N_0 times e, the base of natural logarithms, to the exponent rt.

Pull down the **Model** menu, and select **Exponential.** Using the **Options** and **Run** menus, structure the graph and set population parameters to the same values as for the geometric simulation. Run this simulation. Does the population grow continuously or in a series of steps? What is the population size after 10 time periods? What is the doubling time for this population? Why does the continuous exponential growth model differ from the discrete geometric model?

Populations of animals introduced into extensive, favorable environments sometimes grow for a long time in a manner uninfluenced by their own population density. Using the exponential simulation model, determine r for the following cases given by Brewer (1994, p. 88). Structure the axes of the graph window appropriately for the magnitude of population size and time for these cases, type the initial population size of these populations, and enter what you think to be reasonable estimates of per capita birth and death rates (which lead to an estimate of r) into the simulation parameter set. Run the simulation and see how closely the population comes to that observed in the same time period. Experiment until you obtain a curve closely matching the observed data. What is the value of r? What is the population doubling time?

1. A group of eight ring-necked pheasants were introduced to Protection Island, Washington, in 1898. In 6 years, this population had increased to 1,898 individuals.
2. Eight tule elk were translocated onto Grizzly Island in the Sacramento Delta region near San Francisco, California. In 8 years, this population had grown to 150 individuals.

LOGISTIC POPULATION GROWTH AND LIMITATION

One of the most simple models of population growth under conditions that are limiting is the *logistic model,* which incorporates an environmental *carrying capacity, K.* This value is incorporated in a modified form of the geometric equation so that a density-dependent constraint on growth exists:

$$dN/dt = rN(K-N/K)$$

Some of the best information on the behavior of vertebrate populations in environments of limited carrying capacity has come from studies of white-tailed deer on the George Reserve, a research site of the University of Michigan. Using data from McCullough (1979), we shall examine certain features of the logistic curve.

In 1928, six deer were introduced to the reserve, a fenced, predator-free area. In 7 years, the population grew exponentially to 222 animals, displaying an intrinsic rate of annual increase of 0.675. (You can check this by returning to the **Exponential Population Growth** module.) After this, culling of animals was begun to keep the herd at a level that did not cause severe forage damage.

Long-term studies of the population, which is censused and culled regularly, have suggested that the carrying capacity (K) of the reserve is about 176 animals. Pull down the **Model** menu, and select **Discrete Logistic.** Using an initial population of 6 and an r value of 0.675, what is the approximate population expected after 7 years? (You will need to rescale the axes of the graph and adjust the model parameters to do this.) Does the discrete form of the logistic permit the population to reach 222? How does the introduction of the carrying capacity, K, influence the shape of the population growth curve?

Restructure the simulation so that the initial population is 222. How does the population behave? What do you think would have happened in the years after 1935 if the population of 222 deer had not been culled? Do you think the carrying capacity would have been affected by the high population? How? What do you think permitted the population to grow to a level substantially above the carrying capacity? How might the simple logistic equation be modified to take such an influence into account?

BEHAVIOR OF AGE-STRUCTURED POPULATIONS

The simple models we have just examined consider each of the N individuals of the population to be identical in characteristics such as risk of dying or probability of reproducing. For most plants and animals this is clearly untrue for individuals of different ages. Models are available, however, for specifying these aspects of a species' population ecology and following the behavior of the population through time. These models use matrix algebra to project the behavior of the population through time, and thus are best explored by computer.

Pull down the **Model** menu, and select **Leslie Matrix.** Pull down the **Run** menu and select **Reset Parameter Values.** Note that you can specify the number of age classes, the survival probability of all but the oldest (why not for the oldest?), and the fecundity of individuals of all age classes.

Select **Reset Parameter Values** and create a simulation for four age classes of a short-lived animal such as a meadow vole. Discuss with your colleagues what might be reasonable survival and fecundity values for individuals of different ages. Put these values into the simulation. Run the simulation, which gives the stable proportions of individuals of different age classes that should develop in time. Note these values. Does the age class with the highest survival value necessarily become the most frequent group in the population?

Before abandoning this simulation, assume that a group of 40 individuals of age class #1 and 10 of age class #2 colonized a previously inaccessible island in San Francisco Bay. Following this group through time, using the iteration procedure, determine how many time periods will be required for the population age structure to reach equilibrium. Do the frequencies of different age groups change gradually and progressively toward equilibrium, or are there erratic fluctuations?

INTERSPECIFIC COMPETITION

Two mathematical ecologists, an Italian, Vito Volterra, and an American, A. J. Lotka, in 1926 and 1932 respectively, modified the logistic equation to include competitive effects between individuals of different species. The resulting equations are thus known as the Lotka-Volterra equations, and, in numerous derived forms, still provide the mathematical basis for much theory about interspecific competition, competitive displacement, and resource partitioning. These equations are

$$\text{Species 1} \qquad \frac{dN_1}{dt} = r_1 N_1 \left(\frac{K_1 - N_1 - \alpha N_2}{K_1} \right)$$

$$\text{Species 2} \qquad \frac{dN_2}{dt} = r_2 N_2 \left(\frac{K_2 - N_2 - \beta N_1}{K_2} \right)$$

In these equations, α measures the effect of N_2 on N_1 in terms of the number of individuals of N_1 to which one individual of N_2 is equal. Similarly, β measures the effect of N_1 on N_2 in terms of the number of individuals of N_2 to which one individual of N_1 is equal. The outcome of competition between two species depends on the ratio of α to K_1/K_2 and of β to K_2/K_1. There are four possible outcomes:

1) $\alpha > K_1/K_2$, $\beta > K_2/K_1$ Only one species survives, depending on initial relative abundances

2) $\alpha < K_1/K_2$, $\beta < K_2/K_1$ Both species survive

3) $\alpha < K_1/K_2,\ \beta > K_2/K_1$ Species #1 always wins
4) $\alpha > K_1/K_2,\ \beta < K_2/K_1$ Species #2 always wins

These relationships can be shown on graphs in which the abundance of one species is shown on the vertical axis and of the other on the horizontal axis. Isoclines can then be drawn showing the number of each species that can exist at different numbers of the other. From any point on this graph, representing an initial combination of abundances of the two species, one can project how competition will proceed. If the initial point is below one species' isocline, that species can increase until it reaches the isocline. The species that can increase after the other has reached its isocline will ultimately survive.

Pull down the **Model** menu, and select **Lotka-Volterra Competition.** Note the four cases corresponding to the cases listed above (**a12** = α, and **a21** = β in this listing). Once you understand the meaning of these coefficients, and are able to interpret the graphs, carry out the following simulations:

The blue-winged warbler and the golden-winged warbler are two closely related birds of eastern North America. The blue-winged warbler appears to be replacing the golden-winged warbler in some areas. Suppose that interspecific competition is the mechanism, and explore what might happen in particular locations, given hypothetical values for their population growth, carrying capacity, and competitive interaction. Suppose that the K and r values of these two species are very similar:

	Blue-winged	Golden-winged
K	52 birds/km^2	48 birds/km^2
r	0.35	0.42

Simulation #1

Suppose that $\alpha = 0.9$ and $\beta = 1.3$, and that six blue-winged warblers (three pairs) have become established in an area with 46 golden-winged warblers (23 pairs). Using **Reset Parameter Values,** enter these values and conduct the simulation. Does one, or do both, species survive? What happens if the values of α and β are switched? What is the position of the isocline for the surviving species in each case?

Simulation #2

Suppose that the intensity of interspecific competition is weaker for both species, and that $\alpha = 0.8$ and $\beta = 0.7$. Using the same initial population densities, what happens? What do these values of the competition coefficient imply about the relative intensity of intraspecific and interspecific competition? How do the isoclines for this case differ from those in simulation #1?

Simulation #3

Suppose now that $\alpha = 1.4$ and $\beta = 1.2$. Given the same initial populations, what is the outcome of interaction of the species? Change the initial numbers, and determine whether the outcome is always the same. Does the species with the lower coefficient always survive? Experiment until you find the critical relation that determines the outcome of competition.

PREDATOR-PREY INTERACTION

Lotka and Volterra also devised a simple model of interaction between a predator (P) and a herbivore prey (H). In this model, the change in the prey population depends on its growth rate, r, and the rate of capture, p, by the predator:

$$dH/dt = rH - pH$$

The change in the predator population, in turn, depends on the growth resulting from the capture rate, p, and assimilation rate, a, by the predator, and the decline due to the mortality rate, m, of the predator:

$$dP/dt = apH - mP$$

These equations show that populations of the two species can sometimes cycle indefinitely, the ups and downs of the predator population following those of the prey. When such a cycle is graphed on a two-axis graph with the prey on the horizontal axis and the predator on the vertical axis, the points representing their combined populations through time form a closed loop.

From the **Model** menu, select **Volterra Predation.** Carry out the simulation using the parameters of the base simulation, noting the behavior of the predator and the prey in both graph and results windows. What happens to populations of the two species when their combined populations fall in the lower left quadrant of the graph? In the lower right? In the upper right? In the upper left?

By resetting parameter values, carry out simulations to find out what eventually happens when the predator death rate decreases and when it increases. What happens when the prey growth rate decreases and increases? When capture or assimilation rates of the predator increase or decrease?

✪ SUGGESTED ACTIVITIES

1. Examine the numbers of an invading species, such as the European starling or cattle egret in the western United States or the house finch in the eastern United States, in Christmas bird counts at a particular location. These counts have been published over the years in *American Birds* and its predecessors (*Audubon Field Notes, Audubon Magazine*). Are the patterns of change in abundance fitted well by geometric, exponential, or logistic curves? Are reasonable growth rates or carrying capacities suggested?

2. Solve the Lotka-Volterra equations given earlier for α and β. Locate census data for a pair of species presumed to be close competitors (e.g., blue-winged and golden-winged warblers in a locality in New York State). Using the observed changes in abundance of the species, together with estimates of their r and K values, estimate the value of the competition coefficients.

3. List the assumptions about individuals and their activities that are implicit in the model of predator-prey interaction. List 10 predator-prey systems in nature (e.g., wolf-moose) and check off the assumptions that seem likely to be violated in the real world.

✪ QUESTIONS FOR DISCUSSION

1. What are the implications of the fact that the carrying capacity, K, in the logistic equation and its modifications is not constant through time? What patterns and degrees of change in K are likely to occur during the seasons for species such as annual plants and insects? For vertebrates? What patterns probably occur through the years for vertebrate populations in different ecosystems?

2. What sorts of interactions might occur between predation and competition? Can predation modify the outcome of interspecific competition?

3   Do you think r is necessarily a constant in populations interacting in ecological time? Can you describe a scenario in which changes in r might occur during a seasonal cycle, over the course of ecological succession, or in different units of a metapopulation?

✪ SELECTED BIBLIOGRAPHY

Begon, M., and R. Wall. 1987. Individual variation and competitor coexistence: a model. Functional Ecology 1:237–241.

Brewer, R. 1994. The science of ecology. 2nd ed. Saunders College Publishing, Ft. Worth, Texas, USA.

Dorshner, K. W., S. F. Fox, M. S. Keener, and R. D. Eikenbary. 1987. Lotka-Volterra competition revisited: the importance of intrinsic rates of increase to the unstable equilibrium. Oikos 48:55–61.

Gilpin, M. E. 1992. Demographic stochasticity: a Markovian approach. Journal of Theoretical Biology 154:1–8.

Kingsland, S. 1982. The refractory model: the logistic curve and the history of population biology. Quarterly Review of Biology 57:29–52.

May, R. M. 1974. Biological populations with non-overlapping generations: stable points, stable cycles and chaos. Science 186:645–647.

Murray, B. G., Jr. 1987. The calculation of the rate of increase, *r*, of populations with heterogeneous life histories. Oikos 50:262–266.

Pielou, E. C. 1981. The usefulness of ecological models: a stock-taking. Quarterly Review of Biology 56:17–31.

Schaffer, W. M. 1985. Order and chaos in ecological systems. Ecology 66:93–106.

Stone, L., and A. Roberts. 1991. Conditions for a species to gain advantage from the presence of a competitor. Ecology 72:1964–1972.

Wangersky, P. J. 1978. Lotka-Volterra population models. Annual Review of Ecology and Systematics 9:189–218.

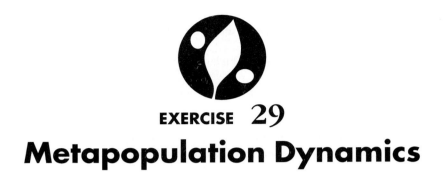

Metapopulation Dynamics

INTRODUCTION

Frequently, the overall population of a species is distributed as a series of semi-isolated subpopulations among which occasional dispersal occurs. Such units constitute a *metapopulation*. The sizes and isolation of the units of a metapopulation, together with the species' population growth and dispersal abilities, influence the regional persistence of the species. This has made metapopulation dynamics of major concern to conservation biologists, who increasingly are dealing with fragmented habitats containing semi-isolated portions of a once-continuous population.

A major feature of small populations is the importance of *stochastic,* or randomly operating, factors on population dynamics. Stochastic factors fall into two categories. *Demographic stochasticity* involves random influences on mortality and reproduction—factors that operate in all populations but assume greater relative importance in small populations. The accidental failure of individuals to mate during a window of fertility, for example, probably would not lead to a major impact in a population of 1,000 pairs, but a few such failures in a population of five pairs might cause a near-catastrophic decline in numbers in the next generation. *Environmental stochasticity* is the effect of chance variation in environmental conditions on the survival and reproduction of members of a population. This sort of influence can vary from the usual variability seen in almost every factor of the environment, ranging from season-to-season variations in weather to the catastrophic impact of random events such as hurricanes, volcanic eruptions, and the like. Because of the relative importance of stochastic influences in small populations, much of the theory of metapopulation dynamics has centered on these influences.

MODELING METAPOPULATION DYNAMICS

We will examine metapopulation dynamics by using *RAMAS/space,* a program that models the dynamics of single populations or multiple populations that constitute a metapopulation (appendix B). RAMAS/space (version 1.31) incorporates environmental and demographic stochasticity, but does not take into account the age structure of the population(s). It estimates the pattern of change in a population or metapopulation over future time, and computes the probabilities of exceeding or falling below threshold sizes (including a size that corresponds to extinction). This program is now being used by a number of investigators and environmental consultants in population viability analysis.

In the program, the size and spacing of metapopulation units can be specified. The finite growth rate, *R,* is used to describe potential population growth, and environmental stochasticity is modeled as random variation in this parameter. (*R* equals 1.0 when the population is stable.) Demographic stochasticity is modeled by random variation in the product of survivorship and fecundity. Density-dependent population behavior relative to a carrying capacity, *K,* can be specified. One can also include the *Allee Effect,* which is simply an optimal population

---

Brewer (1994): 4:114, 19:632–636   Pianka (1994): 9:199–200
Krebs (1994): 16:331–335, 19:402–428   Ricklefs (1993): 16:289–293

size, both above and below which R decreases. Migration among units of the metapopulation is modeled in relation to the size of each unit and the distance between units. A density-dependent pattern of migration can be specified, meaning that as a population unit approaches its K, the per capita probability of migration increases, and as it declines to a low level, the probability of migration decreases to near zero. The RAMAS/space manual describes these relationships in more detail.

Opening and Becoming Familiar with RAMAS/space

To open RAMAS/space from its resident directory, type **SPACE.** You will see the opening screen, and find basic instructions for initial use of the program. RAMAS/space is a challenging program that requires some preliminary study to learn its capabilities and commands. First, therefore, you should work your way through the tutorial section of the manual, so that you become familiar with the following:

- Use of the keys, especially the following:
    F1—Help. **PgDn** and **PgUp** enable movement through help text.
    F2—Fill. Fills **Migration** and **Correlation** matrices with default values.
    F3—Load. Enables recall of a model that exists as a saved file (e.g., **RAIL.SP**).
    F4—Save. Enables a created or modified file to be saved.
    F5—Add. Enables a new population to be added to an existing model.
    F9—View. Displays a graph on screen. **Ctrl-D** while graph is on screen causes graph to be printed.
    F10—Menu. Returns you to the **Input/Results** menu screen.
    Ctrl-C—Reveals **Correlation** matrix.
    Ctrl-M—Reveals **Migration** matrix.
    Ctrl-L—Reveals list of populations in metapopulation.
    Ctrl-S—Reveals map of metapopulation spatial structure.
- Content and use of the **Input** and **Results** screens, particularly the following:
    Input
        General Information
        Populations
        Spatial Structure
        Migration Function
        Migration Matrix
        Correlation Function
        Correlation Matrix
    Results
        Trajectory Summary
        Metapopulation Occupancy
        Interval Extinction Risk
        Terminal Extinction Risk
        Time to Quasi-extinction
- The sample files, especially the following, and the program capabilities they illustrate:
    1LARGE.SP
    5SMALL-C.SP
    5SMALL-I.SP
    5SMALL-M.SP
    5SMALL-D.SP
    METAPOP6.SP

Before actually running simulations with these, you may want to adjust the number of replicates to 10 or so.

Conducting a Simulation

Three simulation files based on real natural populations are included with the RAMAS/space program: **RAIL.SP, OWL.SP,** and **ALBATROS.SP.**

Load the **RAIL.SP** file. This file describes the metapopulation of the light-footed clapper rail, a federally listed endangered species occurring in coastal salt marshes of southern California. The population sizes and parameters for this metapopulation are based on several years of study, so this can be considered a realistic simulation.

Select the general information screen and set replications to **10** and duration to **100.** Return to the main menu screen and select **Run** to carry out the simulation. Examine the **Interval Extinction Risk** graph and print it (remember some peculiarity of your graph, because the printouts as they appear will look very similar). What is the probability that the metapopulation will go extinct some time in the next 100 years under the base conditions (each tick on the probability scale is 2%)?

Examine the **OWL.SP** and **ALBATROS.SP** files in similar fashion. These are also realistic simulations of metapopulations of the California spotted owl and Galapagos waved albatross, respectively.

⊘ SUGGESTED ACTIVITIES

1. Design a modification of the **RAIL.SP** file to explore what changes in coastal marsh habitat, or in assumptions about clapper rail behavior, might improve the survival probability of this species. Assume that we wish to achieve/maintain a 95% probability of survival of the species for 100 years. For example, what if
   - maximum migration distance is greater than 40 km?
   - migration tendency is density dependent?
   - the K levels of certain populations are increased by restoration efforts?
   - a new population is created at Los Angeles (geographical coordinates: $X = 107.2$, $Y = 63.2$)?
   - habitat protection increases the R value of certain populations?
   - the basic degree of correlation of environmental conditions is less than the base value of 0.8?
   - quality of habitat gradually increases over the years by natural recovery?

   If you make changes in basic migration or correlation parameters, you must select the respective matrices and update them by use of the **F2** key, which enters default values that correspond to the altered basic parameters. If you want to make migration density dependent, you would specify 0.05 (a 5% increased tendency at high density), or some such positive value in the population listing for each metapopulation unit. To gradually increase habitat quality over years, indicate a value such as 0.005 for the change in K option (this would be 0.5% improvement in K per year, for 100 years).

   Initially, don't try more than 10 replicates for 100 years. If this goes fast enough, then you may increase the number of replicates. When you have made a change, run the simulation and check the output screens to see what happened. When you get a good one (i.e., with improved survival probability), print the **Interval Extinction Risk** screen, and note on it what you did.

2. Examine the **OWL.SP** model and determine the critical factor for the tendency for the population to decline progressively over the simulation period. Discuss whether or not the various simulation parameters (i.e., R's, K's, migration and correlation parameters) are reasonable. Make adjustments to these parameters, and see what influence they have on survival of the species in southern California. Assume that habitat loss occurs in some of the metapopulation units nearest large cities such as Los Angeles and San Diego (by modifying the **Change in K** parameter). Determine what effect this will have on persistence of the species. Determine if such an effect can be offset by improved habitat conditions in other metapopulation units.

3. Create a metapopulation model for a species (real or hypothetical) in your own area by specifying x- and y-coordinates and filling in subpopulation parameter data on an input screen for each metapopulation unit.

QUESTIONS FOR DISCUSSION

1. Why are some units in metapopulations known as *sources* and some as *sinks*? Can you identify such units in one of your simulations?

2. What sorts of organismal and environmental attributes affect the dispersal of individuals of a species among metapopulation units? What characteristics of corridors linking such units could be important?

3. What sorts of genetic processes occur in small metapopulation units? Are these considered in RAMAS/space? How would you take such factors into account in a metapopulation simulation?

SELECTED BIBLIOGRAPHY

Akçakaya, H. R., and S. Ferson. 1992. Spatially structured population models for conservation biology. Applied Biomathematics, Setauket, New York, USA.

Gilpin, M. E., and I. Hanski, editors. 1991. Metapopulation dynamics. Academic Press, London, England.

Gotelli, N., and W. G. Kelley. 1993. A general model of metapopulation dynamics. Oikos 68:36–44.

Hanski, I. 1994. A practical model of metapopulation dynamics. Journal of Animal Ecology 63:151–162.

———, and M. Gyllenberg. 1993. Two general metapopulation models and the core-satellite species hypothesis. American Naturalist 142:17–41.

Hastings, A. 1991. Structured models of metapopulation dynamics. Biological Journal of the Linnean Society 42:73–88.

Mangel, M., and C. Tier. 1993. Dynamics of metapopulations with demographic stochasticity and environmental catastrophes. Theoretical Population Biology 44:1–31.

Nee, S., and R. M. May. 1992. Dynamics of metapopulations: habitat destruction and competitive coexistence. Journal of Animal Ecology 61:37–40.

EXERCISE 30

Detecting Density-Dependent Change in Populations

INTRODUCTION

Population regulation is the tendency for a population to grow asymptotically toward a certain density, and to return to that density when accidental deviations occur from it. Regulation results from the action of direct density-dependent factors, which are factors that reduce reproductive success or increase mortality rate as population density increases. Direct density-dependent factors are typically biotic relationships such as competition (intraspecific or interspecific, including interference), predation, parasitism, and disease. These factors reduce population growth as the population approaches or exceeds its asymptotic level, but relax their action and permit faster growth when the population drops below this level. Thus, they cause the population growth rate to show a negative relationship to population size (McCullough 1990).

In a sense, density-dependent regulation is responsible for two phenomena: (1) the cessation of population growth at some characteristic density level, and (2) the corrective adjustment of a near-equilibrium population to the small imbalances that frequently result from the accidents of nature.

Density-dependent behavior can be recognized in sequential population censuses. For this, data are needed on population sizes at successive and comparable points in time (the same phase in the annual cycle of species with overlapping generations, or the same developmental stage for species with nonoverlapping generations). Analysis of such data can reveal whether or not density-dependent factors are the dominating influence on a population as it grows in size or fluctuates about some average level (Vickery and Nudds 1991).

DETECTING DENSITY-DEPENDENT TRENDS

Several techniques have been suggested for the detection of density-dependent patterns in census data. Slade (1977) compared the performance of several of the simplest of these. More recent studies of the problem of detecting density dependence have suggested that simulation procedures (Vickery and Nudds 1984) or computerized randomization procedures (Pollard et al. 1987) are more effective. These techniques differ in their accuracy, ease of calculation, and efficiency in detecting density dependence under different circumstances (Gaston and Lawton 1987, Vickery and Nudds 1991). We shall consider two of the simplest techniques (Slade 1977), which are applicable to somewhat different types of data.

Brewer (1994): 4:102–114
Colinvaux (1993): 10:174–202
Ehrlich and Roughgarden (1987): 5:90–101
Krebs (1994): 16:322–348
Odum (1993): 6:151–157
Pianka (1994): 9:188–189

Ricklefs (1990): 17:330–346
Ricklefs (1993): 15:274–280
Smith (1990): 15:373–383
Smith (1992): 13:190–203
Stiling (1992): 15:283–293

Populations Growing Toward Equilibrium

The simplest technique for detecting density dependence in growing populations is calculation of the slope of the standard major axis, b_s, for sets of paired estimates of populations at successive points in time (t and $t + 1$). For a sequence of population censuses, t and $t + 1$ values are constructed for overlapping pairs of census values: $t1$ and $t2$, $t2$ and $t3$, $t3$ and $t4$, etc. These values are then converted to natural logarithms, here designated by the symbol X. The slope of the standard major axis, b_s, for N sets of X_t and X_{t+1} data is calculated by the equation

$$b_s = \left(\frac{\Sigma X_{t+1}^2 - \dfrac{(\Sigma X_{t+1})^2}{N}}{\Sigma X_t^2 - \dfrac{(\Sigma X_t)^2}{N}} \right)^{1/2}$$

A value of 1.0 for b_s indicates that population growth is essentially exponential, showing no tendency to slow down as it should if it is approaching an asymptote and experiencing the effects of density-dependent factors. A value less than 1.0 is evidence of direct density-dependent response. Deviation of a calculated value of b_s from 1.0 is evaluated by a t test, as follows:

$$t = \frac{1.0 - b_s}{s_b} \qquad DF = N - 2 \qquad \text{Where: } s_b = \left(\frac{\text{Residual Mean Square}}{\Sigma X_t^2 - \dfrac{(\Sigma X_t)^2}{N}} \right)^{1/2}$$

The residual mean square is obtained from an analysis of variance table for determining the significance level of a regression coefficient (table 8.2, p. 54). Critical t values are obtained from table A.2, p. 247.

Populations Fluctuating About Equilibrium

Bulmer (1975) developed a technique which is most efficient in detecting density dependence in populations fluctuating about a general equilibrium level. In this procedure, a density-dependence index, R, is calculated by the equation

$$R = V/U \qquad \text{Where: } U = \sum_{t=1}^{N-1} (X_{t+1} - X_t)^2$$

$$V = \sum_{t=1}^{N} (X_t - \bar{X})^2$$

The variables N and X are as defined earlier.

This calculation effectively compares the differences between successive population values to the deviations of these values from the equilibrium level. If deviations from equilibrium are small relative to those between times, most fluctuations must be back and forth across the equilibrium level, as would be expected when density-dependent factors are operating.

Critical R values were determined empirically by computer simulation and are obtained for various probability levels by the equation

$$R_{critical} = 0.25 + (N - 2) X_L \qquad \text{Where: } X_L = 0.0366 \ (\alpha = 0.05)$$
$$0.0248 \ (\alpha = 0.01)$$
$$0.0170 \ (\alpha = 0.001)$$

To indicate a significant degree of density dependence, a calculated value of R must be equal to or less than the critical R.

🔾 SUGGESTED ACTIVITIES

1. Examine breeding or wintering bird census data for a locality that has been studied for many years. Breeding and wintering population censuses for North American localities are published in the *Journal of Field Ornithology* (and formerly, *American Birds, Audubon Field Notes,* and *Audubon Magazine*). Most censuses that span a significant period of years (e.g., 10+) are for mature habitats, and the populations are therefore likely to be fluctuating about a general equilibrium. For such data, significant comparisons can be made between the degree of density dependence for summer vs. winter censuses, total birds vs. individual species, migrant vs. resident species, seed-eaters vs. insect-eaters, small-bodied vs. large species, hole-nesters vs. open-nesters, and many others.

2. Analyze data for a growing population of birds or other animals. Data for species such as the European starling, the cattle egret, or the house finch (in eastern North America) can be obtained from Christmas bird counts published in *American Birds.*

3. Examine data on the growth of the North American population of the whooping crane since 1941 (Doughty 1989 and recent issues of *American Birds*) by calculating the slope of the standard major axis. You may want to try this analysis both with and without inclusion of the experimental Rocky Mountain subpopulation (which did not lead to a viable breeding population).

    Data of these types can be recorded in table 30.1, and calculations of b_s and R also carried out and recorded in this table.

🔾 QUESTIONS FOR DISCUSSION

1. Does recognition of density-dependent fluctuations in a population provide much insight into the actual mechanisms affecting population growth?

2. How does the ability to detect density dependence vary with the importance of density-independent factors and their effects on population size?

3. What do you think the capability of these techniques would be to detect density-dependent fluctuations about an equilibrium level that is gradually changing? In what situations might such a pattern occur?

🔾 SELECTED BIBLIOGRAPHY

Bulmer, M. G. 1975. The statistical analysis of density dependence. Biometrics 31:901–911.

Doughty, R. W. 1989. Return of the whooping crane. University of Texas Press, Austin, Texas, USA.

Gaston, K. J., and J. H. Lawton. 1987. A test of statistical techniques for detecting density dependence in sequential censuses of animal populations. Oecologia 74:404–410.

Holyoak, M. 1993. New insights into testing for density dependence. Oecologia 93:435–444.

McCullough, D. R. 1990. Detecting density dependence: filtering the baby from the bathwater. Transactions of the North American Wildlife and Natural Resources Conference 55:534–543.

Pollard, E., K. H. Lakhani, and P. Rothery. 1987. The detection of density dependence from a series of annual censuses. Ecology 68:2046–2055.

Slade, N. A. 1977. Statistical detection of density dependence from a series of sequential censuses. Ecology 58:1094–1102.

Stenning, M. J., P. H. Harvey, and B. Campbell. 1988. Searching for density-dependent regulation in a population of pied flycatchers *Ficedula hypoleuca* Pallas. Journal of Animal Ecology 57:307–317.

Vickery, W. L. 1991. An evaluation of bias in k-factor analysis. Oecologia 85:413–418.

———, and T. D. Nudds. 1984. Detection of density dependent effects in annual duck censuses. Ecology 65:96–104.

———. 1991. Testing for density-dependent effects in sequential censuses. Oecologia 85:419–423.

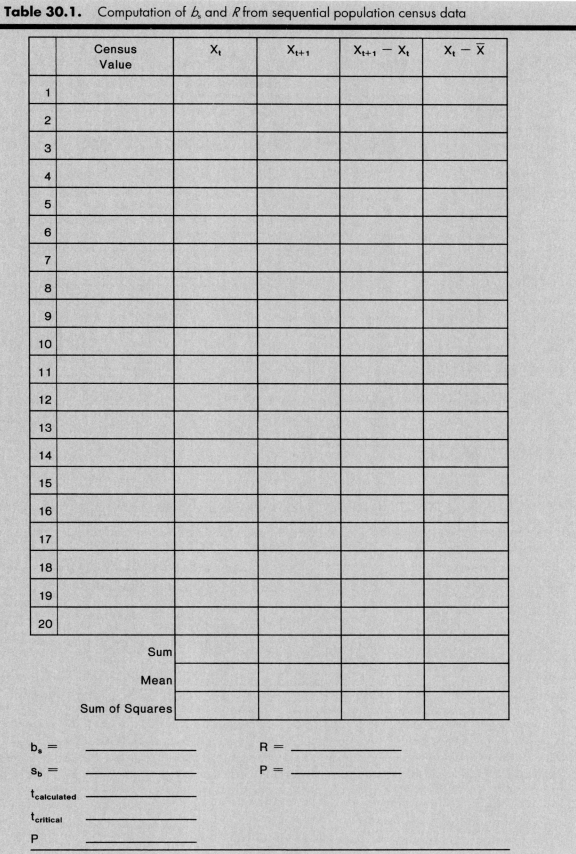

Table 30.1. Computation of b_s and R from sequential population census data

	Census Value	X_t	X_{t+1}	$X_{t+1} - X_t$	$X_t - \overline{X}$
1					
2					
3					
4					
5					
6					
7					
8					
9					
10					
11					
12					
13					
14					
15					
16					
17					
18					
19					
20					
Sum					
Mean					
Sum of Squares					

$b_s =$ _____ $R =$ _____

$s_b =$ _____ $P =$ _____

$t_{calculated}$ _____

$t_{critical}$ _____

P _____

Interspecific Association

INTRODUCTION

Quantifying the degree of spatial or temporal association of individuals of different species is a valuable technique for studying interspecific relationships and community structure. Positive association—co-occurrence more frequently than expected by chance—may indicate, for example, an interaction favorable to one or both species, such as mutualism, commensalism, or a specific herbivore-plant or parasite-host relation. In a heterogeneous environment, a positive association may also indicate similarity of the species in adaptations and responses to habitat conditions. Measures of interspecific association therefore provide an objective method for grouping species with similar habitat relations. Negative associations, or co-occurrence less frequently than expected by chance, may reflect interactions that are detrimental to one or both species, such as interspecific competition, interference, or the avoidance of predators by potential prey. In a heterogeneous environment, negative associations can result from preference of species for different habitat conditions.

This exercise describes methods for measuring the degree of positive or negative association between individuals of different species in a specific habitat or habitat complex. Similar procedures can be used to analyze the co-occurrence of species in different communities, such as those of different islands or other geographical units. The latter situations, however, present some difficulties in defining the expected random co-occurrence pattern (Wilson 1987).

CALCULATING COEFFICIENTS OF INTERSPECIFIC ASSOCIATION

Coefficients of interspecific association can be calculated from unit-area (quadrat), unit-volume, or unit-time sampling data. The data are arranged in a 2 × 2 contingency table. To construct this table (see below), the sampling units are examined for presence or absence of both species (A and B), and the number of samples containing both species (a), only species A (b), only species B (c), and neither (d) are recorded.

		Species B		
		Present	Absent	
Species A	Present	a	b	a + b
	Absent	c	d	c + d
		a + c	b + d	T

Letter designations should be assigned to the species so that species A does not occur in more samples than does species B (i.e., a + b must be equal to or less than a + c). From these data, a coefficient of association (C_7), varying from +1.0 for a condition of maximum possible co-occurrence of species, to −1.0 for a condition of minimum possible co-occurrence, is calculated. A value of 0 for this coefficient indicates that the frequency of co-occurrence is that expected by chance. The exact form of the calculation depends on the numerical relationships of values in the contingency table. The appropriate equation given below should be used for this calculation (Ratliff 1982):

$$\text{When } ad \geq bc \text{ and } c < b \qquad C_7 = \frac{ad - bc}{(a + b)(b + d)}$$

$$\text{When } ad \geq bc \text{ and } c \geq b \qquad C_7 = \frac{ad - bc}{(a + c)(c + d)}$$

$$\text{When } ad < bc \text{ and } a \leq d \qquad C_7 = \frac{ad - bc}{(a + b)(a + c)}$$

$$\text{When } ad < bc \text{ and } a > d \qquad C_7 = \frac{ad - bc}{(b + d)(c + d)}$$

A contingency chi-square can be calculated from the values in the association table, using the procedure described for a 2×2 contingency table in exercise 7.

Hurlbert (1969) found that the C_7 coefficient is somewhat biased, because it is influenced somewhat by the actual frequencies of the species. He presents a modified calculation for a coefficient of association (C_8) that removes this bias. This modification does not, however, affect the validity of the chi-square testing procedure outlined above for recognition of significant positive or negative association between species. Ratliff (1982) provides a correction for the coefficient suggested by Hurlbert (1969). The difference between C_7 and C_8 in most cases is slight. Furthermore, the calculations for C_8 are somewhat involved and are best done by the use of a computer program (see below).

COMPUTER ANALYSIS OF INTERSPECIFIC ASSOCIATION

The software package Ecological Analysis–PC Vol. 1, distributed by Oakleaf Systems (Appendix B), contains an interspecific-association module. This module requires only the input of the frequencies of the four co-occurrence combinations, and gives values of C_7 and C_8, together with the chi-square value required to test the level of significance for the association.

✪ SUGGESTED ACTIVITIES

1. Sample one or more areas of natural vegetation, such as a grassland, shrubland, or forest, using quadrats of appropriate size. If time permits, select two areas that differ in heterogeneity or intensity of disturbance. Obtain random quadrat samples of species (exercise 5). For pairs of common species, determine the numbers of samples with both, each alone, and neither species. Calculate the values of C_7 and C_8, or obtain them by use of the Ecological Analysis software module. Compare association coefficients for particular pairs of species in heterogeneous versus homogeneous, or undisturbed versus disturbed, sites. Results of contingency-table analyses for pairs of species can be recorded in table 31.1.

    Alternatively, sample the desert vegetation map accompanying this manual, using clear plastic squares or circles as quadrats. Obtain one set of samples from the wash area and a second set from the nonwash area. Compare association coefficients for pairs of common species in the wash and nonwash sections of the plot, and also for the combined data from both sections.

2. Select an invertebrate herbivore (e.g., snail, insect) that has different potential plant hosts available in an area of habitat. Devise a plan for random sampling (exercise 5) and record the presence or absence of the herbivore and the various plant species in a series of quadrats. Determine if significant positive or negative associations exist between the herbivore and particular plant species.

Table 31.1. Results of contingency-table analyses of degree of interspecific association for pairs of species

Species Combination	Number of Samples	Coefficient of Association C_7	Coefficient of Association C_8	Chi-Square	Significance Level

3. Visit a habitat with several species of birds, frogs, or calling insects (e.g., grasshoppers, cicadas). Formulate one or more null hypotheses about whether different species should tend to sing simultaneously (e.g., mutually territorial bird species) or avoid simultaneous song (e.g., amphibians or insects with mate-attracting calls). Record the singing or calling activity of these species in short time intervals. Test for positive or negative temporal association of vocalizations by the species.

QUESTIONS FOR DISCUSSION

1. How might the degree of interspecific association be related to the overall dispersion pattern of individuals in a community?
2. What are some examples of heterogeneous environments in which high positive or negative interspecific associations might be expected? What are some specific mechanisms by which strong negative associations might be produced in a homogeneous environment?
3. How does the size of the quadrat or other sampling unit influence the sensitivity of the contingency-table technique for measuring association?

SELECTED BIBLIOGRAPHY

Cole, L. C. 1949. The measurement of interspecific association. Ecology 30:411–424.

Connor, E. F., and D. Simberloff. 1984. Neutral models of species' co-occurrence patterns. Pages 316–331 *in* D. R. Strong, Jr., D. Simberloff, L. G. Abele, and A. B. Thistle, editors. Ecological communities: conceptual issues and the evidence. Princeton University Press, Princeton, New Jersey, USA.

Dice, L. R. 1945. Measures of the amount of ecologic association between species. Ecology 26:297–302.

Greig-Smith, P. 1964. Quantitative plant ecology. 2nd ed. Butterworth and Co., London, England.

Hurlbert, S. H. 1969. A coefficient of interspecific association. Ecology 50:1-9.

Madgewick, H. A. I., and P. A. Desrochers. 1972. Association-analysis and the classification of forest vegetation of the Jefferson National Forest. Journal of Ecology 60:285-292.

Pielou, E. C. 1977. Mathematical ecology. John Wiley & Sons, New York, New York, USA.

Ratliff, R. D. 1982. A correction of Cole's C_7 and Hurlbert's C_8 coefficients of interspecific association. Ecology 63:1605-1606.

Sloan, N. A. 1982. Size and structure of echinoderm populations associated with different coexisting coral species at Aldabra Atoll, Seychelles. Marine Biology 66:67-75.

Strong, D. R. 1982. Harmonious coexistence of hispine beetles on *Heliconia* in experimental and natural communities. Ecology 63:1039-1049.

Wilson, J. B. 1987. Methods for detecting non-randomness in species co-occurrences: a contribution. Oecologia 73:579-582.

EXERCISE 32

Measurement of Species Diversity

INTRODUCTION

Species diversity is a characteristic unique to the community level of organization. The concept of diversity, however, has been applied to three quite different relationships. *Alpha diversity,* the central and perhaps most important concept, refers to the diversity of species within a given habitat. *Beta diversity,* in contrast, describes the degree of change in species from one habitat to another. *Gamma diversity,* finally, relates to the total regional species diversity that results from the number of habitats present, the diversity of species within each, and the degree of turnover of species between habitats. In this exercise we shall confine our attention to alpha diversity.

Alpha diversity, as most ecologists have utilized the concept, combines two distinct aspects of the species composition of communities: *number of species* and *equitability* of their abundance. These aspects of diversity can be described separately, by indices of species richness or of equitability of abundance, or they may be considered together, by indices of *heterogeneity.* Many indices of richness, equitability, and heterogeneity have been proposed. These indices vary widely in their mathematical basis, and often behave quite differently as sample size increases. None of the commonly used indices is independent of sample size, so that most comparisons of diversity indices require samples of equal size.

SPECIES RICHNESS

Species richness refers simply to the number of species in a given area of habitat or in a sample of given size. A number of mathematical indices of species richness, some tending to be independent of sample size, have been suggested (Pielou 1977). These indices, however, assume specific patterns of frequency of common to rare species, and are not accurate if these assumptions are not met. Rarely can a particular pattern be assumed with confidence, so that most ecologists prefer simply to use the number of species itself as the index of species richness.

The number of species can be determined either for a unit area of habitat or for a sample of a certain number of individuals. When number of species is related to area of habitat, the value is best considered to be *species density.* Species density varies widely with the productivity and favorability of the habitat and is important to consider, for example, in the selection of areas for preservation of biotic diversity. *Numerical species richness* is the number of species present in samples of a certain number of individuals. Communities very different in species density might be similar in numerical species richness, and vice versa. To determine numerical species richness, it is best to obtain samples of equal numbers of individuals by direct sampling. However, given samples of different total size, the number of species expected in a sample of certain size can be computed, although the computations can be very tedious.

Begon et al. (1990) 17:615-619   Odum (1993): 3:55-58   Smith (1990): 24:603-609
Brewer (1994): 10:296-302   Pianka (1994): 17:384-400   Smith (1992): 19:301-312
Colinvaux (1993): 16:312-334   Ricklefs (1990): 36:748-775   Stiling (1992): 17:318-335
Krebs (1994): 23:514-542, Appendix IV:704-706   Ricklefs (1993): 24:451-475

INDICES OF HETEROGENEITY AND EQUITABILITY

Interest in diversity indices and concern for the environmental impacts of pollution developed at about the same time, and diversity indices were quickly introduced into analyses of environmental quality. The observation that mature communities of stable environments typically show high species diversity, and those of disturbed or stressed situations less diversity, led some investigators to utilize diversity indices as measures of environmental stress (Wilhm 1970, Zand 1976). Other studies have shown that this must be done with great care (Godfrey 1978). Nevertheless, indices of heterogeneity and equitability remain in general use. The most important of these are discussed below.

Shannon Index

This index (Peet 1974, Pielou 1977) is the most widely used index of heterogeneity. It is derived from a function used in the field of information theory to describe the average degree of uncertainty of occurrence of a particular symbol at a certain point in a message, and consequently, the amount of information conveyed by its presence. As a diversity index for biotic communities, the function describes the average degree of uncertainty of predicting the species of an individual picked at random from the community. This uncertainty increases both as the number of species increases and as the individuals are distributed more and more equitably among the species already present. The general form of this index, H', is

$$H' = - \Sigma\, p_i \log p_i \qquad \text{Where: } p_i = \text{Decimal fraction of individuals belonging to the } i^{th} \text{ species}$$

This index is often calculated with natural logarithms (to base e), but often as well to base 2, the base employed in its use in information theory. Conversion of H' to base 2 from base e or base 10 is easy:

$$H' \text{ (base 2)} = 1.4427\, H' \text{ (base e)}$$
$$H' \text{ (base 2)} = 3.3219\, H' \text{ (base 10)}$$

The index can also be calculated from the numbers of individuals of each species (n_i) and their total (N) by the expression

$$H' = \log N - \frac{1}{N} \Sigma\, n_i \log n_i$$

These calculations are easily carried out with most available hand calculators.

The Shannon index varies from a value of 0 for communities with only a single species to high values for communities having many species, each with a few individuals. For a community containing nine individuals of one species and one of a second, H' (base 2) equals 0.469, and for a community with two species with five individuals each, the index equals 1.000. This index is somewhat more sensitive to the addition or loss of rare species than is the Simpson index, discussed below.

The null hypothesis that two Shannon diversity indices come from communities equal in diversity can be tested by a t test (Zar 1984). This test requires calculation of the variance, s^2, and the standard error of the difference, s_d, for each community by the equations

$$s^2 = \frac{\Sigma\, n_i \log n_i - (\Sigma\, n_i \log n_i)^2/N}{N^2}$$
$$s_d = \sqrt{s_1^2 + s_2^2}$$

For the test itself, t is given by

$$t = \frac{H_1' - H_2'}{s_d}$$

The t value has a number of degrees of freedom, DF, equal to

$$DF = \frac{(s_1^2 + s_2^2)^2}{\frac{(s_1^2)^2}{N_1} + \frac{(s_2^2)^2}{N_2}}$$

A measure of the equitability of abundance of species can be derived, using H'. The simplest such index, J', is simply

$$J' = H'/H'_{max}$$

For this calculation, H'_{max} is the value of H' computed with the same number of species, but equal p_i values.

Simpson Index

Simpson (1949) suggested an index corresponding to the number of randomly selected pairs of individuals that must be drawn from a community to have an even chance of obtaining a pair with both individuals of the same species. This index, D_s, is calculated by the equation

$$D_s = \frac{N(N - 1)}{\Sigma n(n - 1)}$$

Where: N = Total number of individuals of all species

n = Number of individuals of a species

This index increases from a value of 1.0 for a community containing only one species, to infinity for a community in which every individual belongs to a different species. For a community consisting of one species with nine individuals and a second species with one individual, the index equals 1.25. For a community with two species, each having five individuals, the index equals 2.25.

The null hypothesis that two Simpson indices come from communities of equal diversity can be tested by a procedure described by Keefe and Bergerson (1977).

A measure of equitability of abundance of species, V, can also be calculated for the Simpson index by the expression

$$V = N_2/N_{2\ max}$$

For this calculation, $D_{s\ max}$ is D_s calculated for the same number of species and individuals, but equal abundances of each.

The reciprocal of the Simpson index has also been used as a measure of concentration of dominance, C_d, or the degree to which the total abundance (or another quantitative value, such as cover) is concentrated in a few of the total species of the community. This index is

$$C_d = \frac{\Sigma\ n_i(n_i - 1)}{N(N - 1)}$$

Rarefaction and Relative-Abundance Curves

James and Rathbun (1981), among others, have pointed out that any heterogeneity index hides information on the two components of diversity that it combines. They recommend that, in addition to basing diversity measurements on samples of standard numbers of individuals or samples from areas of standard size, curves of rarefaction or relative abundance be provided. Rarefaction curves (figure 32.1) show how the number of species would change as the number of individuals is reduced toward zero by removing individuals at random. The number of species expected for a sample of a certain size is the sum of the probabilities of inclusion of the various species present. Calculations of this relationship are involved, and are best done by computer (see below).

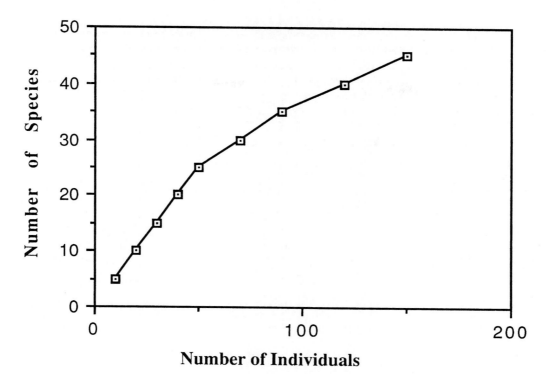

Figure 32.1.
A rarefaction curve plots expected number of species against number of individuals from one to the number observed in a sample from a community

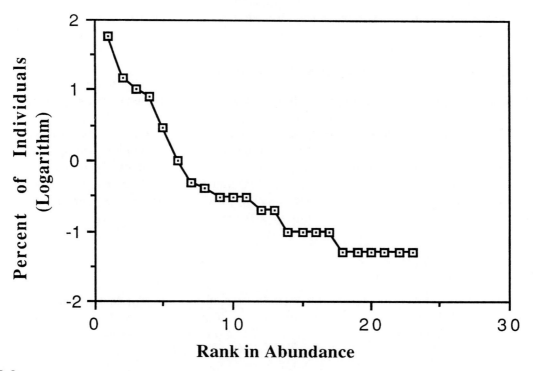

Figure 32.2.
A relative-abundance curve plots the percentage of individuals of each species on a logarithmic scale, ranked from most to least abundant

Table 32.1. Summary of species diversity indices for a series of natural communities

Community Description	Size of Sample Area	Number of Species	Number of Individuals	Simpson Index		Shannon Index	
				D_S	V	H'	J

Relative-abundance curves (figure 32.2) show the percentage abundance of species plotted as a logarithm on the ordinate against rank in abundance, from highest to lowest, on the abscissa. In such a plot, horizontal sections of the resulting curve identify groups of species of nearly equal abundance, and steep sections sets of species with markedly different relative abundances. Such a curve would reveal, for example, if two indices of heterogeneity having the same value were similar in equitability and richness patterns, or whether the similar values were the product of high equitability in one case and high richness in the other.

SOFTWARE FOR SPECIES DIVERSITY COMPUTATION

The EcoStat software package (appendix B) contains a module for calculating indices of heterogeneity and equitability, including H', J', D_s, C_d, and V.

The Ecological Analysis–PC software package (appendix B) also contains a module for diversity index computation. This module provides Simpson indices D_s and C_d as well as Shannon indices H', H_{max}, and J'.

Ecological Measures 1.10 (Appendix B), a set of BASIC software programs, also contains modules for the various Shannon and Simpson indices, as well as for several other diversity and evenness indices.

Simberloff (1978) provides a short FORTRAN IV program, titled SIM, for computing a rarefaction curve for a community containing up to 550 species with up to 6,000 individuals of any single species.

SUGGESTED ACTIVITIES

1. Obtain data on numbers of individuals of different species in two different habitats (1) with samples representing equal areas of habitat and (2) from samples containing equal total numbers of individuals. These may be obtained in field situations, or by sampling the wash and nonwash portions of the vegetation map included in this manual. Work out the various indices of richness, heterogeneity, and equitability for these samples. Compare diversity indices for samples of numbers of individuals with indices computed for the same samples using basal area or canopy-coverage values. Record values in table 32.1.

2. Examine diversity in communities of different stages of succession, at different points on gradients of environmental conditions, or in areas with different degrees of disturbance. Calculate the various indices of richness, heterogeneity, and equitability for these samples.

3. Examine the diversity of bark or rock-inhabiting lichens occupying similar microhabitats in localities differing in air pollution levels. Calculate the various indices of richness, heterogeneity, and equitability for these samples.

QUESTIONS FOR DISCUSSION

1. Does diversity necessarily increase with stability and freedom from disturbance? Do you think a relation exists between diversity and community stability?

2. What relation should exist between diversity and the degree of niche specialization of the member species of a community?

3. How do you think natural selection should influence diversity in a community over evolutionary time?

SELECTED BIBLIOGRAPHY

Godfrey, P. J. 1978. Diversity as a measure of benthic macroinvertebrate community response to water pollution. Hydrobiologia 57:111-122.

Huston, M. A. 1994. Biological diversity: the coexistence of species on changing landscapes. Cambridge University Press, New York, New York, USA.

James, F. C., and S. Rathbun. 1981. Rarefaction, relative abundance, and diversity of avian communities. Auk 98:785-800.

Keefe, T. J., and E. P. Bergersen. 1977. A simple diversity index based on the theory of runs. Water Resources 11:689-691.

Kvalseth, T. O. 1991. Note on biological diversity, evenness, and homogeneity measures. Oikos 62:123-127.

Magurran, A. E. 1988. Ecological diversity and its measurement. Princeton University Press, Princeton, New Jersey, USA.

Mason, C. F. 1977. The performance of a diversity index in describing the zoobenthos of two lakes. Journal of Applied Ecology 14:363-367.

Olson, C. M. 1992. Monitoring species diversity: a sampling approach. Pages 30-35 *in* H. M. Kerner, editor. Proceedings of the symposium on biodiversity of northwestern California. Report No. 29, Wildland Resources Center, University of California, Berkeley, California, USA.

Peet, R. K. 1974. The measurement of species diversity. Annual Review of Ecology and Systematics 5:285-307.

Pielou, E. C. 1975. Ecological diversity. John Wiley & Sons, New York, New York, USA.

————. 1977. Mathematical ecology. 2nd ed. John Wiley & Sons, New York, New York, USA.

Simberloff, D. 1978. Use of rarefaction and related methods in ecology. Pages 150-165 *in* K. L. Dickson, J. Cairns, Jr., and R. J. Livingston, editors. Biological data in water pollution assessment: quantitative and statistical analyses. American Society for Testing and Materials STP 652.

Simpson, E. H. 1949. Measurement of diversity. Nature 163:688.

Wilhm, J. L. 1970. Range of diversity index in benthic macroinvertebrate populations. Journal of the Water Pollution Control Federation 42:221-224.

Zand, S. M. 1976. Indexes associated with information theory in water quality. Journal of the Water Pollution Control Federation 48:2026-2031.

Zar, J. H. 1984. Biostatistical analysis. 2nd ed. Prentice-Hall, Englewood Cliffs, New Jersey, USA.

Community Similarity and Ordination

INTRODUCTION

Community ecologists have sought ways to express degree of similarity of communities to each other in composition and how this similarity relates to habitat conditions. They have invented a number of indices of similarity, and a host of techniques of *ordination*—mathematical procedures for arranging communities in graph space so that those most similar in composition are close together and those most different are far apart.

In this exercise we shall examine several indices of similarity, and illustrate three basic techniques of ordination.

COMMUNITY SIMILARITY INDICES

The simplest indices of similarity of communities are those which consider only the presence or absence of species. *Jaccard's index (J)* and *Sorenson's index (S)* are calculated as follows:

$$J = \frac{2w}{(A + B - w)}$$

$$S = \frac{2w}{(A + B)}$$

Where: A = Number of species in community A
B = Number of species in community B
w = Number of species in common

Jaccard's index varies from 0, for communities with no species in common, to 2.0 for communities with all species in common, whereas Sorenson's index varies from 0 to 1.0. Both indices suffer from the fact that rare species carry the same weight as common species in their influence on index value.

Other indices take into account quantitative measures of the abundance of species, such as density, cover, frequency, or importance value.

The simplest index, the *coefficient of community,* can be calculated either with actual values (e.g., densities) or relative values (e.g., relative densities):

$$C_c = \frac{2w}{(A + B)}$$

Where: A = Sum of values for community A
B = Sum of values for community B
w = Sum of lower of the two values for shared species

This coefficient, like the similar Sorenson's index, ranges between 0 and 1.0.

Begon et al. (1990) : 17:620–628
Brewer (1994) : 10:293–296
Ehrlich and Roughgarden (1987) : 15:323–325
Krebs (1994) : 20:431–457

Ricklefs (1990) : 32:662–667
Ricklefs (1993) : 22:414–416
Smith (1992) : 19:305

If quantitative values are expressed as percentages of their total for each community, *percent similarity* can be expressed simply by computing

$$C_\% = \Sigma(\text{lower \% values for shared species})$$

For data on numbers of individuals of various species in different communities, *Morisita's index (C_M)* is often used. This index is computed as follows:

$$C_M = \frac{2 \Sigma X_i Y_i}{(S_A + S_B)N_A N_B}$$

Where: X_i = Number of species i in community A
Y_i = Number of species i in community B
$N_A = \Sigma X_i$
$N_B = \Sigma Y_i$
$S_A = \dfrac{\Sigma[X_i (X_i - 1)]}{N_A (N_A - 1)}$
$S_B = \dfrac{\Sigma[Y_i (Y_i - 1)]}{N_B (N_B - 1)}$

This index, which ranges from 0 to 1.0, is essentially the probability that single individuals drawn randomly from the two communities will be of the same species, relative to the probability that two individuals drawn from one or the other of the communities will be of the same species.

Horn's index (C_H), based on the Shannon-Weaver function from information theory, is also used frequently to compare community data on numbers of individuals. This index is computed by the equation

$$C_H = \frac{\Sigma[(X_i + Y_i) \log (X_i + Y_i)] - \Sigma(X_i \log X_i) - \Sigma(Y_i \log Y_i)}{[(N_A + N_B) \log (N_A + N_B)] - (N_A \log N_A) - (N_B \log N_B)}$$

Where: X_i = Number of species i in community A
Y_i = Number of species i in community B
$N_A = \Sigma X_i$
$N_B = \Sigma Y_i$

ORDINATION PROCEDURES

Two rather different approaches have been taken to ordination. In one, *indirect gradient analysis,* communities are characterized by their actual similarities of composition. Each community is located on a graph with axes that represent major components of variation in composition. Correlations can then be sought between the major compositional axes of the ordination and gradients of environmental conditions.

Direct gradient analysis includes environmental information from the outset. Community composition is related to measured conditions such as temperature, moisture, and nutrient availability. Implicit in early direct gradient analysis procedures, however, was the assumption that responses of species to such conditions are linear (Peet 1980). This means, for example, that a unit change in moisture is assumed to produce the same effect on the community in relatively wet conditions as in relatively dry conditions—an assumption seldom fulfilled. Recently, methods have been developed which assume a unimodal relation of species' abundances to environmental factors (ter Braak 1987).

Most ordination procedures are mathematically complicated and impractical to carry out by hand. Thus, we shall use procedures programmed for microcomputers.

We shall describe three procedures for ordination: (1) the *Bray-Curtis polar ordination,* (2) *detrended correspondence analysis,* and (3) *canonical correspondence analysis.* For a class exercise, a relatively small number of community samples (e.g., 10–20) should be used, since the purpose of the activity is to understand the principle of ordination, not to wrestle with voluminous input and output files.

Bray-Curtis Polar Ordination

This is an indirect ordination procedure that is based on coefficients of similarity, such as the *coefficient of community*, that are calculated for each possible pair-combination of communities. Coefficients can be calculated from data on the presence or absence of species, or from detailed structural data such as density, dominance, or frequency. The latter values can be expressed either in absolute or relative form, or they can be combined into a single importance value.

The Bray-Curtis procedure then converts similarity coefficients to dissimilarity values, since distance between communities in an ordination represents degree of difference rather than similarity. The position of the communities in a two-axis ordination is then determined by calculating *x*- and *y*-coordinates on two axes. The calculation is done so that the greatest component of community variation is spread along the *x*-axis. The calculation of the *y*-axis coordinate is done so that the greatest component of the remaining variation is spread along this axis. To do this, poorness-of-fit values are calculated for each sample on the *x*-axis. The two communities most dissimilar in this measure, and located within a certain distance of each other along the *x*-axis, are found. These communities form the ends of the *y*-axis spread, and the remaining samples are located with reference to them.

Bray-Curtis ordination is available in a number of software packages. Ecological Analysis–PC, Vol. 3 (see below) provides a user-friendly module for this analysis. Data can be entered on screen for up to 200 species and 100 communities. Spreadsheet data files can also be imported. The analysis and output of results are carried out by selection from on-screen menus. The output contains a table of *x*- and *y*-coordinates for the communities.

The direct distance between any two communities in the ordination can be calculated from the equation

ordination interval $= \sqrt{dx^2 + dy^2}$ Where: dx = Difference between communities
on the *x*-axis
dy = Difference between communities
on the *y*-axis

The degree to which the spacing of the communities on the two-axis ordination accounts for the total variation in community composition can be estimated by correlating the ordination interval with calculated dissimilarity values (e.g., $1 - coefficient\ of\ community$) for community pairs. If ordination interval is designated *x* and dissimilarity value *y*, the correlation coefficient, *r*, for these two values can be calculated as described in exercise 8. For this analysis 20–25 sets of *x* and corresponding *y* values should be chosen randomly.

The value of this ordination lies in the insight that may be gained into environmental factors most important in controlling community composition. After the ordination has been constructed, therefore, environmental factors, singly or in combination, correlated with the *x*- and *y*-axis variables should be sought.

Advanced Ordination Procedures Using CANOCO

The CANOCO software package (see below), is best developed for IBM-compatible microcomputers (although a Macintosh version exists, as described below) and supports a variety of indirect and direct ordination procedures. CANOCO is capable of handling up to 500 samples with up to 500 species.

The CANOCO software package (Version 3.12) includes three programs: (1) the ordination program, CANOCO itself; (2) a data file editor, CEDIT; and (3) a graphics preparation program, CanoDraw LITE.

Data files can be prepared for CANOCO by using the CEDIT program. CEDIT can import data from various sources and prepare them for use by CANOCO, but we shall describe only a simple on-screen input procedure. A single data file with values for species' abundances is required for indirect ordination procedures. Separate data files for species' abundances and environmental measurements are required for direct ordination procedures.

Type **CEDIT** to open it from the CANOCO directory on the hard disk. Respond to the various questions: type * to indicate a new file, type the number of rows and columns, type names for rows and columns (give rows the same names in both species and environmental files). Choose **Ed** to edit the file, and **IR** for data input one row at a time. Type in the individual values. When finished, type **Qu** to leave the editor and **Wr** to save it. Provide a file name ending in **.SPE** for species' values or **.ENV** for environmental data, and select **Sp** or **En** to identify the file as species or environmental data, respectively. Give the file a descriptive title and indicate the numerical format of your data by typing in two numbers (e.g., **6 2,** meaning that the values have a maximum of 6 numerals with two after a decimal point). Type **Y** (if the format appears correct) and **Qu** to quit CEDIT.

To use CANOCO proper from its directory on a hard disk, type **CANOCO** and answer a sequence of questions about the desired analysis. We shall describe the procedure for conducting two of the analyses.

Detrended Correspondence Analysis

Detrended correspondence analysis, or *DCA* (also known as *DECORANA*), is an advanced indirect ordination procedure that removes some of the computational artifacts that affect Bray-Curtis ordination and other procedures (Hill and Gauch 1980). In particular, it eliminates the tendency for communities to be arranged in an "arch-like" pattern in ordination space.

For a basic DCA analysis, default or **0** selections can be made for most of the questions. One must provide the name of the data file (with the suffix **.SPE**), select **7** for the DCA analysis, and specify any transformation desired on data entering the analysis. One must also specify the name of a printer output file (ending in **.OUT**) and an answer file (ending in **.CON**) for use by CanoPlot LITE. Once the analysis is complete, a **Summary** table can be viewed. Among other things, this shows the cumulative percent of total variation in community composition that is accounted for by ordination axes 1–4. Requesting **more information** and selecting choice **6 6** leads to the display of tables of species' positions and stand positions on ordination axes, as well as the storage of data in the output file. After CANOCO is closed, the command **PRINT XXXXX.OUT** will give the tabular results.

Canonical Correspondence Analysis

Canonical correspondence analysis (CCA) is actually a form of direct gradient analysis, with communities being positioned in ordination space on the basis of their overall relationships with a set of environmental variables measured for the community sites. The first ordination axis shows the greatest component of variation that can be accounted for, the second axis the next greatest amount, and so on. A linear relationship between species' abundances and environmental factors is not assumed in this analysis.

CCA requires a community-composition input file like that for DCA, but also requires an environmental-variable data file structured in exactly the same manner, but named with the suffix **.ENV.** To conduct a CCA ordination, open CANOCO and make selections in response to on-screen queries, supplying the names of both input files and selecting **5** for the CCA procedure. When the analysis is finished, one can view the matrix of correlations among environmental variables, a table of means and standard deviations of environmental variables, a summary table of the percentages of variation in species composition and species-environmental variable relations accounted for by the ordination axes, and the locations of species and stands on ordination axes. After CANOCO is closed, this output can be printed as for DCA.

With the CanoDraw LITE program, the results of DCA or CCA analyses can be viewed graphically. Type **CDLITE** to open CanoDraw LITE. Double-click on **Files in** the drop-down menu and select the **.CON** file you specified in the analysis. Then, under the **Scatter** selection, plots of **Species** and **Sites** along the ordination axes can be viewed for DCA, and **Env. Variables** also viewed for CCA. This plot shows, by the direction and length of the arrows, the closeness and strength of correlation of environmental factors with ordination axes. By selecting **Create Virtual Image,** you can create a file for printing these graphics.

SOFTWARE FOR COMMUNITY SIMILARITY AND ORDINATION ANALYSIS

The EcoStat software package (appendix B) includes a module that allows calculation of the Jaccard, Sorenson, percent similarity, Morisita, and Horn indices. Ecological Analysis–PC, Vol. 3 (appendix B) contains a module for calculating Jaccard, Sorenson, percent similarity, Morisita, Horn, and coefficient of community indices. This package also includes a module for Bray-Curtis polar ordination that can handle up to 100 data sets and up to 200 total species.

CANOCO and the companion programs CEDIT and CanoDraw LITE are available for IBM microcomputers and compatibles (see appendix B). A Macintosh version of CANOCO is also available, but lacks CEDIT and has a CanoPlot module for graphical display of ordination results. This version requires data input as a file with a very specific FORTRAN format. Both versions of CANOCO support DCA, CCA, and a variety of other ordination procedures. A separate CanoDraw program with broader capabilities than CanoDraw LITE is also available for IBM microcomputers.

The Cornell Ecology Programs (see appendix B) consist of a variety of programs for community analysis, including Bray-Curtis polar ordination (as part of the ORDIFLEX program) and DECORANA. These programs require data input in a file with a very specific FORTRAN format. ORDIFLEX produces a Bray-Curtis dissimilarity matrix, tabulated values for the x-axis coordinates (identified as **ORDN NO. 1**) and y-axis coordinates (**ORDN NO. 2**). The positions of the samples on these axes (**X**), poorness-of-fit values (**E**), dissimilarity values with end samples (**D1** and **D2**), and positions rescaled to a spread of 100 units (**XREL**) are given in the tabled x- and y-axis output sets. A schematic graph of the two-dimensional ordination is also given. The DECORANA program produces ordinations of both species and samples, giving their positions on four axes.

🐾 SUGGESTED ACTIVITIES

1. Select 5–10 sets of data collected by class members or taken from a published data set in the literature. Use density, cover, frequency, or importance values of species (or other quantitative values from published material) to compare community similarity by different indices, and to carry out ordinations.

2. Sample several different sites of the wash and bajada sections of the vegetation map accompanying this manual. Use summarized data on density, cover, frequency, or importance values of species (or other quantitative values from published material) to compare community similarity by different indices, and to carry out ordinations.

3. Examine breeding-bird census data published in the *Journal of Field Ornithology* (or earlier in *American Birds* and *Audubon Field Notes*). Select a series of areas censused in the same region and year (e.g., the California desert, Point Reyes National Seashore). Calculate similarity indices and perform one or more ordinations on data on numbers of breeding pairs.

🐾 QUESTIONS FOR DISCUSSION

1. What pattern of correlation should be shown between environmental factors and position of communities in an ordination, assuming the individualistic hypothesis of community structure to be correct? The organismal hypothesis?

2. What would a low r value in the correlation of dissimilarity values and ordination intervals indicate about the nature of the environmental factors controlling community composition? What pattern should be shown by the samples from the wash and bajada sections of the lab manual vegetation map? Does this pattern support either an organismal or an individualistic hypothesis of community structure?

3. What are the relative merits of direct versus indirect gradient analysis? Which do you think is likely to give greatest insight into patterns of compositional variation in communities? Into community relationships with habitat conditions?

☯ SELECTED BIBLIOGRAPHY

Austin, M. P. 1985. Continuum concept, ordination methods, and niche theory. Annual Review of Ecology and Systematics 16:39–61.

Gauch, H. G., Jr., and W. M. Scruggs. 1979. Variants of polar ordination. Vegetatio 40:147–153.

———, and T. R. Wentworth. 1976. Canonical correlation analysis as an ordination technique. Vegetatio 33:17–22.

Goodall, D. W., and R. W. Johnson. 1982. Non-linear ordination in several dimensions. Vegetatio 48:197–208.

Hill, M. O., and H. G. Gauch, Jr. 1980. Detrended correspondence analysis: an improved ordination technique. Vegetatio 42:47–58.

Hobbs, R. J., and J. Grace. 1981. A study of pattern and process in coastal vegetation using principal components analysis. Vegetatio 44:137–153.

Ihm, P., and H. van Groenewoud. 1975. A multivariate ordering of vegetation data based on Gaussian type gradient response curves. Journal of Ecology 63:767–777.

Johnson, R. W., and D. W. Goodall. 1979. A maximum likelihood approach to non-linear ordination. Vegetatio 41:133–142.

Krebs, C. J. 1989. Ecological methodology. Harper & Row, New York, New York, USA.

Palmer, M. W. 1993. Putting things in even better order: the advantages of canonical correspondence analysis. Ecology 74:2215–2230.

Peet, R. K. 1980. Ordination as a tool for analyzing complex data sets. Vegetatio 42:171–174.

Shipley, B., and P. A. Keddy. 1987. The individualistic and community-unit concepts as falsifiable hypotheses. Vegetatio 69:47–55.

Spellerberg, I. F. 1991. Monitoring ecological change. Cambridge University Press, Cambridge, England.

ter Braak, C. J. F. 1987. Ordination. Pages 91–173 *in* R. H. Jongman, C. J. F. ter Braak, and O. F. R. van Tongeren, editors. Data analysis in community ecology. Pudoc, Wageningen, The Netherlands.

———, and I. C. Prentice. 1988. A theory of gradient analysis. Advances in Ecological Research 18:272–317.

van der Maarel, E., editor. 1980. Classification and ordination. Vegetatio 42:1–185. (This special volume contains 22 papers on ordination procedures and their application.)

EXERCISE 34

Mechanisms of Biotic Succession

INTRODUCTION

Biotic succession is the change in composition and structure of a biotic community through time, with external controlling factors of the ecosystem, such as climate, remaining constant. A century ago, pioneering studies of succession in the U.S. Midwest and Great Plains provided a major stimulus to ecology in North America. Early studies of succession were synthesized by the plant ecologist Frederick C. Clements into a dynamic theory of the process that dominated ecology until recently. Clements's ideas, which may be termed the *facilitation hypothesis,* were that the first colonists of new or disturbed habitats modified physical conditions in ways that favored the invasion of other species. This modification, termed *reaction,* was continued by the next group of invaders, favoring the establishment of still other species. Thus, species succeeded each other in groups, each modifying the habitat to an additional degree, until a set of species appeared that could reproduce indefinitely under the existing conditions but effect no further change in the habitat. This was the *climax community.* This hypothesis gained support from studies of biotic succession in several terrestrial environments, such as the sand dunes of Lake Michigan and areas released by glacial retreat in Glacier Bay, Alaska. These studies demonstrated that major changes in soil physical and chemical conditions occurred through successional time.

Ecologists have now studied succession in many different environments—deserts, lakes, rocky intertidal shores—and realize that not all changes in community composition occur through the same mechanisms. In fact, the major emphasis in successional studies now is determining the mechanisms that are operating to bring about change.

HYPOTHESES OF SUCCESSION

Connell and Slatyer (1977) and Sousa (1979) proposed three alternate hypotheses of successional dynamics. To these we shall add a fourth model—a successional sequence due only to different dispersal and colonization abilities. These four alternate hypotheses are

I. Differential colonization hypothesis

Under this hypothesis, species have different probabilities of dispersing to and colonizing an area during a given time interval. As time passes, the density of individuals of all species increases, but change in the community structure reflects only the accumulation of individuals of the various species according to their different colonization rates. Thus, no change occurs in their relative abundances.

II. Facilitation hypothesis

This hypothesis assumes that only certain colonizing species can become established under initial habitat conditions. Occupation and modification of habitat conditions by these species (1) reduces the

Begon et al. (1990) 17:628–647
Brewer (1994): 13:379–403
Colinvaux (1993): 20:418–443
Ehrlich and Roughgarden (1987): 19:397–400, 408–416

Krebs (1994): 22:483–513
Odum (1993): 7:187–203
Pianka (1994): 4:70–73, 16:359–362
Ricklefs (1990): 33:677–707

Ricklefs (1993): 23:427–450
Smith (1990): 25:634–661
Smith (1992): 20:320–336
Stiling (1992): 18:341–349

favorability of the habitat for their own reproduction and establishment, and (2) improves the habitat for other species, facilitating their establishment. Once later invaders become established, they further modify the habitat to the disfavor of colonizing species.

III. Competition (tolerance) hypothesis

This hypothesis proposes that any species might become established under initial conditions, but that certain pioneer species appear in abundance first because of their efficient dispersal and establishment abilities. Subsequently, with the arrival of other species, community composition changes as some species displace others due to superior competitive abilities.

IV. Survival (inhibition) hypothesis

According to this hypothesis, any species can become established under initial conditions. Subsequent change in community composition reflects the differential abilities of species to tolerate inhibiting or damaging influences such as herbivore feeding, severe weather, disease, and other stresses (interspecific competition being unimportant).

These different hypotheses lead to alternative predictions that can form the basis for observations and experiments designed to discriminate among them. Although manipulative experiments are clearly the best way to test these alternatives, certain differences are predicted in the pattern of change in species composition through successional time. These differences may allow preliminary tests of the alternate hypotheses by observational data.

🌊 SUGGESTED ACTIVITIES

1. Succession in rock outcrop plant communities.
    Examine biotic succession in communities of saxicolous lichens, mosses, and vascular plants growing on rock outcrops. First, locate a site with rock surfaces that have been exposed for varying lengths of time and possess communities ranging from early stages of colonization to late in succession. Then proceed as follows:
    A. Survey the species present on the rock surfaces to be studied. Unless one is an expert in the taxonomy of lichens and mosses, identification to species may not be practical. Arbitrary names can usually be formulated without much difficulty, however, based on characteristics such as color, shape, texture, and other general features.
    B. Select a series of rock surfaces of differing age or successional status but similar mineralogy and exposure.
    C. Determine an appropriate procedure and quadrat size for sampling the percent cover of plants on the rock surfaces. Clear plastic sheets 15 × 15 cm in size and gridded into 100 cells, for example, allow estimates of percent cover to be obtained quickly.
    D. Each student should obtain a specified number of samples (e.g., 10) from the selected rock exposures, locating sample quadrats as directed by the instructor. For each sample, record the percent cover for each species and the total percent cover for the community in table 34.1.
    E. After samples have been taken, each student should fill in the overtopping matrix (table 34.2) by observing contacts between colonies or clusters of as many pairs of species as possible and recording which species is able to overgrow the other.

    Combine the samples of all class members or groups for analysis. First, rank samples from pioneer to climax on the basis of information on actual or relative age of the rock surfaces. If no direct information is available on age, rank the samples on the basis of total percent cover (*pioneer* = least cover, *climax* = greatest cover). Distinguish pioneer and climax sample groups. These should be the samples from the youngest and oldest rock surfaces, respectively, or those with the least and greatest total percent cover (e.g., 0–20% for pioneer and 80–100% for climax). For each species, average the percent cover values for pioneer and climax sample groups and record the values in table 34.3.

Table 34.1. Percent cover for species of rock-inhabiting organisms in quadrat samples along a successional gradient

Species	Sample												
Total													

Testing the differential colonization hypothesis

A test of the differential colonization hypothesis should be conducted first to determine whether successional change in relative abundance of species is occurring. This hypothesis postulates that the cover of a species should be directly proportional to the time it has occupied the area. Thus, the correlation between percent cover of n species in pioneer and climax communities should be very strong and positive, and the correlation coefficient, r, of these values should not differ significantly from 1.0. The calculated r should be transformed (see exercise 8) to a z' value by the equation

$$z' = 0.5 \ln [(1 + r)/(1 - r)]$$

A standard z test (see exercise 8) is then conducted:

$$Z = \frac{3.80 - z'}{\sigma_z} \qquad \text{Where: } \sigma_z = \sqrt{\frac{1}{n - 3}}$$

In this calculation, the value 3.80 is the z' value for an r of 0.999, since a real value for r of 1.000 cannot be calculated. Calculated z values of 1.96 and 2.58 lead to rejection of the null hypothesis at $\alpha = 0.05$ and 0.01, respectively.

If the hypothesis of random colonization is rejected, we may examine the remaining three hypotheses. These hypotheses are tested by comparing the rankings of species in the climax community (highest rank values assigned to species with highest percent cover) to the rankings expected by predictions based on each hypothesis. In assigning ranks, species with tied values are given the average value of the ranks for which they are tied. Rankings are then compared by determining the difference, d, between rank values for each of the n species, and calculating the Spearman rank correlation coefficient, r_s, by the equation

$$r_s = 1 - \frac{6\Sigma d^2}{n^3 - n}$$

The calculated r_s value is compared to critical tabled values (table A.7, p. 253) for α and n. If the calculated value exceeds the tabled critical value, the null hypothesis (no correlation) is rejected.

Testing the facilitation hypothesis

This hypothesis predicts that the species most important in the climax community should be those most responsive to modification of initial habitat conditions. If we assume that the degree of habitat modification is proportional to the total fraction of the rock surface covered by organisms, *responsiveness* of a species may be calculated as the difference between maximum and minimum (> 0) cover percentages for that species, divided by the maximum difference in total percent cover (all species) in those same plots. If a species has 0 cover in one or more plots, the minimum difference in total cover between plots with 0 and that with maximum cover of the species should be used to calculate responsiveness. If a species' cover declines as total cover increases, responsiveness will be negative. Calculate responsiveness values (r) for each species and rank the species by this value (highest ranks assigned to highest r values). Then calculate r_s for the rankings in importance in the climax community and in responsiveness, and determine whether this correlation is significant. A significant positive correlation supports this hypothesis.

Testing the competition hypothesis

The competition hypothesis predicts that the most abundant species in the climax community should be those with the greatest competitive ability. Among sessile organisms, the ability of one species to overgrow the other is an expression of competitive ability. Using the overtopping matrices developed by class members (table 34.2), determine the linear ranking of species in overtopping ability. Assign the highest rank in *competitiveness, C,* to the species with greatest overtopping ability. In table 34.3, calculate r_s for overtopping rank and rank in importance in the climax community, and determine whether this correlation is significant. A significant positive correlation supports this hypothesis.

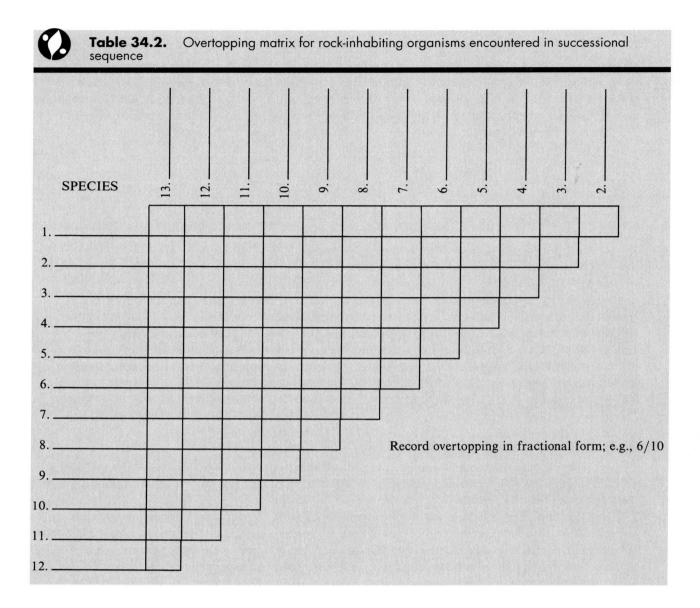

Table 34.2. Overtopping matrix for rock-inhabiting organisms encountered in successional sequence

Record overtopping in fractional form; e.g., 6/10

Survival hypothesis

This hypothesis predicts that the most abundant members of the climax community are those with the greatest survival ability. We can define survival ability as the maximum range of total percent cover over which the species exists. For each species, find the samples with the highest and lowest total percent cover for all species. Using this difference as a measure of *survival ability, S,* rank the species (highest ranks assigned to species with greatest difference). Calculate r_s in table 34.3 for survival ability and importance in the climax community, and determine whether this correlation is significant. A significant positive correlation supports this hypothesis.

2. Succession on sites exposed by melting glaciers. Analyze Viereck's (1966) data on plant succession on a glacial outwash substrate in Alaska. This study presents cover data for plant members of tundra communities on sites released from glacial coverage at five periods ranging from 25–30 years earlier to 5,000–9,000 years before present.

Examine table 3 of Viereck's study, and restrict your attention to those species that show cover classes of 1 (5% or less) to 5 (75–100%) in at least one stage (30 species).

1. Test the differential colonization hypothesis by calculating the correlation coefficient, *r,* between percent abundances of species in pioneer and climax communities.

Table 34.3. Rankings and tests of predictions of different hypotheses of biotic succession (R = Responsiveness, C = Competitive Overtopping, S = Survival)

Species	RANDOM Percent Cover		FACILITATION			TOLERANCE			INHIBITION		
	Pioneer	Climax	R	R Rank	d	Climax Rank	C Rank	d	Climax Rank	S Rank	d	Climax Rank
r or r_s												
P												

2. Rank the species in the climax community from highest to lowest cover.

3. Using midpoint values of the five cover class ranges, compute the total plant cover in each successional stage (assign the "+" cover class a value of 0.5%). Calculate responsiveness values and rank them as described above. Test the facilitation hypothesis by calculating r_s for the rankings in importance in the climax community and in responsiveness.

4. Consult manuals of lichen, moss, and vascular plant floras that cover the Alaskan region, and determine the characteristics of the various plant species. Using the available information, construct an index of "potential competitiveness." You might, for example, group the species in five height classes (e.g., 1 = prostrate, 2 = 5-25 cm, 3 = 25-50 cm, 4 = 50-100 cm, 5 = > 100 cm) and five permanence classes (e.g., 1 = seasonal herbaceous, 2 = evergreen nonvascular, 3 = evergreen vascular, 4 = deciduous woody, 5 = evergreen woody) and characterize the potential competitiveness of each species as the product of values for the two classes. The species can then be ranked in competitiveness, and the competition hypothesis tested by calculating r_s for overtopping rank and rank in importance in the climax community.

5. For each species, find the stages with the highest and lowest total percent cover for all species (assign the "+" cover class a value of 0.5%). Rank species according to this difference, S (highest ranks assigned to species with greatest difference). To test the survival hypothesis, calculate r_s for survival ability and importance in the climax community.

3. Succession in rocky intertidal plant and invertebrate communities

Examine biotic succession in a community of intertidal plants and sessile invertebrates, using a procedure similar to that outlined above for rock outcrop plants.

COMPUTER SOFTWARE FOR DATA ANALYSIS

The E-Z Stat software package for IBM PCs and compatibles (appendix B) contains modules for simple linear correlation and Spearman rank correlation. The Ecological Analysis—PC (Vol. 1) software package (appendix B), also for IBM PCs and compatibles, likewise contains a correlation module that carries out both simple linear correlation and Spearman rank correlation.

QUESTIONS FOR DISCUSSION

1. Which hypothesis of community development is supported best by the various comparisons? If the process does not appear to be random colonization, what precise mechanisms might cause facilitation, competition, or differential inhibition?

2. Can you develop other predictions from the four models of community development and succession? How would you test them? What sorts of experimental approaches might be taken to these alternative hypotheses?

3. Can you propose still other models of succession, or modified forms of the models presented in this exercise? Can a successional process combine elements of two or more of these models at the same time, or in different phases of the successional process?

SELECTED BIBLIOGRAPHY

Armesto, J. J., and S. T. A. Pickett. 1986. Removal experiments to test mechanisms of plant succession in old fields. Vegetatio 66:85-93.

Armstrong, R. A. 1982. Competition between three saxicolous species of *Parmelia* (Lichens). New Phytologist 90:67-72.

Connell, J. H., and R. O. Slatyer. 1977. Mechanisms of succession in natural communities and their role in community stability and organization. American Naturalist 111:1119–1144.

Connell, J. H., I. R. Noble, and R. O. Slatyer. 1987. On the mechanisms producing successional change. Oikos 50:136–137.

Del Moral, R., and L. C. Bliss. 1993. Mechanisms of primary succession: insights resulting from the eruption of Mount St. Helens. Advances in Ecological Research 24:1–66.

Farrell, T. M. 1991. Models and mechanisms of succession: an example from a rocky intertidal community. Ecological Monographs 61:95–113.

Gray, A. J., M. J. Crawley, and P. J. Edwards, editors. 1987. Colonization, succession, and stability. Blackwell Scientific Publications, Palo Alto, California, USA.

McCook, L. J. 1994. Understanding ecological community succession: causal models and theories, a review. Vegetatio 110:115–147.

McCormick, P. V., and R. J. Stevenson. 1991. Mechanisms of benthic algal succession in lotic environments. Ecology 72:1835–1848.

Pentecost, A. 1980. Aspects of competition in saxicolous lichen communities. Lichenologist 12:135–144.

Pickett, S. T. A., S. L. Collins, and J. J. Armesto. 1987. Models, mechanisms and pathways of succession. Botanical Review 53:335–371.

Sousa, W. P. 1979. Experimental investigations of disturbance and ecological succession in a rocky intertidal algal community. Ecological Monographs 49:227–254.

Turner, T. 1983. Facilitation as a successional mechanism in a rocky intertidal community. American Naturalist 121:729–738.

Van Andel, J., J. P. Bakker, and A. P. Grootjans. 1993. Mechanisms of vegetation succession: a review of concepts and perspectives. Acta Botanica Neerlandica 42:413–433.

Viereck, L. A. 1966. Plant succession and soil development on gravel outwash of the Muldrow Glacier, Alaska. Ecological Monographs 36:181–199.

Walker, L. R., and F. S. Chapin, Jr. 1987. Interactions among processes controlling successional change. Oikos 50:131–135.

EXERCISE 35

Species-Area Curves

INTRODUCTION

The number of species in a patch of habitat is a balance between two processes: the appearance of new species by dispersal and colonization (and, over longer periods, by speciation), and the loss of species by extinction. In some areas, such as those in early stages of secondary biotic succession, this number may be rapidly changing. In other, more mature situations, it may be a near-equilibrium number that reflects basic biogeographic relationships.

The number of species is also a function of area, either the size of the habitable area, as in the case of an island, or the area arbitrarily chosen for examination, as in the case of a sample plot. Typically, as the size of the area examined increases, the number of species encountered also increases. This increase tends to be rapid at first, but then slows as the number of species approaches the number in the region as a whole. This species-area relation is described quite well by a simple equation:

$$S = bA^z$$

Where: S = Number of species
A = Area of sample
b = Coefficient varying with group of organisms and biogeographic region
z = Slope of species-area relationship

In this equation, the exponent, z, varies with several factors of ecological and biogeographic interest. These include (1) the degree of isolation of the areas sampled, (2) the degree of stability of the environment, and (3) the dispersal capabilities of the organisms.

Where areas are isolated, as in the case of oceanic islands, extinctions of species are not quickly countered by reinvasion of the same species from other areas. In addition, extinction is frequent in small populations (e.g., those on tiny islands). The result is that the species-area relation is a steep one, with a high z value which indicates that the rate of decline of species number with decrease in size of occupied area is very rapid. Much the same is true of fragmented habitats, such as woodlots in agricultural landscapes, in which the species occupy patches of varying size. In a homogeneous environment, the number of species tends to decline more slowly with decrease in size of the area examined. This slower decline largely reflects the fact that accidental disappearance of a population in one location is quickly offset by reinvasion from surrounding areas of similar habitat.

In unstable environments, fluctuation of environmental conditions tends to perturb the smaller islands or individual patches most intensely. This is likely to cause local extinction of species that exist in small populations. Thus, a high degree of instability often produces species-area relations with high z values.

The dispersal ability of organisms obviously relates to the species-area relationship, as well. If the species disperse effectively, local extinctions are more likely to be offset quickly, and the species-area relation will have

Begon et al. (1990): 22:768–789
Brewer (1994): 10:302–306, 19:622–625
Colinvaux (1993): 21:447–456
Ehrlich and Roughgarden (1987): 19:397–400, 408–416

Krebs (1994): 25:587–597
Pianka (1994): 2:27–34
Ricklefs (1990): 34:721–726
Ricklefs (1993): 22:423–426

Smith (1990): 24:611–620
Smith (1992): 19:312–317
Stilling (1992): 18:349–357

a flat slope (low z value). If the species involved disperse only with difficulty, the species-area relation will be steep (high z value).

Gould (1979) has shown that the coefficient, b, of the species-area relation also has an ecological meaning. When the slope, z, of this relation is the same for two sets of data, the ratio of b values indicates the relative richness in species of the regions from which the two sets of data are taken (or of the different taxa that are compared in a single region).

The slope and intercept of species-area relations are influenced by a variety of other factors, and, as Martin (1981) has noted, slope is to a degree influenced by the sizes of the areas used in computing the species-area relation. This means that caution should be used in interpreting slopes and coefficients of species-area curves that span different ranges of area size.

COMPUTING THE SPECIES-AREA RELATION

The species-area relation can be calculated by linear regression analysis of data on number of species and size of area, both expressed as logarithms (see exercise 6). In this analysis, the logarithm of the number of species (S) is the dependent variable, the logarithm of area (A) is the independent variable, b is the intercept of the regression line, and z is the slope of the species-area relationship.

COMPUTER SOFTWARE FOR SPECIES-AREA ANALYSIS

The software package Ecological Analysis—PC Vol. 2, for IBM PCs and compatibles, contains a module for calculating the species-area relation (appendix B). Any software that can perform logarithmic data transformation and simple linear regression computation (e.g., the linear regression module in E-Z Stat for the PC; appendix B) can also be used. Most basic statistical and graph-making software packages have this capability.

🌐 SUGGESTED ACTIVITIES

1. Collect original data on species-area relationships in some actual field situation. Many small-scale systems show insular features. Some of these island "analogues" are
   A. Woodlots of varying size in agricultural areas
   B. Patches of chaparral, grassland, desert scrub, or other vegetation types in suburban areas
   C. Small vegetated islands in rivers
   D. Rocks or rock outcrops bearing moss, lichen, and rock-loving vascular plant communities
   E. Patches of disturbed soil in areas of perennial grassland
   F. Rocks or rock outcrops on sandy beach areas of lakes or the ocean
   G. Vacant lots in urban areas
   H. Temporary ponds
      Pick one or more comparisons that you think will give insight into factors controlling the numbers of species that exist in a given area. Calculate the species-area relationships. Compare the z values obtained to those given for various groups of organisms in published sources.
2. Examine published data on numbers of species in areas or samples of different sizes. Published data are available for many groups of organisms. One particularly good source is the journal *American Birds,* which publishes breeding and wintering bird censuses from all of North America. Data on the vertebrate faunas of oceanic islands of varying sizes are available from various sources. In this case, it is useful if actual species lists are used, since this allows species-area relations to be determined for different ecological groupings of species.

3. Examine species-area relationships in noninsular situations. Compare areas differing in one or more of the following features:
    A. Homogeneity vs. heterogeneity of habitat conditions
    B. Early successional vs. late successional status
    C. Disturbed (grazing, burning, cultivating) vs. undisturbed conditions
        Or, make comparisons of species-area relationships for different groups of organisms, such as
    D. Groups of organisms differing in some basic ecological or behavioral characteristic (e.g., annual vs. perennial plants, birds vs. mammals, migratory vs. resident birds, flying vs. flightless insects, wind- vs. animal-dispersed plants)
    E. Groups of organisms differing in trophic status (e.g., herbivores, insectivores, and carnivores; fungi and mosses)
        Calculate the species-area relationships. Compare the z values obtained to those given for various groups of organisms in published sources.

QUESTIONS FOR DISCUSSION

1. What range of values do you think z might show? Does the actual diversity of organisms in a region influence the value of z?
2. In the situation you examined, would you expect the z relation to vary much from season to season, or from year to year?
3. What happens to the species-area relation when an area of continuous habitat is fragmented into islandlike remnants, as in the case of much of the eastern deciduous forests? What are the implications of this process for conservation, and for the design and maintenance of parks and refuges?

SELECTED BIBLIOGRAPHY

Armesto, J. J., and L. C. Contreras. 1981. Saxicolous lichen communities: nonequilibrium systems? American Naturalist 118:597–604.

Connor, E. F., and E. D. McCoy. 1979. The statistics and biology of the species-area relationship. American Naturalist 113:791–833.

Dunn, C. P., and C. Loehle. 1988. Species-area parameter estimation: testing the null model of lack of relationship. Journal of Biogeography 15:721–728.

East, R. 1981. Species-area curves and populations of large mammals in African savanna reserves. Biological Conservation 21:111–126.

Kitchener, D. J., A. Chapman, and B. G. Muir. 1980. The conservation value for mammals of reserves in the western Australian wheatbelt. Biological Conservation 18:179–207.

Lawrey, J. D. 1992. Natural and randomly-assembled lichen communities compared using the species-area curve. The Bryologist 95:137–141.

———. 1991. The species-area curve as an index of disturbance in saxicolous lichen communities. The Bryologist 94:377–382.

MacArthur, R. H., and E. O. Wilson. 1967. The theory of island biogeography. Princeton University Press, Princeton, New Jersey, USA.

Martin, T. E. 1981. Species-area slopes and coefficients: a caution on their interpretation. American Naturalist 118:823–837.

Rey, J. R., E. D. McCoy, and D. R. Strong, Jr. 1981. Herbivore pests, habitat islands, and the species-area relation. American Naturalist 117:611–622.

Ryti, R. T. 1984. Perennials on rock islands: testing for patterns of colonization and competition. Oecologia 64:184–190.

Schmiegelow, F. K. A., and T. D. Nudds. 1987. Island biogeography of vertebrates in Georgian Bay Islands National Park. Canadian Journal of Zoology 65:3041–3043.

Wilcox, B. A., D. D. Murphy, P. R. Ehrlich, and G. T. Austin. 1986. Insular biogeography of the montane butterfly faunas in the Great Basin: comparison with birds and mammals. Oecologia 69:188–194.

Wright, S. J. 1981. Intra-archipelago vertebrate distributions: the slope of the species-area relation. American Naturalist 118:726–748.

———. 1988. Patterns of abundance and the form of the species-area relation. American Naturalist 131:401–411.

Terrestrial Primary Production

INTRODUCTION

This exercise describes methods for estimating net primary production (*NPP*) in terrestrial ecosystems in which the producer organisms are annual or perennial herbaceous plants. The basic method was used by Wiegert and Evans (1964) in measuring *NPP* in an old field in Michigan. The estimate obtained by this method is corrected for loss of shoot material by death and decomposition, but is uncorrected for similar losses of root material and for losses of shoot or root material as a result of grazing by herbivores. The method utilizes measurements of the standing crops of living and dead plant materials and assumes continuous gradual change in standing crop values between periods of measurement. The most accurate estimates of *NPP* are, therefore, obtained if measurements are made during a portion of the growing season in which a continuous increase in biomass of living plant material is occurring, and in situations in which herbivore grazing is minimal. A modification of the Wiegert/Evans method, devised by Lomnicki et al. (1968), is also described.

Cox and Waithaka (1989) adapted the Wiegert/Evans technique to estimate grazing harvest by large herbivores. This procedure, which requires measurements inside and outside of grazing exclosures, is also described.

WIEGERT AND EVANS ESTIMATE OF *NPP*

If herbivore action is negligible, the *NPP* of shoot tissue during a period of time is equal to the increase in standing crop of live tissue, corrected for the loss of live material by mortality. Estimates of mortality of shoot tissue are difficult to obtain directly, but can be obtained indirectly from measurements of the change in standing crop of dead plant material and of the disappearance (decomposition) rate of dead material. A rough estimate of the net production of root material may be obtained by determining the increase in standing crop of root tissue during the period of measurement. This estimate will be low, since no correction is made for mortality of root tissue. The raw data needed for an estimate of *NPP* according to the above theory consist of measurements of the standing crops of live shoot material, dead shoot material, and total root material on sample plots at the beginning and end of a production period. Production periods of between two weeks and a month should probably be used if measurements are made during a favorable portion of the growing season.

Begon et al. (1990): 18:648–670
Brewer (1994): 11:316–324
Colinvaux (1993): 24:499–511
Krebs (1994): 26:604–632
Odum (1993): 4:79–83
Pianka (1994): 16:355–358

Ricklefs (1990): 11:190–198, 14:261–263
Ricklefs (1993): 6:106–111
Smith (1990): 11:211–218
Smith (1992): 22:362–369
Stilling (1992): 19:372–378

Establishment of Plots and Initial Sampling

The size of sample plots should be determined on the basis of density of vegetation and the time and facilities available for processing samples. For most herbaceous communities, samples of 0.1 m² should be adequate.

Each individual or class group selects three adjacent plots that are as similar in plant density and composition as possible. These plots are marked with stakes. From plot #1, the living and dead plant materials are collected. The living shoot material is clipped off at the ground surface, any associated dead material removed, and the living material placed in a bag. All dead shoot material from on and above the ground surface is collected and placed in a second bag. The soil then is removed from the plot to a depth of 25 cm or to the maximum depth of root penetration. The bulk of the soil may be separated from the root material by placing the sample on a sieve of fine window screening and washing it with water. Root material sorted out in this manner is placed in a third bag.

From plot #2, all of the living shoot material is removed, but all of the dead plant material is left in place. The living shoot material from this quadrat may be kept as a replicate sample of the initial standing crop of live shoots. This plot will furnish an estimate of the rate of disappearance of dead shoot material during the production period. If it is likely that the dead material will be lost from the plot by processes other than normal decomposition (e.g., wind), or if addition of material from neighboring areas may occur, the plot should be covered with a hardware-cloth cage during the measurement period.

In the laboratory, the root material may be cleaned by washing and flotation procedures. All of the plant material samples then are dried at 100° C for 48 hours and weighed to the nearest centigram. These dry biomass values are recorded in table 36.1.

Final Sampling

At the end of the production period, the dead shoot material is collected from plot #2 and the living shoot material, dead shoot material, and total root material from plot #3. This material is sorted, oven dried, and weighed as for the initial samples. Dry biomasses of the samples should be recorded in table 36.1.

Calculation of Net Primary Production

The instantaneous rate of disappearance of dead shoot material (r) in grams per gram per day is calculated from the biomasses of dead shoot material from plot #1 (W_0) and plot #2 (W_1), using the equation

$$r \text{ (in grams/gram/day)} = \frac{\text{Log}_e \ (W_0/W_1)}{t}$$
Where: t = Length of production period in days

An estimate of the quantity of dead material disappearing from the undisturbed plot (x) during the production period then is determined from the standing crops of dead material at the beginning and end of the period and the rate of disappearance of dead material. This estimate is given by the equation

$$x \text{ (in grams per plot)} = \frac{a_0 + a_1}{2} \ rt$$
Where: a_0 = Weight of dead shoot material from plot #1
a_1 = Weight of dead shoot material from plot #3

The mortality of live shoot material (d) during the production period then is obtained from the change in standing crop of dead material and the quantity of dead material disappearing during the production period, using the equation

$$d \text{ (in grams per plot)} = x + (a_1 - a_0)$$

Table 36.1. Oven-dry weights in grams for live shoot material, dead shoot material, and total root material collected from sample plots at beginning and end of productivity measurement period

Plot Location	Measurement Date ____ Plot #1			Measurement Date ____ Plot #2			Measurement Date ____ Plot #3		
	Live Shoot Material	Dead Shoot Material	Total Root Material	Live Shoot Material	Dead Shoot Material	Total Root Material	Live Shoot Material	Dead Shoot Material	Total Root Material

The net production of shoot material (y) during the period then is calculated from the change in standing crop of live shoot material during the period and the estimate of shoot mortality:

$$y \text{ (in grams per plot)} = (b_1 - b_0) + d$$

Where b_0 = Weight of live shoot material
from plot #1
b_1 = Weight of live shoot material
from plot #3

The net production of shoot material is combined with the increase in root biomass during the production period to give an estimate, in grams dry biomass per plot, of total *NPP*. *NPP* can also be expressed in gram-calories of energy if estimates of the caloric value of the plant materials are obtained. Values for a variety of plant materials are given by Golley (1961) and Weigert and Evans (1964). Although significant variation does occur in the caloric value of different types of plant material, most herbaceous plant tissues have values near 4,000 gram-calories per gram of dry biomass. To aid in comparison of results with those of other workers, the *NPP* values may be converted to gram-calories per square meter per day. Calculations of *NPP* may be carried out and recorded in table 36.2.

SIMPLIFIED PROCEDURE FOR ESTIMATING *NPP*

Lomnicki et al. (1968) suggested a simplified procedure that assumes no significant decomposition of plant material dying during the interval between the initial and the final sampling. These authors feel that for weedy field or grassland communities such an assumption is valid for periods of up to about a month.

This procedure requires only two plots. On plot #1 the dead shoot material (a_0) is removed and measured at the start of the period, but all live shoot material is left standing. On plot #2 the living shoot material (b_0) and root material are removed and measured at the start of the period.

At the end of the period the living shoot material (b_1), the dead shoot material produced during the interval (d), and the root material, are collected and weighed. From these data *NPP* is given by

$$\frac{\text{Net Primary}}{\text{Production}} = (b_1 - b_0) + d + \text{Change in root biomass}$$

ESTIMATION OF GRAZING HARVEST

To extend these measurements to estimation of grazing harvest of shoot material by large herbivores, the Wiegert/Evans procedure must be carried out on sets of plots located inside and outside exclosures (Cox and Waithaka 1989). In this case, the plots inside the exclosure give estimates of r, x, d, and y in the absence of grazing, as described above. From y, b_0, and b_1, the production rate, p, in grams per gram per day can be estimated by the equation

$$p = y/[(b_0 - b_1)/2](t)$$

If it is assumed that the decomposition rate, r, and the production rate, p, prevailing on the ungrazed plot also prevail on the grazed plot, values of total decomposition, x_g, and shoot net production, y_g, on the grazed plot can be estimated by the equations

$$x_g = [(W_0 + W_g)/2]rt$$
$$y_g = [(b_0 + b_g)/2pt$$

Where: W_g and b_g are final weights of dead and live material on the grazed plots

Grazing harvest, g, is then given by the equation
$$g = b_0 + W_0 + y_g - b_g - W_g$$

Table 36.2. Calculation of net primary production in grams dry biomass and gram-calories of energy for a simple terrestrial ecosystem

Plot Location	Dis. Rate for Dead Mat. (g/g/day) r	Dis. Dead Mat. During Interval (g/plot) x	Change in Standing Crop of Live Mat. (g/plot) $b_1 - b_0$	Change in Standing Crop of Dead Mat. (g/plot) $a_0 - a_1$	Mortality of Live Mat. (g/plot) d	Tot. Shoot Growth During Interval (g/plot) y	Increase in Standing Crop of Roots (g/plot)	Tot. Net Prod. During Interval (g/plot)	Net Prod. During Interval (g-cal./m²)	Net Prod. Rate (g-cal./m²/day)

☯ SUGGESTED ACTIVITIES

1. Carry out *NPP* measurement in a grassland or old field ecosystem in which large grazing herbivores are absent or excluded, using either the Wiegert and Evans (1964) or the Lomnicki et al. (1968) technique. Replicate each set of plots at least 10 times. Allow 4-6 weeks between initial and final sampling.

2. Simulate a field *NPP* study with plants grown in pots in a greenhouse. A month in advance of the first sampling period, plant a series of 30 pots with equal quantities of seeds and cover the soil surface with a measured mass (e.g., 5 grams) of straw or other dead herbaceous plant material. Designate the pots as #1, #2, and #3 samples, and intersperse them in the greenhouse array. Allow the plants to germinate and grow for 4-6 weeks under conditions in which all the pots receive equal watering and light. Conduct the initial harvest of materials from pots #1 and #2 and process the materials. Allow pots #2 and #3 to remain for an additional 4-6 weeks. Conduct the final harvests, process the samples, and calculate *NPP*.

3. Simulate the measurement of *NPP* in a system subject to grazing by medium to large mammals, using the technique described immediately above. Establish two sets of pots as described, but between the initial and final harvests, clip a fraction of the shoot biomass from each of the #3 pots, keeping track of the individual quantity taken. Dry, weigh, and record these "herbivore harvests." Conduct the final harvest, process the materials, and compute *NPP* and grazing harvest. Compare the individual and average grazing harvests based on the pot samples with the actual amounts clipped.

☯ QUESTIONS FOR DISCUSSION

1. What factors are likely to influence the accuracy of estimates of net shoot and net root production? In what ways might the estimate of net root production be improved?

2. What kinds of herbivores other than large grazing animals might influence estimates of net root and shoot production? How could the net production estimates be corrected for losses due to herbivory by these animals?

3. What information would be needed for an estimate of gross primary production? What difficulties would be involved in obtaining this information for the community studied?

☯ SELECTED BIBLIOGRAPHY

Cox, G. W., and J. M. Waithaka. 1989. Estimating aboveground net production and grazing harvest by wildlife on tropical grassland range. Oikos 54:60-66.

Glooschenko, W. A., and N. S. Harper. 1982. Net aerial production of a James Bay, Ontario, salt marsh. Canadian Journal of Botany 60:1060-1067.

Golley, F. B. 1961. Energy values for ecological materials. Ecology 42:581-584.

Groenendijk, A. M. 1984. Primary production of four dominant salt-marsh angiosperms in the SW Netherlands. Vegetatio 57:143-152.

Hogeland, A. M., and K. T. Killingbeck. 1985. Biomass, productivity and life history traits of *Juncus militaris* Bigel. in two Rhode Island (U.S.A.) freshwater wetlands. Aquatic Botany 22:335-346.

Hopkinson, C. S., J. G. Gosselink, and R. T. Parrondo. 1978. Aboveground production of seven marsh plant species in coastal Louisiana. Ecology 59:760-769.

Houghton, R. A. 1985. The effect of mortality on estimates of net above-ground production by *Spartina alterniflora*. Aquatic Botany 22:121-132.

Linthurst, R. A., and R. J. Reimold. 1978. An evaluation of methods for estimating the net aerial production of estuarine angiosperms. Journal of Applied Ecology 15:919-931.

———. 1978. Estimated net aerial primary production for selected angiosperms in Maine, Delaware, and Georgia. Ecology 59:945-955.

Lomnicki, A., E. Bandola, and K. Jankowska. 1968. Modification of the Wiegert-Evans method for estimation of net primary production. Ecology 49:147-149.

Marinucci, A. C. 1982. Trophic importance of *Spartina alterniflora* production and decomposition to the marsh-estuarine ecosystem. Biological Conservation 22:35-58.

Sasser, C. E., and J. G. Gosselink. 1984. Vegetation and primary production in a floating freshwater marsh in Louisiana. Aquatic Botany 20:245-255.

White, D. A., T. E. Weiss, J. M. Trapani, and L. B. Thien. 1978. Productivity and decomposition of the dominant salt marsh plants in Louisiana. Ecology 59:751-759.

Wiegert, R. G., and F. C. Evans. 1964. Primary production and the disappearance of dead vegetation on an old field in southeastern Michigan. Ecology 45:49-63.

EXERCISE 37

Primary Production in Aquatic Systems

INTRODUCTION

Several techniques have been used to measure primary production in aquatic ecosystems. Principal among these are techniques which measure the oxygen by-product of photosynthesis, the uptake of carbon dioxide or radiocarbon in photosynthesis, the change in pH that accompanies carbon dioxide depletion of water, the concentration of chlorophyll in the water (in relation to light intensity and temperature), and direct change in the biomass of aquatic plants. Some of these techniques, such as that for measuring uptake of ^{14}C in photosynthesis (Wetzel and Likens 1991), require specialized equipment and complicated procedures. Techniques that measure the net release of oxygen during periods of active photosynthesis, and its uptake during periods when respiration predominates, are easier, and can be used to measure gross and net primary production in aquatic systems in either the field or the laboratory.

LIGHT AND DARK BOTTLE TECHNIQUE

This technique is based on the fact that the respiration of organisms, including plants, animals, and microorganisms, removes oxygen from water, whereas photosynthesis releases oxygen into the water. Changes in oxygen concentration in the water thus reflect the net effect of these two processes. By using clear glass containers, which permit light to enter (*light* bottles), and containers painted or wrapped to exclude light (*dark* bottles), these two processes can be separated. In the dark bottle, only respiration occurs, and the decrease in oxygen concentration is thus a measure of respiration by all organisms. In the light bottle, both respiration and photosynthesis occur, and the oxygen concentration increases, or at least decreases less than in the dark bottle, depending on the rate of photosynthesis. The difference between the final oxygen concentrations in the light and dark bottles is thus a measure of the total photosynthesis, or gross primary production, over the period.

This technique obviously makes the assumption that respiration is actually occurring in the light bottle at the same rate as in the dark bottle; in actuality this does not seem to be exactly the case (Yallop 1982). For simple experiments designed to illustrate the light and dark bottle technique, however, this assumption does not create major difficulties.

In practice, the technique is also influenced by the fact that samples of water, containing organisms, must be enclosed in containers. All glass containers modify the light transmitted to the inside somewhat. Often they also produce some distortion of normal temperature relations. To the extent that the organisms enclosed do not represent a perfect cross section of those in the natural system, still another bias is introduced.

In dark and light bottle studies, the period of exposure of the bottles should be chosen so that the changes of oxygen concentration are small—only a few parts per million. If left too long, the oxygen in the dark bottle

---

Begon et al. (1990): 18:648–670
Brewer (1994): 11:316–324
Colinvaux (1993): 24:504–518
Krebs (1994): 26:604–621

Odum (1993): 4:79–80
Pianka (1994): 16:355–358
Ricklefs (1990): 11:190–199
Ricklefs (1993): 6:106–111

Smith (1990): 11:211–218
Smith (1992): 22:362–369
Stiling (1992): 19:372–380

may be exhausted, and respiration will thus cease, giving an underestimate of respiration in the natural ecosystem. In the light bottle, on the other hand, abnormally high levels of oxygen may cause some oxygen to come out of solution, leading to an underestimate of photosynthesis.

Lake Studies with Light and Dark Bottles

In a large pond or lake, photosynthesis and respiration can be measured in a water column from surface to bottom. For such a study, water samples are obtained from several depths, placed in 250 ml light and dark bottles, and resuspended under normal light conditions at the depths from which they were taken. At the end of 24 hours, or some other selected interval, the bottles are retrieved and their oxygen content determined.

In setting up this experiment, water samples should be taken from depths chosen to reflect the expected variation at different levels of the water column. It is most convenient if samples are spaced so that they furnish estimates of community metabolism in sections of the water column an even number of meters in length. The following procedure can be used:

1. Collect water samples with a Van Dorn water sampler or similar device from the midpoint of each section of the water column, e.g., 0.5 m (0-1), 1.5 m (1-2), 2.5 m (2-3), 4.5 m (3-6), 8 m (6-10), 15 m (10-20). These samples should contain, as nearly as possible, a normal complement of planktonic plant and animal life. Introduce the samples into glass-stoppered, 250 ml light and dark bottles with as little turbulence as possible, so that the original oxygen content of the water is maintained. No air bubbles should remain in these bottles after stoppering. The dark bottles should be painted black, covered with black tape, or completely enclosed in aluminum foil. Cover the stopper and neck of painted or taped bottles with aluminum foil to ensure that no light can enter.

2. Obtain measurements of the initial oxygen concentration of water samples from the various depths at the time the bottles are filled. Readings can be made quickly with a dissolved oxygen meter and probe (see below).

3. Resuspend the light and dark bottles at the depths from which the water samples were obtained. Place at least two or three light and dark bottles at each depth. At the end of the measurement period, recover the bottles and measure the oxygen content with the dissolved oxygen meter and probe.

The values of oxygen concentration in milligrams per liter (ppm) for the initial oxygen samples (I), the final light bottle samples (L), and the final dark bottle samples (D) are used to calculate the various measures of community metabolism in milligrams carbon per cubic meter of water per hour by the following equations:

$$\text{community respiration} = (375.36)\ (I - D)\ (RQ)$$

$$\text{gross primary production} = \frac{(375.36)\ (L - D)}{PQ}$$

Where: PQ = Photosynthetic quotient (molecules of O_2 produced/molecules of CO_2 taken up)

RQ = Respiratory quotient (molecules of CO_2 produced/molecules of O_2 taken up)

$$\text{net community production} = \text{gross primary production} - \text{community respiration}$$

The value 375.36 represents a conversion factor from the mass of the 12 atoms of oxygen ($6O_2$) to that of the six atoms of carbon (0.37536) in the basic photosynthetic equation, and the conversion from 1 l to 1 m^3 (1,000 l).

Appropriate PQ and RQ values must be selected for calculation of respiration and photosynthesis. Values of 1.0 may be used for both on the assumption that only carbohydrates are being produced by photosynthesis or broken down in respiration. Where the bulk of the community metabolism is due to phytoplankton, however, a *PQ* of 1.2 and an *RQ* of 1.0 is probably more accurate (Strickland and Parsons 1972).

To obtain the total metabolism beneath a square meter of lake surface, community metabolism measurements, in milligrams of carbon per cubic meter, should be multiplied by the length of the water column they represent, and these values totaled for samples from all depths. These calculations and results can be recorded in table 37.1.

Shallow-Water Studies with Bell Jars

A somewhat simpler approach can be used to measure community metabolism in shallow-water areas such as temporary ponds, pond and lake margins, or shallow, soft-bottomed streams. In these situations, light and dark bell jars can be placed over small areas of the natural ecosystem, enclosing a column from near the surface to the bottom.

Large bell jars without screw-cap openings at the top can be used for this experiment. In this case a rubber or plastic tube must also be buried under the lip of the bell jar to permit samples of water to be obtained from inside the bell jar at the end of the experiment.

Smaller bell jars can be made from screw-cap liter bottles by cutting off the bottom of the bottle. The screw cap makes it easy to obtain oxygen readings at the end of the experiment. If the cap is left slightly above the water surface, an oxygen probe can be inserted directly into the bell jar, or a rubber tube can be used to siphon water from the bell jar into a beaker or chemical fixation bottle.

To allow calculations of productivity to be made, two additional items of information are needed. These are *V,* the volume of the water mass enclosed in the bell jar, and *A,* the cross-sectional area of the bell jar. These should be determined as follows:

V = Bell jar volume in cubic centimeters. While the bell jar is in position, measure the distance from the top of the bell jar to the surface of the sediment within the jar. After the bell jar is removed, it should be inverted and the volume of water required to fill it to the same level determined.

A = Bell jar cross section in square centimeters. The inside diameter of the bell jar should be measured in centimeters, and the circular area calculated.

Using data on oxygen concentration in milligrams per liter (ppm) for the initial oxygen concentration (*I*), the final light bell jar sample (*L*), and the final dark bell jar sample (*D*), community metabolism estimates can be calculated in milligrams carbon per square meter per hour by the following equations:

$$\text{community respiration} = \frac{3.7536\ V}{A}(I - D)\ (RQ)$$

$$\text{gross primary production} = \frac{(3.7536\ V)\ (L - D)}{(A)\ (PQ)}$$

$$\text{net community production} = \text{gross primary production} - \text{community respiration}$$

In these calculations, *RQ* and *PQ* are the respiratory and photosynthetic quotients described earlier. The value 3.75 is a composite conversion factor incorporating the conversion from weight of oxygen to that of carbon, together with conversions involving volume and area of the bell jar (adjusting area to square meters and calculating the actual number of milligrams of oxygen produced in the actual water volume of the bell jar).

Conversions to Biomass and Energy

The community metabolism measurements can be converted from milligrams of carbon per square meter of lake surface to grams dry weight of biomass by assuming that dry-weight biomass is about 50% carbon. From dry-weight biomass, the values can be converted to calories of energy, since the caloric value of tissue of planktonic forms is about 4.9 kcal and of other algae about 4.5 kilocalories per gram dry weight.

Table 37.1. Community metabolism beneath one square meter of surface in an aquatic ecosystem as determined by the light and dark bottle technique

1. Sample Depth (m)	2. Length of Water Column Represented (m)	3. Initial Oxygen Concentration (mg/l)	4. Final Light Bottle Oxygen (mg/l)	5. Final Dark Bottle Oxygen (mg/l)	6. Community Respiration (mg C/m³)	7. Columns 2 × 6	8. Net Community Production (mg C/m³)	9. Columns 2 × 8	10. Gross Primary Production (mg C/m³)	11. Columns 2 × 10	12. P/R Ratio

Totals   MgC/m²

Grams dry weight biomass/m²

Gram-calories/m²

From the gross primary production and community respiration values, *P/R* (production/respiration) ratios can be calculated. The *P/R* ratio can be determined for various depths and for the water column as a whole.

INSTRUMENTATION FOR OXYGEN MEASUREMENT

A YSI Dissolved Oxygen Meter equipped with a Model 5905 BOD (Biological Oxygen Demand) Probe (appendix B) is one of the most convenient devices for measuring oxygen in glass-stoppered bottles or small bell jars. This probe is slender enough to be inserted directly into the bottle or jar, so that readings can be taken without any disturbance of the water.

 ## SUGGESTED ACTIVITIES

1. Measure photosynthesis and respiration at various levels in a lake or large pond over a 24-hour period. Follow the procedures outlined above for a lake ecosystem study.
2. Select a shallow-water ecosystem with a sand or mud bottom and a good, relatively uniform growth of submerged plants. Prepare a series of light and dark screw-cap bell jars. Select a site with water about 3–5 cm shallower than the height of the bell jars. Submerge the bell jars so that they do not contain any trapped air. Place each jar over a section of the submerged plant community, burying the bottom lip deep (3–5 cm) enough in the substrate to prevent water exchange between inside and outside, but leaving the top of the neck rim just above the water surface. Using an oxygen meter and probe, obtain measurements of the initial concentration of oxygen in the water at the time the bell jars are put in place.

   Allow the jars to remain in place for 3–6 hours. At the end of this period, carefully unscrew the caps of the bell jars, making sure that water is not exchanged between the inside and the outside of the jar. Immediately obtain measurements of oxygen concentration with an oxygen meter by inserting the probe into the bell jar, or by testing water siphoned with as little turbulence as possible into a beaker. Calculate measures of community metabolism as described earlier in the exercise.
3. Set up one or more aquaria with a bottom substrate of sand about 5 cm deep. Prepare a series of light and dark screw-cap bell jars. Cut sections of an aquatic plant such as *Elodea* to a standard length that corresponds to the height of the bell jar. Place equal numbers of plant sections in each bell jar, submerge the bell jars so that no air is trapped, and bury their bottom edges in the aquarium substrate. Make sure that the bell jars are fully submerged. Take measurements of the oxygen concentration of the water at the time the bell jars are put into the aquaria.

   If enough aquaria and bell jars are available, place aquaria containing both light and dark jars in locations differing in light intensity or temperature. Allow the jars to remain in place for 3–6 hours. At the end of this period, remove water from the aquaria to a level that allows the screw caps of the bell jars to be removed and oxygen determinations made without contamination from the aquarium water outside the jars. Calculate measures of community metabolism as described earlier in the exercise.

QUESTIONS FOR DISCUSSION

1. What is the general trend in the *P/R* ratio from the surface to the bottom of a lake? What becomes of the net community production occurring in the surface layers? How would the *P/R* ratio for the lake as a whole change during the year? What would the *P/R* ratio probably be for the lake as a whole over the entire year?
2. What modifications in procedure would need to be made to include the littoral areas of the lake in measurements of community metabolism?
3. How could net primary production for the lake ecosystem be measured? How does the gross primary production in this study compare with that observed for other ecosystems?

✪ SELECTED BIBLIOGRAPHY

Bott, T. L., J. T. Brock, C. S. Dunn, R. J. Naiman, R. W. Ovink, and R. C. Petersen. 1985. Benthic community metabolism in four temperate stream systems: an inter-biome comparison and evaluation of the river continuum concept. Hydrobiologia 123:3–45.

Busch, D. E., and S. G. Fisher. 1981. Metabolism of a desert stream. Freshwater Biology 11:301–307.

Graham, J. M., and L. E. Graham. 1987. Growth and reproduction of *Bangia atropurpurea* (Roth) C. Ag. (Rhodophyta) from the Laurentian Great Lakes. Aquatic Botany 28:317–331.

Graham, J. M., J. A. Krantzfelder, and M. T. Auer. 1985. Light and temperature as factors regulating the seasonal growth and distribution of *Ulothrix zonata* (Ulvophyceae). Journal of Phycology 21:228–234.

Kemp, W. M., M. R. Lewis, and T. W. Jones. 1986. Comparison of methods for measuring production by the submersed macrophyte, *Potamogeton perfoliatus* L. Limnology and Oceanography 31:1322–1334.

Lieth, H., and R. H. Whittaker, editors. 1975. Primary production of the biosphere. Ecological Studies, Vol. 14. Springer-Verlag, New York, New York, USA.

Strickland, J. D. H., and T. R. Parsons. 1972. A practical handbook of seawater analysis. Bulletin of the Fisheries Research Board of Canada 167:1–310. 2nd ed.

Vollenweider, R. A., editor. 1974. Primary production in aquatic environments. 2nd ed. IBP Handbook No. 23. Blackwell Scientific, Oxford, England.

Wetzel, R. G., and G. E. Likens. 1991. Limnological analyses. 2nd ed. Springer-Verlag, New York, New York, USA.

Yallop, M. L. 1982. Some effects of light on algal respiration and the validity of the light and dark bottle technique for measuring primary production. Freshwater Biology 12:427–433.

Ecosystem Simulation

INTRODUCTION

Systems ecology emphasizes the development of predictive mathematical models of ecological systems. If we consider the general goal of science as seeking natural principles and laws, this goal is a traditional one. Rather than focusing on individual cause and effect relationships, however, systems ecology seeks to understand complex phenomena that involve many cause-effect relations.

A *system* is a set of components united by regular patterns of interaction. In ecology, these may be living organisms and nonliving components of the environment. The complexity of ecological systems can be very great because they may have many abiotic components and numerous species, each with many individuals that differ in age, sex, and other attributes. The overall behavior of such a system depends on interactions among all of these abiotic and biotic components. To complicate matters, interactions often show thresholds, discontinuities, and time lags in their behavior. As a result, even if we understand the basic nature of the specific cause-effect relations in the system, it is very difficult to predict, intuitively, how the total system will respond. The system may be *counterintuitive;* it may not respond in the way we think it should, based on our experience with simple cause-effect relationships of everyday life.

Nevertheless, humans must manage such complex systems, and for this, *simulation models* are proving increasingly useful. A simulation model consists of two parts. First, it contains a set of mathematical equations describing how major relationships within the system are believed to operate. Second, it includes mathematical "bookkeeping" techniques that track quantitative features of the system as they change through time.

This exercise utilizes *STELLA II* (Version 3.0.2), a modeling software program designed for Macintosh microcomputers, to illustrate the nature of ecosystem simulation models (see appendix B). For this exercise the recommended hardware is a Macintosh computer with a math coprocessor, operating system 7.0.1, 4 mb RAM, a color monitor, and STELLA II software installed on the hard disk. The description of STELLA II procedures below is limited to essentials. A full description is given in the user's guide to this software (Patterson and Richmond 1993). In addition, the exercise assumes that students have had an introduction to standard Macintosh keyboard and mouse procedures such as opening and closing windows, selecting and opening items by clicking, typing and deleting commands, opening and choosing menus, and moving icons or symbols by dragging.

THE SILVER SPRINGS ECOSYSTEM

The simulation utilizes the data on ecosystem energy flow from Silver Springs, Florida (Odum 1957). It does not pretend to simulate this ecosystem realistically, however, but merely to use the data to illustrate the characteristics of ecosystems models. The model describes the quantity and the flux of energy for major biotic

Begon et al. (1990): 18:670–680	Odum (1993): 4:91–99	Smith (1990): 3:28–30, 11:208–244
Brewer (1994): 11:307–328	Pianka (1994): 16:355–358	Smith (1992): 23:374–394
Colinvaux (1993): 2:23–26	Ricklefs (1990): 10:173–188, 11:189–209, 14:259–276	Stilling (1992): 19:368–372
Krebs (1994): 26:603–621, 27:633–658	Ricklefs (1993): 6:103–122, 9:162–174	

components of Silver Springs, a large clear-water spring in central Florida. The producers in this ecosystem are mainly eelgrass, attached algae, and some emergent macrophytes. The herbivores include fish, turtles, and invertebrates; carnivores comprise other fish and invertebrates. The top carnivores include largemouth bass and gar. Decomposers include bacteria and crayfish, which are generalized scavengers. Odum (1957) measured the standing crop biomasses of these organisms, and determined the energy equivalent of their biomasses. He also obtained estimates of the rate of energy flow through these components of the ecosystem. This study was one of the first to examine energy flow through an entire ecosystem.

The Silver Springs model includes five *state variables:* the quantitative features of the system whose variation through time is described by the model. These state variables are the quantities of energy in the standing crop biomass of producers, herbivores, carnivores, top carnivores, and decomposers. A series of 19 flux rates characterize the input of energy to various components of the system, the transfer of energy between components within the system, and the loss of energy from the system by respiration and outflow (in the stream flowing from the spring). These state variables and flux rates are summarized graphically in figure 38.1.

Structure of the Model

Mathematically, the Silver Springs model consists of five differential equations that compute the change in value of the state variables during a defined period of time. These changes are used to adjust the values of the state variables, starting with the initial values and continuing over the desired number of time periods.

The five equations (see figure 38.1 for definition of symbols) are

Producers
$$\frac{dX1}{dt} = Z01 - (Z10 + Z12 + Z15 + Z16)$$

Herbivores
$$\frac{dX2}{dt} = (Z12 + Z72) - (Z20 + Z23 + Z25 + Z26)$$

Carnivores
$$\frac{dX3}{dt} = Z23 - (Z30 + Z34 + Z35 + Z36)$$

Top Carnivores
$$\frac{dX4}{dt} = Z34 - (Z40 + Z45 + Z46)$$

Decomposers
$$\frac{dX5}{dt} = (Z15 + Z25 + Z35 + Z45) - (Z50 + Z56)$$

The equations use very simple relationships based on the energy values of the standing crops of the five groups of organisms. The various flux rates (Zs) are represented in the model by a coefficient (P) multiplied against an expression involving one or more state variables (see figure 38.1). One input, that of bread fed to herbivorous fish and turtles by tourists, is defined in simple energetic terms. These expressions are extremely simplified—much more so than in a realistic simulation of an ecosystem.

USING STELLA II TO MODEL ECOSYSTEM ENERGY FLOW

STELLA II is a sophisticated program and has capabilities that cannot be mastered with just a brief introduction. Before attempting to create an ecosystem model, one should therefore work through at least the first three sections of the tutorial. The tutorial describes how a STELLA II model can be used to explore the implications of different management options or environmental changes. It also illustrates the procedure for model construction and explains the use of the various tools. This should perhaps be an activity carried out during the first of a pair of laboratory sessions using this program.

To create a model of the Silver Springs ecosystem, open STELLA II by double-clicking on its icon. This reveals the high-level **Mapping Layer** window. In this window, the major elements of the model can be created

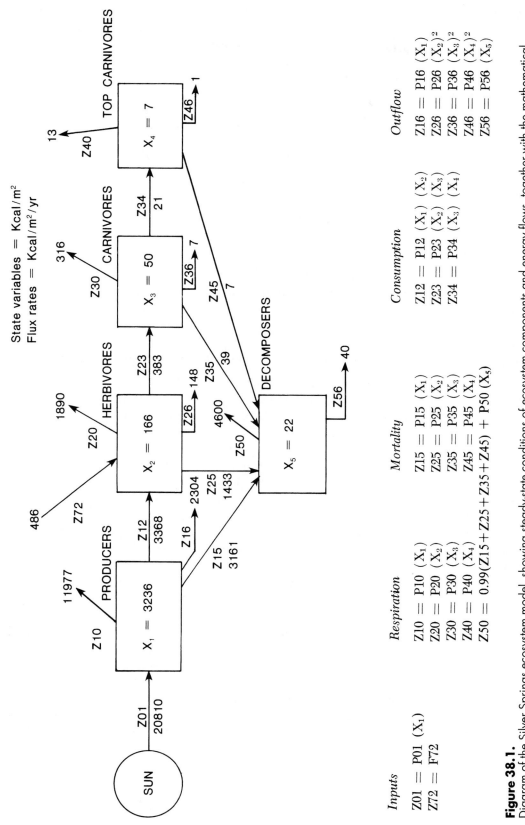

Figure 38.1.
Diagram of the Silver Springs ecosystem model, showing steady-state conditions of ecosystem components and energy flows, together with the mathematical expressions by which flows are calculated. Symbols are as follows: Z = flow rate, P = coefficient (table 38.1), F = externally defined flow.

and positioned. For Silver Springs, these consist of the **Producers, Herbivores, Carnivores, Top Carnivores,** and **Decomposers.** Use the hand tool to capture the large **Process Frame** icon at the left end of the menu bar and place it in the model window. Type in the name of the trophic group. Repeat this procedure and position process frames roughly as in figure 38.1.

Click on the down arrow in the title bar of the producer icon to move to the model **Construction Layer.** The tools along the top of the menu bar of this window will be used to build a graphical model of the ecosystem. First, locate the tools and review their function. To the left are four model components:

1. Rectangle. A *stock* or *state variable,* which changes in value as energy flows in and out. The energy value of the producer standing crop is an example.
2. Flow arrow with balloon. An energy flow channel to, from, or between system components. Herbivore energy intake by feeding on producers is an example.
3. Converter balloon. A site where a window can be opened to create an equation describing a model relationship. A balloon appears automatically when a flow is created, but can also be placed manually for use with a converter arrow. The energy value of the bread input to herbivores and the equation defining herbivore feeding on producers are examples of relationships defined in converter balloons.
4. Connector arrow. A connecting element that designates a controlling relationship. The influence of producer biomass on the export of producers from the system is an example.

In the center of the menu bar are four manipulation tools, two of which are essential:

5. Hand. Used to capture and position model elements.
6. Dynamite stick. Use to eliminate model elements.

Four steps should be followed to create a graphic model of Silver Springs:

1. Place a state variables rectangle in the **Producer** process frame, title it appropriately, and insert its initial value. Repeat this procedure for the remaining four state variables in their respective process frames.
2. Locate and name the inflows and outflows for the **Producer** state variable. Capture a flow arrow and move it to the position of the sun in figure 38.1. Click, hold, and drag the symbol until it touches the producer box, and then release. This produces a flow channel from an external "cloud" to the producer state variable. Delete the **noname** label and type a real, unique label (e.g., **Z01**) in the name space. Create and name outflows for respiration and export in a similar fashion, but begin the flow inside the **Producer** box and end it outside. Create and name a flow channel to the **Herbivore** box by beginning the flow in the **Producer** box and ending it in the **Herbivore** box. Create and name a flow to the **Decomposer** box in the same manner. Repeat this procedure for the flows into and out of the **Herbivore, Carnivore, Top Carnivore,** and **Decomposer** boxes. Remember, in this or any other steps in building the model, an item can be deleted by capturing the dynamite symbol and clicking on the item.
3. Click on the globe symbol in the left margin of the window to switch from the mapping to the modeling mode in the construction window. Consult the equations given in figure 38.1, and note the state variables that influence the value of each flow. Use the connector tool to create loops that connect the state variable box to the balloons of all the flows that it influences (e.g., the **Herbivore** box to the balloon of the **Z12** flow). To use the decomposer respiration equation stated in figure 38.1, connectors must be placed between converter balloons of flows to decomposers and that of the decomposer respiration outflow. Use the click-and-hold technique to drag the connector arrow from the state variable box to the converter balloon (or one converter balloon to another).
4. Note that all the converter balloons now have **?** in their center. Double-click on each of these and provide values or expressions to complete the equation in the dialog window. To do this, you will need to type or click in values from the table of coefficients (table 38.1) and click on the **required inputs** given in the dialog window. In these expressions, arithmetic operations are indicated by $+$, $-$, $/$, and $*$

Table 38.1. Steady-state conditions for the Silver Springs ecosystem model

State Variables (kcal/m²)		Coefficients		
Producers	3236	P01 = 6.43078	P30 = 6.32000	
Herbivores	166	P10 = 3.70117	P34 = 0.06000	
Carnivores	50	P12 = 0.00627	P35 = 0.78000	
Top Carnivores	7	P15 = 0.97682	P36 = 0.00280	
Decomposers	22	P16 = 0.71199	P40 = 1.85714	
		P20 = 11.3855	P45 = 1.00000	
Bread Input (Kcal/m²/yr)		P23 = 0.04614	P46 = 0.02041	
To Herbivores		P25 = 8.63253	P50 = 0.29090	
F72 = 486		P26 = 0.00537	P56 = 1.81820	

symbols. When the equation is completed, click on the **OK** box. This procedure clears the **?** from the converter balloon. The completed model should be free of all **?** symbols in converter balloons.

In preparation for running the model, the variables must be scaled. Open the **Time Specs** window under the **Run** menu and select years as the time unit, **0** to **20** as the length of simulation, and **0.05** as **DT,** the interval over which model calculations will be done. Click on **OK** and close the window. Open the **Range Specs** window to scale the five state variables so that values can range from a minimum of 0 to a maximum of 2–3 times their initial values. Do this by typing in minimum and maximum values, and clicking on **Set.** Close the window by selecting **OK.**

The output of the model can be viewed and printed as either a graph or a table. To create a graph, click on the **Graph** icon on the menu bar, and drop it onto an open space on the edge of the model window. Double-click on this icon to open it, and on the graph to open a **Define graph** dialog window. Double-click on the variables to be graphed and select the **Time series** option. Select **OK** to close the window.

The model is now ready to run. Select **Run** under the run menu. Values of the selected variables will be graphed. If the model has been constructed correctly, all variables will project as nearly level lines across the graph at about their initial (steady state) values.

Tabular output can also be obtained by capturing the table icon from the menu bar and placing it on the model window. It can be opened like the graph window, and the content defined by selecting the variables to be listed and the time interval at which they are printed.

The **File** menu permits the model, a graph of a model run, or a table of a model run to be printed, simply by selecting the appropriate commands.

SUGGESTED ACTIVITIES

1. Once a working program has been obtained, the model can be manipulated. Initial state conditions, external inputs, or coefficients can be changed to simulate some perturbation of the ecosystem. These changes are made by (1) changing initial biomass values in the state variable boxes, (2) changing coefficients or variable relationships in equations in converter balloons associated with the flow channels, or (3) eliminating certain state variables (e.g., top carnivores). Until some appreciation is gained of the magnitude of response of the model to various types of changes, only one or a few changes should be made. Changing state variables by 50–200% or coefficients by 10–20% might be tried at the outset, for example. For example, suppose it were possible to replace the existing herbivores with forms having a 20% greater ability to harvest plant tissues. To simulate this, coefficient *P12* might be changed from 0.00627 to 0.00752, and the model run with this single change.

To explore other changes, think of some environmental change or human manipulation that might occur in Silver Springs. Translate this change into a change in initial state conditions, coefficient values, or external inputs. Write down a prediction of how you think the system will respond; then run the model under the changed conditions. Some of the following questions might be examined:

1. What would happen if the solar radiation input to the ecosystem were reduced by 20% due to air pollution or increased cloudiness? Would the equilibrium biomasses of all groups of organisms be reduced by 20%?
2. How would the system change if DDT magnification along food chains caused the top carnivores to be eliminated?
3. What would happen if thermal pollution increased the temperature of the water enough to increase the respiratory rate of the various organisms by 10%?
4. What would happen if the carnivore abundance was increased by stocking fish in a quantity equal to that already present?
5. What would occur if herbivores were replaced by larger-bodied species that had a respiration level 20% lower than that of those originally present?
6. What would happen if tourism increased or decreased, so that 10% increase or decrease in bread input occurred?
7. What would happen if nutrient pollution increased photosynthetic production by 10%?

2. Replace one or more of the controls on energy flows between trophic groups with a graphical expression that contains a density-dependent relationship (e.g., predation on herbivores), or a time lag (e.g., a reduced recovery rate for producers depleted below a certain level).

3. Develop a model for a different ecosystem, such as a Georgia salt marsh (Teal 1962), the high-arctic tundra (Bliss 1975), or a Tennessee deciduous forest (Harris et al. 1975).

✪ QUESTIONS FOR DISCUSSION

1. Give examples of counterintuitive responses of the Silver Springs model, based on your exploration of its behavior. What relationships in the model do you think are least and most realistic?
2. How would you modify the model to accommodate the human harvest of one or more components of the ecosystem, as by fishing? How might you modify the model to take into account the fact that primary production might be influenced by the biomass of plants present and by the activity of decomposers in regenerating nutrients, as well as by the amount of light radiation available?
3. To what do you think the simple coefficients used in this model would correspond in a realistic ecosystem model?

✪ SELECTED BIBLIOGRAPHY

Bliss, L. C. 1975. Truelove lowland—A high arctic ecosystem. Pages 51–85 *in* T. W. M. Cameron and L. W. Billingsley, editors. Energy flow—Its biological dimensions. Royal Society of Canada, Ottawa, Canada.

Bratley, P., B. L. Fox, and L. E. Schrage. 1987. A guide to simulation. Springer-Verlag, New York, New York, USA.

Hall, C. A. S., and J. W. Day, Jr., editors. 1977. Ecosystem modeling in theory and practice: an introduction with case histories. John Wiley & Sons, New York, New York, USA.

Harris, W. F., P. Sollins, N. T. Edwards, B. E. Dinger, and H. H. Shugart. 1975. Analysis of carbon flow and productivity in a temperate deciduous forest ecosystem. Pages 116–122 *in* D. E. Reichle, J. F. Franklin, and D. W. Goodall, editor. Productivity of world ecosystems. National Academy of Sciences, Washington, D.C., USA.

Jakeman, A. J., M. B. Beck, and M. J. McAleer, editors. 1993. Modelling change in environmental systems. John Wiley & Sons, London, England.

Jeffers, J. N. R. 1978. An introduction to systems analysis: with ecological applications. University Park Press, Baltimore, Maryland, USA.

Jorgensen, S. E. 1986. Fundamentals of ecological modelling. Elsevier, Amsterdam, The Netherlands.

Jorgensen, S. E., editor. 1993. Theoretical modeling aspects. Ecological Modelling 68:1-118.

Law, A. M., and W. D. Kelton. 1991. Simulation modeling and analysis. 2nd ed. McGraw-Hill, New York, New York, USA.

Mitsch, W. J., M. Straskraba, and S. E. Jorgensen, editors. 1988. Wetland modelling. Elsevier, New York, New York, USA.

Odum, H. T. 1957. Trophic structure and productivity of Silver Springs. Ecological Monographs 27:55-112.

———. 1983. Systems ecology: an introduction. John Wiley & Sons, New York, New York, USA.

Peterson, S., and B. Richmond. 1993. STELLA II. Technical documentation. High Performance Systems, Hanover, New Hampshire, USA.

Richter, O. 1990. Parameter estimation in ecology: the link between data and models. VCH, Berlin, Germany.

Swartzman, G. L., and S. P. Kaluzny. 1987. Ecological simulation primer. Macmillan Publishing Company, New York, New York, USA.

Teal, J. M. 1962. Energy flow in the salt marsh ecosystem of Georgia. Ecology 43:614-624.

Wiegert, R. G. 1975. Simulation models of ecosystems. Annual Review of Ecology and Systematics 6:311-338.

EXERCISE *39*

Geographic Information Systems

INTRODUCTION

A *geographic information system (GIS)* consists of computer hardware and software designed for the input, storage, manipulation, and output of data referenced by a common system of spatial coordinates. The function of a GIS is to permit data on many sets of variables over a specified geographic area to be combined or manipulated to answer questions about spatial relationships. The GIS enables the results of such analyses, as well as any of the data sets themselves, to be printed out in map form. For example, a GIS might combine data on climate, soils, and slope steepness to create a map of land suitability for crop cultivation.

In principle, the spatial scale of a GIS can vary widely. A GIS might be developed for data on microsite conditions and vegetation characteristics of a small area of grassland, using spatial coordinates measured in centimeters. Most GISs are being developed, however, for areas such as parks, national forests, or political units such as cities, counties, or states. The United Nations Environment Program maintains a global GIS database, the *Global Environmental Monitoring System.*

Many types of data can be included in a GIS. For ecological applications, data might include many sorts of physical or chemical characteristics of the environment, and many biotic characteristics, including the presence or abundance of individual species or communities. Cultural variables, such as land ownership or type of human use, can also be included. Data on each variable are stored in a separate computer file, or *layer,* using a common system of coordinates, enabling values of any combination of variables to be determined for a given location.

Two somewhat different systems of storing spatial information exist, and these are combined in more sophisticated GISs. The *raster* format divides the spatial area into cells or pixels, with conditions in each cell being specified by a value or index. Data on plant biomass might be recorded in this manner, for example, using a scale of values from zero to some maximum. The *vector* format uses lines, their positions referenced by x- and y-coordinates, to show the boundaries of areas of similar conditions. The distribution of soil types might, for example, be recorded in this fashion.

A GIS FOR MACINTOSH MICROCOMPUTERS: macGIS

We shall illustrate the nature and capabilities of GIS systems by exploring *macGIS 2.0,* a raster-format GIS system developed for Macintosh computer systems by David Hulse and Kit Larson (1992). For this exercise, a Macintosh II computer with 4MB of RAM and color monitor, or a more advanced member of the Macintosh family, is recommended. Copies of the macGIS program used in this exercise are subject to the conditions noted in appendix B. Before undertaking this exercise, students should acquaint themselves with the basics of the Macintosh graphical interface: icons, menus, and mouse operations such as highlighting, selecting, opening windows, scrolling, and control of window size.

Brewer (1994) 21:701–703

A simple GIS database is included with the macGIS program. This database is for Mt. Pisgah, Oregon, and consists of 18 data layers. The land area in this database covers 3,600 hectares, each hectare being represented by one pixel in the GIS database system.

Layers of the Mt. Pisgah GIS

The objective of this activity is to become familiar with the nature of the data layers and how they can be displayed.

1. Open the program by double-clicking on the **macGIS** icon. This reveals the **COMMAND LOG** window that simply records the instructions you give during your macGIS session.
2. Select **OPEN** from the **FILE** menu. Highlight and select **ALTITUDE.** Enlarge the **ALTITUDE** window to show the entire Mt. Pisgah landscape, which shows elevational ranges in shades of color. Check the elevational-range index window. You will have to scroll down the index window to see all the color shades that might be used to code altitudes. What is the most frequent elevational range?
3. From the **DISPLAY** menu, select **SHOW WIREFRAME.** This opens a new window showing this same landscape in three dimensions. Enlarge this window to show the entire landscape.
4. Investigate the tools menu at the left. Click and drag the topmost **POSITION CUBE** icon and observe how the landscape can be rotated and tilted. Click on the **STOP** icon (second row, left side) as the image is being drawn. What happens? Click on the **POSITION CUBE** icon. What happens? Click on the **KEY** icon. A menu of parameters for the wireframe view is revealed. Change **vertical exaggeration** to **4.** What happens? Return **vertical exaggeration** to its original value. Note the other parameters that can be modified.

    Click on the **ENLARGE** and **REDUCE** mountain peak icons to see how the size of the image can be changed. Note the **PERSPECTIVE** icons (fourth row) that allow the image to be redrawn from a closer or more distant perspective.

    Click on the **BOUNDARY/FILL** icon (left side, second from bottom) and note the three types of views of the landscape that can be obtained. Click on the **LABEL** icon (right side, second from bottom) to see how labels can be omitted or added. Using the **BOUNDARY/FILL** tool, create a landscape outline without the color-coded elevations. Click on the **SETTINGS** icons (third row from bottom) to see how these affect the level of detail shown on the landscape.

    Click on the **DRAPE** icon (left side, bottom row) to obtain a menu of the other layers of data that can be superimposed on the landscape. Highlight **SOIL DEPTH** and click on **OPEN.** This recodes the landscape with colors representing three soil depths ($<20''$, 20–$40''$, $>40''$). Rotate the landscape and determine on which side of the mountain the soils tend to be shallowest (green).
5. Close the **ALTITUDE** window, and quit macGIS from the **FILE** menu.

Creating Synthetic Data Layers

The objective of this activity is to see how data layers of a GIS can be combined to create a new layer that displays a combination of information that can answer specific ecological questions.

1. Open the program by double-clicking on the **macGIS** icon.
2. From the **FILE** menu, open **WILDLIFE.** From the **POINT** menu, select **RECODE.** Fill in the blanks to assign the value **0** to **0** thru **4** with the legend **other,** clicking on the **DO IT** button when you finish. Repeat this procedure to assign the value **0** to **6** thru **7** with the legend **other.** Now assign the value **1** to **5.** This creates a new **WILDLIFE-A** layer in which old-growth Douglas-fir forest is coded as **1.**
3. From the **FILE** menu, open **ASPECT.** From the **POINT** menu, select **RECODE.** Fill in the blanks to assign the value **9** to **0** with the legend **level,** clicking on the **DO IT** button when you finish.

4. From the **ARITHMETIC** menu, select **MULTIPLY.** Highlight the file name **WILDLIFE-A,** and click on the **SELECT** and **DO IT** buttons in sequence. In the **Label** column of the resulting map legend box, click on the space to the right of **Value 1, Count** 4 and label it **North.** Work downward to fill in **Northeast, East, Southeast, South, Southwest, West, Northwest,** and **Level.** What slope aspects have the most extensive old-growth Douglas-fir stands?

5. From the **DISPLAY** menu, click and hold on **PATTERN SET** and drag down to **SYSTEM PALETTE.** Then double-click on the color code spaces for aspects which have old-growth forest, to view the color wheel. Select a color for each such aspect and click the **OK** button. Click on aspect color boxes for those aspects with no old-growth units, select white at the color wheel center, and click the **OK** button. This leaves only the old-growth stands in color. Are level areas represented in all old-growth stands? Is any of the four individual stands uniform in its aspect?

Analysis of Spatial Relationships

1. From the **FILE** menu, open **VEGETATION.**
2. From the **POINT** menu, select **RECODE.** Fill in the blanks to assign the value **5** to **5** thru **10** with the legend **forest** (remember to click on **DO IT**). From the **POINT** menu, select **ISOLATE.** Fill in the blanks to assign the value **1** to **5** with the legend **forest.**
3. From the **NEIGHBORHOOD** menu, select **CLUMP.** Specify the distance as **3** cells, which specifies that any forest units separated by no more than three grid spaces will be considered to be part of a single clump of forest. The resulting **Val** column lists the individual clumps identified, and the **count** column gives their size. How many forest clumps exist? What is their range of sizes?
4. Close the **VEGETATION** window, and quit macGIS from the **FILE** menu.

Identification of Ecotones

1. From the **FILE** menu, open **WATERBODIES.**
2. With the **POINT** menu, isolate permanent streams by assigning the value **1** to **1** thru **2.** (Remember to click on **DO IT.**)
3. From the **NEIGHBORHOOD** menu, select **SPREAD,** and accept the default value of **4** cells.
4. From the **POINT** menu, isolate the value **1** as applied to **1** thru **1,** and label it **Ecotone.** (Be sure to **DO IT.**)
5. Note that the map now shows only the areas within one grid unit of permanent streams.
6. Close the modified **WATERBODIES-A** window. Trash any new GIS layers (identified by -A suffixes), and quit macGIS from the **FILE** menu.

SUGGESTED ACTIVITIES

Now that you have familiarized yourself with the basic tools of macGIS, you might want to carry out some sort of original analysis as an individual special study or class project.

1. Make a list of the data layers in the Mt. Pisgah GIS. Review the options available under the **ARITHMETIC** and **NEIGHBORHOOD** menus. Pose a question and carry out an analysis that combines or manipulates data from two or more layers. For example, use the layers of **Slope, Soil depth,** and **Erosion** to create an index of land vulnerability to physical disturbance.
2. Create a new layer for the Mt. Pisgah database. Each data layer is a 60 × 60 matrix, each cell of which contains the value for one hectare. Assume, for example, some pattern of acidity of precipitation due to

pollution from a point source that affects a portion of the area. This can be mapped onto one of the existing files and the file saved with a new name.

To do this, open a file such as **Dry** (which shows only two levels of the variable in question). From the **EDIT** menu, drag the cursor down to **Edit Map** and release, toggling the label to **View Map.** Then drag the cursor down to **Expose Non-zero Cells** and release, toggling this label to **Protect Non-zero Cells.** Set up a scale of pH ranges covering a scale from 1 to 9. Double-click on the label spaces opposite the 1–9 scale values and type in the corresponding pH ranges. Then double-click on the color code spaces opposite the 1–9 scale values and choose colors from the palette for the 9 scale values. Click on a color and move to the map, where the cursor position appears as crosshairs. With the mouse button depressed, paint areas to create a zone of hypothetical precipitation pH (e.g., assuming some source in or just outside the mapped area). Do this until the map is filled. From the **FILE** menu, choose **Save as** and label the new layer something like **Acidity.** Close the **Dry** layer and DO NOT save any of the changes.

Using this new layer, determine what fractions of different vegetation types would be affected by various levels of acidity.

3. Create a GIS database for an area of ecological interest for which soils, vegetation, and other kinds of data are available. This need not be a large geographical area, and might be a several-hectare university research area or a portion of campus with diverse ecological conditions. Refer to the section of the macGIS manual on *Creating a Data Base,* and follow the procedure outlined. Or, choose a square sector that covers the area, and divide it into a 60×60 matrix. Create basic data layers by coding available information for each pixel. Collect information on the distribution of selected plant and animal species by field surveys, and create layers for these species.

✪ Questions for Discussion

1. Choose an area of ecological interest near your university. What are the layers of data that you think should be included in a GIS database for this area?
2. How can a GIS system be used to quantify and display changes in conditions through time?
3. How can GIS databases be used to identify species that are not adequately protected in a series of ecological preserves?

✪ Selected Bibliography

Davis, F. W., D. M. Stoms, J. E. Estes, and J. Scepan. 1990. An information systems approach to the preservation of biological diversity. International Journal of Geographical Information Systems 4:55–78.

Hulse, D., and K. Larson. 1992. MacGIS 2.0. A geographic information system for the Macintosh. Harper-Collins, New York, New York, USA.

Johnson, L. B. 1990. Analyzing spatial and temporal phenomena using geographic information systems. Landscape Ecology 4:31–43.

———. 1993. Ecological analyses using geographic information systems. Pages 27–38 *in* S. B. McLaren and J. K. Braun, editors. GIS applications in mammalogy. Oklahoma Museum of Natural History, Norman, Oklahoma, USA.

Maguire, D. J., M. F. Goodchild, and D. W. Rhind. 1991. Geographical information systems: principles and applications. Vol. 1, 2. John Wiley & Sons, New York, New York, USA.

Reinhardt, R. D. 1992. Geographic information systems (GIS)—a global perspective. Cutter Information Corp., Arlington, Massachusetts, USA.

Scott, J. M., F. Davis, B. Csuti, R. Noss, B. Butterfield, C. Groves, H. Anderson, S. Caicco, F. D'Erchia, T. C. Edwards, Jr., J. Ulliman, and R. G. Wright. 1993. Gap analysis: a geographic approach to protection of biological diversity. Wildlife Monographs 123:1–41.

Star, J. L. and J. E. Estes. 1990. Geographical information systems: an introduction. Prentice-Hall, Englewood Cliffs, New Jersey, USA.

Soil Texture Analysis

INTRODUCTION

The particle size composition of soil samples can be analyzed by a hydrometer technique. This technique is based on the facts that (1) particles settle out of a soil-water suspension in order of decreasing size, and (2) the specific gravity of a soil suspension is directly related to the total quantity of soil material still in suspension. In carrying out the analysis, the individual soil particles are first dispersed mechanically by stirring or shaking and are kept from aggregating by a chemical dispersing agent that confers identical charges on the particles. The individual particles are then free to settle out under the force of gravity. The overall settling rate is related to temperature, which affects the viscosity of the water. Detailed descriptions of this method are given by Gee and Bauder (1986) and the American Society for Testing and Materials (1988).

PROCEDURE

The equipment required for this analysis consists of a mechanical dispersing apparatus, sedimentation cylinders, and a soil hydrometer. The recommended dispersing apparatus is a mechanical mixer similar to that used in making milkshakes, and consists of an overhead motor unit with a stirring rod that extends down into a metal cup in which the soil-water mixture is placed. The metal cup should be equipped with sets of baffle rods on the inside to increase turbulence during the dispersion process. The sedimentation cylinders should be glass cylinders calibrated for a volume of 1,000 ml and of such diameter that the 1,000 ml mark is between 34 and 38 cm above the inside bottom of the cylinder. The soil hydrometer (ASTM 152H) is calibrated in grams of soil per liter of soil-water suspension at 68° F (20° C). Apparatus specifically designed for this analysis can be obtained from most major suppliers of scientific equipment.

Preparation of Soil Samples for Analysis

Soil samples should be air-dried in the laboratory for several days prior to the texture analysis. The entire sample should be weighed and then spread out on a flat, newspaper-covered surface. The lumps of soil should be broken up with a wooden roller. Care should be taken during this procedure that rocks and coarse soil particles are not broken or crushed. The sample should then be passed through a 2 mm (#10) sieve. The rock material retained by the sieve is weighed and the percent of the total sample weight consisting of particles greater than 2

Brewer (1994): 3:60-63
Colinvaux (1993): 25:526-541
Ehrlich and Roughgarden (1987): 20:472-477
Odum (1993): 5:137-139
Pianka (1994): 4:70-73

Ricklefs (1990): 13:231-235
Ricklefs (1992): 4:79-86
Smith (1990): 9:175-175
Smith (1992): 9:144-152

mm in diameter determined. This value, along with the total sample weight, can be recorded in table 40.1. The hydrometer analysis is carried out in the <2 mm soil fraction.

A 20–25 g sample of the <2 mm soil should be weighed to the nearest centigram and oven-dried at 105–115° C for at least 24 hours, and the oven-dry weight determined as a percentage of the air-dry weight. The percentages can be recorded in table 40.1 and used to calculate the oven-dry weights of samples used in the hydrometer analysis.

A sample of 100 g, for coarse-textured soils such as sands or sandy loams, or 50 g, for fine-textured soils such as loams and clays, should be weighed to the nearest centigram for the hydrometer analysis. The sample should be placed in a glass container of about 250 ml capacity, and 125 ml of sodium hexametaphosphate, $(NaPO_3)_6$, solution (40 g/l) added. This mixture should be stirred and allowed to slake for at least 16 hours before the mechanical dispersion.

The sample should then be transferred to the metal dispersing cup. Distilled water should be used to wash any residue from the glass container into the metal cup. The sample should then be mixed with the mechanical stirrer. Sandy soils should be mixed for about 5 minutes, soils of intermediate texture for about 10 minutes, and soils with a very high clay content for about 15–20 minutes.

Determination of Hydrometer Correction

A set of correction factors must be determined for the hydrometer used in reading the soil suspensions. This hydrometer has been calibrated by the manufacturer with known soils in pure water at 68° F and is designed to be read at the bottom of the meniscus. With actual soil suspensions, however, it is usually necessary to read the top of the meniscus. Deviation of temperatures from 68° F and addition of the chemical dispersing agent produce further variations for which correction must be made. To obtain a composite correction for these factors, 125 ml of dispersing solution is placed in a sedimentation cylinder and the cylinder filled to the 1,000 ml mark with distilled water. Hydrometer readings at the meniscus top are then taken over the range of temperatures expected in the soil-suspension measurements. The difference between this reading and 0 at a particular temperature constitutes a correction factor to be added to (if the reading is less than 0) or subtracted from (if the reading is greater than 0) the readings made for soil suspensions at that temperature.

Approximate Analysis

The following procedure gives approximate percentages of sand, silt, and clay fractions in the sample. These values are sufficiently accurate for general ecological work. The procedure assumes a value of 2.65 for the specific gravity of soil particles and requires that the sedimentation be carried out at or near 68° F. For the latter condition, a water bath or constant temperature room may be needed.

The dispersed sample is washed into a sedimentation cylinder with distilled water. The cylinder is then filled to the 1,000 ml mark with distilled water. The cylinder contents are then mixed by placing one hand over the cylinder mouth and inverting the cylinder about 60 times (about 1 minute). The cylinder is immediately placed on a flat surface and the time noted.

Hydrometer readings are taken at 40 seconds, 60 minutes, and 120 minutes after the start of sedimentation. The hydrometer should be inserted carefully about 20–25 seconds before the time of the reading. If excessive foam threatens to make a reading impossible, it can be dissipated with a squirt or two of methyl alcohol. The temperature of the suspension should be noted at the time of each reading. The hydrometer should be washed with distilled water between readings. Hydrometer readings, adjusted with the correction factors determined previously, can be recorded in table 40.1.

Table 40.1. Data sheet for soil texture analysis by the hydrometer method

Sample	Total Sample Weight (g)	Over 2 mm Fraction %	Determination of Oven-Dry Weight Correction Factor			Weight of 2 mm Texture Analysis Sample		Corrected Hydrometer Readings			Particle-Size Analysis					
			Air-Dry Weight (g)	Oven-Dry Weight (g)	Corr. Factor	Air Dry (g)	Oven Dry (g)	40 Sec.	60 Min.	120 Min.	Sand %	Silt Old %	Silt New %	Clay Old %	Clay New %

Table 40.2. Values of correction Factor, *a,* for different specific gravities of soil particles*

Specific Gravity	Correction Factor, *a*	Specific Gravity	Correction Factor, *a*	Specific Gravity	Correction Factor, *a*
2.45	1.05	2.65	1.00	2.85	0.96
2.50	1.03	2.70	0.99	2.90	0.95
2.55	1.02	2.75	0.98	2.95	0.94
2.60	1.01	2.80	0.97		

*Reprinted with permission of the American Society for Testing and Materials from ASTM D422-63.

The corrected hydrometer readings can be used to determine the percent by weight of various particle-size classes by the following equations:

sands (2.0–0.05 mm)
$$= 100 - \frac{(\text{corrected 40-second reading}) (100)}{\text{oven-dry sample weight}}$$

old conventional clay (<0.005 mm)
$$= \frac{(\text{corrected 60-minute reading}) (100)}{\text{oven-dry sample weight}}$$

new conventional clay (<0.002 mm)
$$= \frac{(\text{corrected 120-minute reading}) (100)}{\text{oven-dry sample weight}}$$

old silt fraction (0.05–0.005 mm) $= 100 - (\text{sands} + \text{old conventional clay})$

new silt fraction (0.05–0.002 mm) $= 100 - (\text{sands} + \text{new conventional clay})$

Detailed Analysis

A more precise analysis may be obtained by the following procedure, in which corrections are made for deviation of the specific gravity of the soil particles from a value of 2.65 and for differences in settling rate due to deviation of the soil-water suspension temperatures from 68° F. In addition, a correction is made for change in the depth range of the suspension measured by the hydrometer as soil material settles out and the hydrometer sinks to a greater depth.

For this analysis, the sample is transferred to a sedimentation cylinder and mixed as described earlier. Hydrometer and temperature readings are then taken at as many intervals as needed to obtain a smooth curve of change in suspended material (e.g., 40 seconds; 2, 5, 15, and 30 minutes; 1, 2, 4, and 12 hours). These hydrometer readings are corrected as described earlier. Although it is possible to correct for deviation of the suspension temperature from 68° F, the temperature must be held nearly constant at some level for the entire sedimentation period.

The quantity of soil remaining in suspension at any time, as a percentage of the oven-dry sample weight, is given by the equation

$$\text{percent in suspension} = \frac{(\text{corrected hydrometer reading}) (a) (100)}{\text{oven-dry sample weight}}$$

Where: a = Correction factor for deviation of soil specific gravity from value of 2.65 (table 40.2).

Table 40.3. Values of *K* for use in equation for computing diameter of particles in hydrometer analysis*

Temp. (°C)	Specific Gravity								
	2.45	2.50	2.55	2.60	2.65	2.70	2.75	2.80	2.85
16	0.01510	0.01505	0.01481	0.01457	0.01435	0.01414	0.01394	0.01374	0.01356
17	0.01511	0.01486	0.01462	0.01439	0.01417	0.01396	0.01376	0.01356	0.01338
18	0.01492	0.01467	0.01443	0.01421	0.01399	0.01378	0.01359	0.01339	0.01321
19	0.01474	0.01449	0.01425	0.01403	0.01382	0.01361	0.01342	0.01323	0.01305
20	0.01456	0.01431	0.01408	0.01386	0.01365	0.01344	0.01325	0.01307	0.01289
21	0.01438	0.01414	0.01391	0.01369	0.01348	0.01328	0.01309	0.01291	0.01273
22	0.01421	0.01397	0.01374	0.01353	0.01332	0.01312	0.01294	0.01276	0.01258
23	0.01404	0.01381	0.01358	0.01337	0.01317	0.01297	0.01279	0.01261	0.01243
24	0.01388	0.01365	0.01342	0.01321	0.01301	0.01282	0.01264	0.01246	0.01229
25	0.01372	0.01349	0.01327	0.01306	0.01286	0.01267	0.01249	0.01232	0.01215
26	0.01357	0.01334	0.01312	0.01291	0.01272	0.01253	0.01235	0.01218	0.01201
27	0.01342	0.01319	0.01297	0.01277	0.01258	0.01239	0.01221	0.01204	0.01188
28	0.01327	0.01304	0.01283	0.01264	0.01244	0.01225	0.01208	0.01191	0.01175
29	0.01312	0.01290	0.01269	0.01249	0.01230	0.01212	0.01195	0.01178	0.01162
30	0.01298	0.01276	0.01256	0.01236	0.01217	0.01199	0.01182	0.01165	0.01149

*Reprinted with permission of the American Society for Testing and Materials from ASTM D422-63.

The maximum diameter of the particles remaining in suspension at a particular time is given by the equation

$$\text{maximum diameter } K\sqrt{\frac{L}{T}}$$

Where: K = Factor relating specific gravity of soil particles and soil suspension temperature (table 40.3)

L = Effective depth of soil suspension measured by hydrometer (table 40.4)

T = time in minutes from beginning of sedimentation

With these calculations, values for the percentage of the sample remaining in suspension can be graphed against maximum particle diameter. The curve formed by the points on this graph may be used as an overall description of the particle-size composition of the sample. From this curve, the percent of the sample composed of particles of any desired size range can also be determined.

SUGGESTED ACTIVITIES

1. Examine soil textural relationships along an environmental gradient, such as a gradient from hillside residual soils to floodplain alluvial soils in a stream valley, or a gradient inland from the shore of the ocean or a large lake. Correlate differences in texture with the fine-root morphology of plants collected at points along this gradient.

2. Examine soil textural relationships at different depths, using samples collected in a fresh roadcut or excavation. Discuss the factors that may be responsible for the differences found.

3. Collect sediment samples along a transect of increasing depth from a lake shoreline outward to a point 200–500 meters from shore. Analyze textural relations, and correlate these with the way that benthic invertebrates find shelter at the various locations.

Table 40.4. Effective depth (L) in centimeters of specific-gravity measurements of soil suspensions for various hydrometer readings (for hydrometer corresponding to ASTM 152H specifications)*

Hydrometer Reading	Effective Depth	Hydrometer Reading	Effective Depth	Hydrometer Reading	Effective Depth
0	16.3	21	12.9	41	9.6
1	16.1	22	12.7	42	9.4
2	16.0	23	12.5	43	9.2
3	15.8	24	12.4	44	9.1
4	15.6	25	12.2	45	8.9
5	15.5	26	12.0	46	8.8
6	15.3	27	11.9	47	8.6
7	15.2	28	11.7	48	8.4
8	15.0	29	11.5	49	8.3
9	14.8	30	11.4	50	8.1
10	14.7	31	11.2	51	7.9
11	14.5	32	11.1	52	7.8
12	14.3	33	10.9	53	7.6
13	14.2	34	10.7	54	7.4
14	14.0	35	10.6	55	7.3
15	13.8	36	10.4	56	7.1
16	13.7	37	10.2	57	7.0
17	13.5	38	10.1	58	6.8
18	13.3	39	9.9	59	6.6
19	13.2	40	9.7	60	6.5
20	13.0				

*Excerpted with permission of the American Society for Testing and Materials from ASTM D422–63.

QUESTIONS FOR DISCUSSION

1. How does soil texture correlate with the water-holding capacity of a soil? How does it affect the availability of soil water to plants?
2. How does percolation of water affect the vertical profile of particle size in soils under humid climates? How are rainfall and runoff likely to affect soil textures in desert environments?
3. What factors affect the texture of sediments in aquatic ecosystems?

SELECTED BIBLIOGRAPHY

American Society for Testing and Materials. 1988. Standard method for particle-size analysis of soils. Pages 87–93 *in* 1988 annual book of ASTM standards. Vol. 04.08. Soil and rock, building stones; geotextiles. American Society for Testing and Materials, Philadelphia, Pennsylvania, USA.

Gee, G. W., and J. W. Bauder. 1986. Particle-size analysis. Pages 383–411 *in* A. Klute, editor. Methods of soil analysis. Part 1. Physical and mineralogical methods. 2nd ed. American Society of Agronomy, Madison, Wisconsin, USA.

APPENDIX A

Statistical Tables

Table A.1.   Random numbers

855	346	070	675	915	842	966	401	542	766	892	514	973	957	143	469
343	854	630	428	722	031	568	266	641	829	850	008	319	216	532	349
407	225	993	012	353	005	283	254	701	168	128	250	623	839	160	852
315	707	853	288	883	003	600	302	943	581	467	369	087	607	635	091
272	637	089	545	414	864	938	307	435	776	556	420	800	254	029	642
743	118	528	574	096	282	924	248	560	561	608	270	494	747	651	399
547	599	368	560	104	761	855	860	005	399	440	305	163	575	715	405
624	247	292	942	013	687	586	571	047	045	562	309	872	389	501	739
405	757	932	822	026	152	326	042	076	212	982	994	014	049	734	647
464	195	737	644	647	308	845	733	340	159	830	107	460	913	679	605
307	176	137	494	544	706	571	894	477	885	104	414	837	553	655	677
080	899	161	702	822	366	265	393	945	094	683	360	171	694	270	341
488	015	003	567	186	188	564	131	527	002	147	210	432	210	359	868
707	592	718	999	538	290	576	971	230	540	170	998	234	805	422	531
239	593	768	352	842	437	459	869	678	261	597	613	932	247	910	369
136	332	304	096	577	705	755	321	234	449	333	282	805	744	143	096
766	506	079	821	470	950	458	884	318	292	678	436	562	199	215	157
541	177	296	914	241	649	006	622	691	356	416	434	571	070	360	663
605	883	930	615	274	605	420	237	013	190	917	223	732	138	408	850
157	432	718	366	821	249	027	368	370	978	294	322	391	789	826	979
558	539	479	547	563	323	535	001	817	266	092	669	355	960	035	595
355	332	275	029	221	211	951	621	282	316	636	196	393	656	292	350
821	385	788	388	261	245	312	958	366	130	572	095	967	311	133	687
325	679	673	868	520	586	837	094	248	132	986	489	824	952	909	138
113	297	200	296	960	639	973	621	274	557	070	641	246	149	609	361
267	101	306	980	542	827	930	550	793	818	260	624	237	615	504	706
728	862	052	154	852	758	325	246	481	179	180	965	995	285	503	697
460	105	027	764	807	281	507	855	016	485	938	118	025	316	246	609
409	677	370	180	400	693	827	831	304	910	924	864	940	013	698	077
358	078	882	266	457	836	452	749	853	308	698	510	348	002	996	156
614	949	535	934	996	423	334	523	651	680	982	634	818	301	489	219
086	912	979	073	169	045	521	949	112	283	594	822	214	704	242	282
437	487	160	760	213	669	141	666	347	999	828	379	479	250	279	671
199	150	527	643	453	098	051	855	242	124	430	860	776	720	636	403
142	010	174	906	358	273	333	568	441	821	788	143	343	371	938	677
735	347	111	183	761	996	738	390	283	132	734	932	513	119	143	441
047	162	002	755	794	838	364	257	727	445	699	924	393	653	853	617
896	712	795	431	751	760	789	823	577	535	687	767	654	684	206	255
158	786	535	065	095	899	824	467	356	553	188	731	800	936	883	351
523	899	538	911	561	908	822	471	338	497	972	356	241	407	832	722
635	507	617	898	211	850	636	528	373	906	993	349	087	115	885	107
990	838	186	376	287	114	061	995	636	045	860	021	527	820	472	253
077	949	161	537	988	531	508	665	154	734	885	471	979	596	730	834
738	438	523	209	009	402	500	896	951	815	337	487	897	011	998	905

Table A.2.   Critical values of the *t* distribution*

		0.50	0.25	0.10	0.05	0.025	0.01	0.005
α (2–tailed):		0.25	0.125	0.05	0.025	0.0125	0.005	0.0025
α (1–tailed):								
DF	1	1.000	2.414	6.314	12.706	25.452	63.657	127.320
	2	0.816	1.604	2.920	4.303	6.205	9.925	14.089
	3	0.765	1.423	2.353	3.182	4.176	5.841	7.453
	4	0.741	1.344	2.132	2.776	3.495	4.604	5.598
	5	0.727	1.301	2.015	2.571	3.163	4.032	4.773
	6	0.718	1.273	1.943	2.447	2.969	3.707	4.317
	7	0.711	1.254	1.895	2.365	2.841	3.500	4.029
	8	0.706	1.240	1.860	2.306	2.752	3.355	3.832
	9	0.703	1.230	1.833	2.262	2.685	3.250	3.690
	10	0.700	1.221	1.812	2.228	2.634	3.169	3.581
	11	0.697	1.214	1.796	2.201	2.593	3.106	3.497
	12	0.695	1.209	1.782	2.179	2.560	3.054	3.428
	13	0.694	1.204	1.771	2.160	2.533	3.012	3.372
	14	0.692	1.200	1.761	2.145	2.510	2.977	3.326
	15	0.691	1.197	1.753	2.132	2.490	2.947	3.286
	16	0.690	1.194	1.746	2.120	2.473	2.921	3.252
	17	0.689	1.191	1.740	2.110	2.458	2.898	3.222
	18	0.688	1.189	1.734	2.101	2.445	2.878	3.197
	19	0.688	1.187	1.729	2.093	2.433	2.861	3.174
	20	0.687	1.185	1.725	2.086	2.423	2.845	3.153
	21	0.686	1.183	1.721	2.080	2.414	2.831	3.135
	22	0.686	1.182	1.717	2.074	2.406	2.819	3.119
	23	0.685	1.180	1.714	2.069	2.398	2.807	3.104
	24	0.685	1.179	1.711	2.064	2.391	2.797	3.090
	25	0.684	1.178	1.708	2.060	2.385	2.787	3.078
	26	0.684	1.177	1.706	2.056	2.379	2.779	3.067
	27	0.684	1.176	1.703	2.052	2.373	2.771	3.056
	28	0.683	1.175	1.701	2.048	2.368	2.763	3.047
	29	0.683	1.174	1.699	2.045	2.364	2.756	3.038
	30	0.683	1.173	1.697	2.042	2.360	2.750	3.030
	40	0.681	1.167	1.684	2.021	2.329	2.704	2.971
	60	0.679	1.162	1.671	2.000	2.299	2.660	2.915
	120	0.676	1.156	1.658	1.980	2.270	2.617	2.860
	∞	0.674	1.150	1.645	1.960	2.241	2.576	2.807

*Modified from Maxine Merrington, "Table of Percentage Points of the *t* Distribution," *Biometrika* 32(1942):300. Used with permission of the editors of *Biometrika*.

Table A.3. Critical values of the *F* distribution*

α = 0.05 (1-tailed)
α = 0.10 (2-tailed)

DF for denominator	DF for numerator													
	1	2	3	4	5	6	7	8	9	10	20	30	60	∞
1	161.45	199.50	215.71	224.58	230.16	233.99	236.77	283.88	240.54	241.88	248.01	250.09	252.20	254.32
2	18.513	19.000	19.164	19.247	19.296	19.330	19.353	19.371	19.385	19.396	19.446	19.462	19.479	19.496
3	10.128	9.5521	9.2766	9.1172	9.0135	8.9406	8.8868	8.8452	8.8123	8.7855	8.6602	8.6166	8.5720	8.5265
4	7.7086	6.9443	6.5914	6.3883	6.2560	6.1631	6.0942	6.0410	5.9988	5.9644	5.8025	5.7459	5.6878	5.6281
5	6.6079	5.7861	5.4095	5.1922	5.0503	4.9503	4.8759	4.8183	4.7725	4.7351	4.5581	4.4957	4.4314	4.3650
6	5.9874	5.1433	4.7571	4.5337	4.3874	4.2839	4.2066	4.1468	4.0990	4.0600	3.8742	3.8082	3.7398	3.6688
7	5.5914	4.7374	4.3468	4.1203	3.9715	3.8660	3.7870	3.7257	3.6767	3.6365	3.4445	3.3758	3.3043	3.2298
8	5.3177	4.4590	4.0662	3.8378	3.6875	3.5806	3.5005	3.4381	3.3881	3.3472	3.1503	3.0794	3.0053	2.9276
9	5.1174	4.2565	3.8626	3.6331	3.4817	3.3738	3.2927	3.2296	3.1789	3.1373	2.9365	2.8637	2.7872	2.7067
10	4.9646	4.1028	3.7083	3.4780	3.3258	3.2172	3.1355	3.0717	3.0204	2.9782	2.7740	2.6996	2.6211	2.5379
20	4.3513	3.4928	3.0984	2.8661	2.7109	2.5990	2.5140	2.4471	2.3928	2.3479	2.1242	2.0391	1.94642	1.8432
30	4.1709	3.3158	2.9223	2.6896	2.5336	2.4205	2.3343	2.2662	2.2107	2.1646	1.9317	1.8409	1.7396	1.6223
60	4.0012	3.1504	2.7581	2.5252	2.3683	2.2540	2.1665	2.0970	2.0401	1.9926	1.7480	1.6491	1.5343	1.3893
∞	3.8415	2.9957	2.6049	2.3719	2.2141	2.0986	2.0096	1.9384	1.8799	1.8307	1.5705	1.4591	1.3180	1.0000

α = 0.025 (1-tailed)
α = 0.05 (2-tailed)

DF for denominator	DF for numerator													
	1	2	3	4	5	6	7	8	9	10	20	30	60	∞
1	647.79	799.50	864.16	899.58	921.85	937.11	948.22	956.66	963.28	968.63	993.10	1001.4	1009.8	1018.3
2	38.506	39.000	39.165	39.248	39.298	39.331	39.355	39.373	39.387	39.398	39.448	39.465	39.481	39.498
3	17.443	16.044	15.439	15.101	14.885	14.735	14.624	14.540	14.473	14.419	14.167	14.081	13.992	13.902
4	12.218	10.649	9.9792	9.6045	9.3645	9.1973	9.0741	8.9796	8.9047	8.8439	8.5599	8.4613	8.3604	8.2573
5	10.007	8.4336	7.7636	7.3879	7.1464	6.9777	6.8531	6.7572	6.6810	6.6192	6.3285	6.2269	6.1225	6.0153
6	8.8131	7.2598	6.5988	6.2272	5.9876	5.8197	5.6955	5.5996	5.5234	5.4613	5.1684	5.0652	4.9589	4.8491
7	8.0727	6.5415	5.8898	5.5226	5.2852	5.1186	4.9949	4.8994	4.8232	4.7611	4.4667	4.3624	4.2544	4.1423
8	7.5709	6.0595	5.4160	5.0526	4.8173	4.6517	4.5286	4.4332	4.3572	4.2951	3.9995	3.8940	3.7844	3.6702
9	7.2093	5.7147	5.0781	4.7181	4.4844	4.3197	4.1971	4.1020	4.0260	3.9639	3.6669	3.5604	3.4493	3.3329
10	6.9367	5.4564	4.8256	4.4683	4.2361	4.0721	3.9498	3.8549	3.7790	3.7168	3.4186	3.3110	3.1984	3.0798
20	5.8715	4.4613	3.8587	3.5147	3.2891	3.1283	3.0074	2.9128	2.8365	2.7737	2.4645	2.3486	2.2234	2.0853
30	5.5675	4.1821	3.5894	3.2499	3.0265	2.8667	2.7460	2.6513	2.5746	2.5112	2.1952	2.0739	1.9400	1.7867
60	5.2857	3.9253	3.3425	3.0077	2.7863	2.6274	2.5068	2.4117	2.3344	2.2702	1.9445	1.8152	1.6668	1.4822
∞	5.0239	3.6889	3.1161	2.7858	2.5665	2.4082	2.2875	2.1918	2.1136	2.0483	1.7085	1.5660	1.3883	1.0000

α = 0.01 (1-tailed) α = 0.02 (2-tailed)

DF for denominator	DF for numerator													
	1	2	3	4	5	6	7	8	9	10	20	30	60	∞
1	4052.2	4999.5	5403.3	5624.6	5763.7	5859.0	5928.3	5981.6	6022.5	6055.8	6208.7	6260.7	6313.0	6366.0
2	98.503	99.000	99.166	99.249	99.299	99.332	99.356	99.374	99.388	99.399	99.449	99.466	99.483	99.501
3	34.116	30.817	29.457	28.710	28.237	27.911	27.672	27.489	27.345	27.229	26.690	26.505	26.316	26.125
4	21.198	18.000	16.694	15.977	15.552	15.207	14.976	14.799	14.659	14.546	14.020	13.838	13.652	13.463
5	16.258	13.274	12.060	11.392	10.967	10.672	10.456	10.289	10.158	10.051	9.5527	9.3793	9.2020	9.0204
6	13.745	10.925	9.7795	9.1483	8.7459	8.4661	8.2600	8.1016	7.9761	7.8741	7.3958	7.2285	7.0568	6.8801
7	12.246	9.5466	8.4513	7.8467	7.4604	7.1914	6.9928	6.8401	6.7188	6.6201	6.1554	5.9921	5.8236	5.6495
8	11.259	8.6491	7.5910	7.0060	6.6318	6.3707	6.1776	6.0289	5.9106	5.8143	5.3591	5.1981	5.0316	4.8588
9	10.561	8.0215	6.9919	6.4221	6.0569	5.8018	5.6129	5.4671	5.3511	5.2565	4.8080	4.6486	4.4831	4.3105
10	10.044	7.5594	6.5523	5.9943	5.6363	5.3858	5.2001	5.0567	4.9424	4.8492	4.4054	4.2469	4.0819	3.9090
20	8.0960	5.8489	4.9382	4.4307	4.1027	3.8714	3.6987	3.5644	3.4567	3.3682	2.9377	2.7785	2.6077	2.4212
30	7.5625	5.3904	4.5097	4.0179	3.6990	3.4735	3.3045	3.1726	3.0665	2.9791	2.5487	2.3860	2.2079	2.0062
60	7.0771	4.9774	4.1259	3.6491	3.3389	3.1187	2.9530	2.8233	2.7185	2.6318	2.1978	2.0285	1.8363	1.6006
∞	6.6349	4.6052	3.7816	3.3192	3.0173	2.8020	2.6393	2.5113	2.4073	2.3209	1.8783	1.6964	1.4730	1.0000

α = 0.005 (1-tailed) α = 0.01 (2-tailed)

DF for denominator	DF for numerator													
	1	2	3	4	5	6	7	8	9	10	20	30	60	∞
1	16211	20000	21615	22500	23056	23437	23715	23925	24091	24224	24836	25044	25253	25465
2	198.50	199.00	199.17	199.25	199.30	199.33	199.36	199.37	199.39	199.40	199.45	199.47	199.48	199.51
3	55.552	49.799	47.467	46.195	45.392	44.838	44.434	44.126	43.882	43.686	42.778	42.466	42.149	41.829
4	31.333	26.284	24.259	23.155	22.456	21.975	21.622	21.352	21.139	20.967	20.167	19.892	19.611	19.325
5	22.785	18.314	16.530	15.556	14.940	14.513	14.200	13.961	13.772	13.618	12.903	12.656	12.402	12.144
6	18.635	14.544	12.917	12.028	11.464	11.073	10.786	10.566	10.391	10.250	9.5888	9.3583	9.1219	8.8793
7	16.236	12.404	10.882	10.050	9.5221	9.1554	8.8854	8.6781	8.5138	8.3803	7.7540	7.5345	7.3088	7.0760
8	14.688	11.042	9.5965	8.8051	8.3018	7.9520	7.6942	7.4960	7.3386	7.2107	6.6082	6.3961	6.1772	5.9505
9	13.614	10.107	8.7171	7.9559	7.4711	7.1338	6.8849	6.6933	6.5411	6.4171	5.8318	5.6248	5.4104	5.1875
10	12.826	9.4270	8.0807	7.3428	6.8723	6.5446	6.3025	6.1159	5.9676	5.8467	5.2740	5.0705	4.8592	4.6385
20	9.9439	6.9865	5.8177	5.1743	4.7616	4.4721	4.2569	4.0900	3.9564	3.8470	3.3178	3.1234	2.9159	2.6904
30	9.1797	6.3547	5.2388	4.6233	4.2276	3.9492	3.7416	3.5801	3.4505	3.3440	2.8230	2.6278	2.4151	2.1760
60	8.4946	5.7950	4.7290	4.1399	3.7600	3.4918	3.2911	3.1344	3.0083	2.9042	2.3872	2.1874	1.9622	1.6885
∞	7.8794	5.2983	4.2794	3.7151	3.3499	3.0913	2.8968	2.7444	2.6210	2.5188	1.9998	1.7894	1.5325	1.0000

*Condensed from Maxine Merrington and Catharine M. Thompson, "Tables of Percentage Points of the Inverted Beta (*F*) Distribution," *Biometrika* 33(1943):73–88. Used with permission of the editors of *Biometrika*.

Table A.4. Critical Mann-Whitney *U* values for the 0.05 and 0.01 levels of significance in two-sample rank-sum tests (two-tailed) involving samples with 20 or fewer items (To show a significant difference, observed values must be less than or equal to the tabled values.)*

α = 0.05

N' (Larger Sample)

N	1	2	3	4	5	6	7	8	9	10	11	12	13	14	15	16	17	18	19	20
1	—	—	—	—	—	—	—	—	—	—	—	—	—	—	—	—	—	—	—	—
2		—	—	—	—	—	—	0	0	0	0	1	1	1	1	1	2	2	2	2
3			—	—	0	1	1	2	2	3	3	4	4	5	5	6	6	7	7	8
4				0	1	2	3	4	4	5	6	7	8	9	10	11	11	12	13	13
5					2	3	5	6	7	8	9	11	12	13	14	15	17	18	19	20
6						5	6	8	10	11	13	14	16	17	19	21	22	24	25	27
7							8	10	12	14	16	18	20	22	24	26	28	30	32	34
8								13	15	17	19	22	24	26	29	31	34	36	38	41
9									17	20	23	26	28	31	34	37	39	42	45	48
10										23	26	29	33	36	39	42	45	48	52	55
11											30	33	37	40	44	47	51	55	58	62
12												37	41	45	49	53	57	61	65	69
13													45	50	54	59	63	67	72	76
14														55	59	64	67	74	78	83
15															64	70	75	80	85	90
16																75	81	86	92	98
17																	87	93	99	105
18																		99	106	112
19																			113	119
20																				127

α = 0.01

N' (Larger Sample)

N	1	2	3	4	5	6	7	8	9	10	11	12	13	14	15	16	17	18	19	20
1	—	—	—	—	—	—	—	—	—	—	—	—	—	—	—	—	—	—	—	—
2		—	—	—	—	—	—	—	—	—	—	—	—	—	—	—	—	—	0	0
3			—	—	—	—	—	—	0	0	0	1	1	1	2	2	2	2	3	3
4				—	—	0	0	1	1	2	2	3	3	4	5	5	6	6	7	8
5					0	1	1	2	3	4	5	6	7	7	8	9	10	11	12	13
6						2	3	4	5	6	7	9	10	11	12	13	15	16	17	18
7							4	6	7	9	10	12	13	15	16	18	19	21	22	24
8								7	9	11	13	15	17	18	20	22	24	26	28	30
9									11	13	16	18	20	22	24	27	29	31	33	36
10										16	18	21	24	26	29	31	34	37	39	42
11											21	24	27	30	33	36	39	42	45	48
12												27	31	34	37	41	44	47	51	54
13													34	38	42	45	49	53	56	60
14														42	46	50	54	58	63	67
15															51	55	60	64	69	73
16																60	65	70	74	79
17																	70	75	81	86
18																		81	87	92
19																			93	99
20																				105

*Modified from D. Auble, "Extended Tables for the Mann-Whitney Statistics," *Bulletin of the Institute of Educational Research, Indiana University,* 1(2)(1953):1–39. Used with permission of the author and publisher.

Table A.5. Critical Wilcoxon *T* values for one- and two-tailed tests with α equal to 0.05, 0.01, 0.001, and 0.0025

α =	One-tailed Test 0.05	0.01	0.0025	Two-tailed Test 0.05	0.01	0.001
n = 6	2					
7	3			2		
8	5	1		3		
9	8	3		5	1	
10	10	5	1	8	3	
11	13	7	3	10	5	
12	17	9	5	13	7	1
13	21	12	7	17	9	2
14	25	15	9	21	12	4
15	30	19	12	25	15	6
16	35	23	15	29	19	8
17	41	27	19	34	23	11
18	47	32	23	40	27	14
19	53	37	27	46	32	18
20	60	43	32	52	37	21
21	67	49	37	58	42	25
22	75	55	42	65	48	30
23	83	62	48	73	54	35
24	91	69	54	81	61	40
25	100	76	60	89	68	45
26	110	84	67	98	75	51
27	119	92	74	107	83	57
28	130	101	82	116	91	64
29	140	110	90	126	100	71
30	151	120	98	137	109	78
35	213	173	146	195	159	120
40	286	238	204	264	220	172
45	371	312	272	343	291	233
50	466	397	350	434	373	304
60	690	600	537	648	567	476
70	960	846	767	907	805	689
80	1276	1136	1039	1211	1086	943
90	1638	1471	1355	1560	1410	1240
100	2045	1850	1714	1955	1779	1578

Source: Selected values from McCornack, R. L. 1965. "Extended tables of the Wilcoxon matched pair signed rank statistic." Journal of the American Statistical Association 60:864–871.

Table A.6. Critical percentage points of the chi-square distribution*

DF	1%	5%	25%	50%	75%	90%	95%	97.5%	99%	99.5%
1	0.0001	0.004	0.102	0.455	1.323	2.706	3.841	5.024	6.635	7.879
2	0.020	0.103	0.575	1.386	2.772	4.605	5.991	7.378	9.210	10.597
3	0.115	0.352	1.212	2.366	4.108	6.251	7.815	9.348	11.345	12.838
4	0.297	0.711	1.922	3.357	5.385	7.779	9.488	11.143	13.277	14.860
5	0.554	1.146	2.675	4.351	6.625	9.236	11.070	12.832	15.086	16.750
6	0.872	1.635	3.455	5.348	7.841	10.645	12.592	14.449	16.812	18.548
7	1.239	2.167	4.255	6.346	9.037	12.017	14.067	16.013	18.475	20.278
8	1.646	2.733	5.071	7.344	10.219	13.362	15.507	17.535	20.090	21.955
9	2.088	3.325	5.899	8.343	11.389	14.684	16.919	19.023	21.666	23.589
10	2.558	3.940	6.737	9.342	12.549	15.987	18.307	20.483	23.209	25.188
11	3.054	4.575	7.584	10.341	13.701	17.275	19.675	21.920	24.725	26.757
12	3.571	5.226	8.438	11.340	14.845	18.549	21.026	23.337	26.217	28.300
13	4.107	5.892	9.299	12.340	15.984	19.812	22.362	24.735	27.688	29.819
14	4.660	6.571	10.165	13.339	17.117	21.064	23.685	26.119	29.141	31.319
15	5.229	7.261	11.036	14.339	18.245	22.307	24.996	27.488	30.578	32.810
16	5.812	7.962	11.912	15.338	19.369	23.542	26.296	28.845	32.000	34.267
17	6.408	8.672	12.792	16.338	20.449	24.769	27.587	30.191	33.409	35.718
18	7.015	9.390	13.675	17.338	21.605	25.989	28.869	31.526	34.805	37.156
19	7.633	10.117	14.562	18.338	22.718	27.204	30.144	32.852	36.191	38.582
20	8.260	10.851	15.452	19.337	23.828	28.412	31.410	34.170	37.566	39.997
21	8.897	11.591	16.344	20.337	24.935	29.615	32.670	35.479	38.932	41.401
22	9.542	12.338	17.240	21.337	26.039	30.813	33.924	36.781	40.289	42.796
23	10.196	13.090	18.137	22.337	27.141	32.007	35.172	38.076	41.638	44.181
24	10.856	13.848	19.037	23.337	28.241	33.196	36.415	39.364	42.980	45.558
25	11.524	14.611	19.939	24.337	29.339	34.382	37.652	40.646	44.314	46.928
30	14.954	18.493	24.478	29.336	34.800	40.256	43.733	46.979	50.892	53.672
40	22.164	26.509	33.660	39.335	45.616	51.805	55.758	59.342	63.691	66.766
50	29.707	34.764	42.942	49.335	56.334	63.167	67.505	71.420	76.154	79.490
60	37.485	43.188	52.294	59.335	66.981	74.397	79.082	83.298	88.379	91.952
70	45.442	51.739	61.698	69.334	77.577	85.527	90.531	95.023	100.425	104.215
80	53.540	60.392	71.144	79.334	88.130	96.578	101.879	106.629	112.329	116.321
90	61.754	69.126	80.625	89.334	98.650	107.565	113.145	118.136	124.116	128.299
100	70.065	77.930	90.133	99.334	109.141	118.498	124.342	129.561	135.807	140.169
200	156.479	168.337	186.172	199.778	213.102	226.021	233.994	241.058	249.445	255.264
500	429.382	449.146	478.323	499.778	520.950	540.930	553.127	563.852	576.493	585.207
1000	898.912	929.595	969.484	999.778	1029.790	1057.724	1074.679	1089.531	1106.969	1118.948

*Modified from Catharine M. Thompson, "Table of Percentage Points of the x^2 Distribution," *Biometrika* 32(1941):187–191. Used with permission of the editors of *Biometrika*.

Table A.7.   Critical values of the Spearman rank correlation coefficient

α (2-tailed)		0.05		0.01
α (1-tailed)	0.05		0.01	
n = 5	0.900	1.000	1.000	———
6	0.829	0.886	0.943	1.000
7	0.714	0.786	0.893	0.929
8	0.643	0.738	0.833	0.881
9	0.600	0.700	0.783	0.833
10	0.564	0.648	0.745	0.794
11	0.536	0.618	0.709	0.755
12	0.503	0.587	0.678	0.727
13	0.484	0.560	0.648	0.703
14	0.464	0.538	0.626	0.679
15	0.446	0.521	0.604	0.654
20	0.380	0.447	0.520	0.570
25	0.337	0.398	0.466	0.511
30	0.306	0.362	0.425	0.467
40	0.264	0.313	0.368	0.405
50	0.235	0.279	0.329	0.363

Table A.8. Critical 0.05 and 0.01 values of the Kruskal-Wallis H statistic for 3, 4, and 5 ANOVA classes (k's). To locate values, rank the number of replicates (n's) in the classes in descending order. Locate the numbers corresponding to the first ($k - 1$) n's along the left side of the appropriate table, and the last n over the columns to the right. For designs with more k's or n's, critical values are approximated by F values with $k - 1$ and $N - k$ degrees of freedom ($N = \Sigma n$), with F calculated as $9H/2(11 - H)$

THREE CLASSES

Last n

First 2 n's	1 — 0.05	1 — 0.01	2 — 0.05	2 — 0.01	3 — 0.05	3 — 0.01	4 — 0.05	4 — 0.01	5 — 0.05	5 — 0.01
32	....	....	4.714	....						
42	....	....	5.333	....						
52	5.000	....	5.160	6.533						
33	5.143	....	5.361	6.250	3.600	7.200				
43	5.208	....	5.444	6.444	5.791	6.745				
53	4.960	....	5.251	6.909	5.648	7.079				
44	4.967	6.667	5.455	7.036	5.598	7.144	5.692	7.654		
54	4.985	6.955	5.273	7.025	5.656	7.445	5.657	7.760		
55	5.127	7.309	5.338	7.338	5.666	7.823	5.666	7.823	5.780	8.000

FOUR CLASSES

Last n

First 3 n's	1 — 0.05	1 — 0.01	2 — 0.05	2 — 0.01	3 — 0.05	3 — 0.01	4 — 0.05	4 — 0.01
331	6.333	....						
421	5.833	....						
431	6.178	7.067						
441	5.945	7.909						
222	5.679	....	6.167	6.667				
322	5.833	....	6.333	7.133				
332	6.244	7.200	6.527	7.636				
422	6.133	7.000	6.545	7.391				
432	6.309	7.455	6.621	7.871				
442	6.386	7.886	6.731	8.346				
333	6.600	7.400	6.727	8.105	7.000	8.538		
433	6.545	7.758	6.795	8.333	6.984	8.659		
443	6.635	8.231	6.874	8.621	7.038	8.876		
444	6.725	8.588	6.957	8.871	7.142	9.075	7.235	9.287

FIVE CLASSES

Last n

First 4 n's	1 — 0.05	1 — 0.01	2 — 0.05	2 — 0.01	3 — 0.05	3 — 0.01
2221	6.750	....				
3211	6.583	....				
3221	6.800	7.600				
3311	7.111	....				
3321	7.200	8.073				
3331	7.576	8.424				
2222	7.133	7.533	7.418	8.291		
3222	7.309	8.127	7.682	8.682		
3322	7.591	8.576	7.910	9.115		
3332	7.769	9.051	8.044	9.505		
3333	8.000	9.451	8.200	9.876	8.333	10.200

Source: Selected values from R. L. Iman, D. Quade, and D. A. Alexander. 1975. "Exact probability levels for the Kruskal-Wallis test." Pages 329-384 *in* H. L. Harter and D. B. Owen, editors. "Selected tables in mathematical statistics, Vol. III." American Mathematical Society, Providence, Rhode Island, USA.

APPENDIX B

Sources of Software and Hardware

Vendors of the following software and hardware should be contacted for up-to-date information on products. Prices are indicated for general reference purposes only, and are those quoted in late 1994.

GENERAL SOFTWARE PACKAGES

E-Z Stat, EcoStat, and EcoSim

These three software packages, designed specifically for use in ecology courses, are distributed by

Trinity Software
Order Dept. M–49
P.O. Box 960
Compton, NH 03223
Telephone: (800) 352–1282 FAX: 603–726–3781

All three packages are designed for IBM PC, PS/2, or compatible computers with 512K, CGA, and a compatible graphics monitor. Output can be printed on dot matrix or HP Laserjet printers. Versions of these packages for Macintosh computers are under development. Single versions of each package (with reference manual), on either 3.5″ or 5.25″ diskettes, are priced at $95.00. Labpacks that allow duplication of four copies of each package are available for $220.00, and multi-user licenses for larger numbers of copies or workstations are also available according to a cost schedule.

E-Z Stat, Version 2.1, is a menu-driven set of basic statistical analyses that are conducted interactively with keyboard or mouse commands. The program package features pull-down menus and data, graph, and output windows. The following tests are performed:

- Parametric descriptive statistics
- Parametric paired and unpaired t tests
- Nonparametric Mann-Whitney two-sample test
- Nonparametric paired Wilcoxon signed-rank test
- Parametric correlation analysis
- Nonparametric Spearman rank correlation test
- Simple parametric regression analysis
- Parametric one-way analysis of variance
- Nonparametric Kruskal-Wallis analysis of variance
- Nonparametric goodness-of-fit tests
- Nonparametric contingency tests
- Kolmogorov-Smirnov test
- Random number generator

EcoStat, Version 1.0, is similar in format to E-Z Stat. The following analyses are included:

- Petersen, Schnabel, and Jolly-Seber mark/recapture analyses
- Density/cover analysis
- Importance value computation
- Population dispersion analyses
- Species diversity indices
- Community similarity indices
- Life-table analysis
- Descriptive statistics

EcoSim, Version 1.0, is also similar in format and includes the following analyses and simulations:

- Exponential and geometric population growth
- Logistic and discrete logistic population growth
- Leslie matrix analysis
- Lotka-Volterra competition simulation
- Volterra predator-prey simulation
- MacArthur predation simulation

Ecological Analysis–PC

The Ecological Analysis–PC software series for IBM microcomputers and compatibles is produced by

Oakleaf Systems
P.O. Box 472
Decorah, Iowa 52101
Telephone and FAX (319) 382–4320

This series now contains four volumes, each with several analysis modules:
Vol. 1

- Life-table analysis
- Interspecific association indices
- Community similarity indices
- Descriptive statistics
- Mark-release recapture analysis
- Regression and correlation analysis

Vol. 2

- Community similarity analysis
- Intrapopulation dispersion analysis
- Species-area curve computation
- Stepwise multiple regression

Vol. 3

- Community similarity indices
- Bray-Curtis ordination
- Cluster analysis
- Diversity indices

Vol. 4

- Descriptive statistics
- Population dispersion analysis

Single copies of each of these programs are priced at $69.95, but site license arrangements can be made for classroom use at reduced cost. A number of other ecology simulation packages are also marketed by Oakleaf Systems.

Ecological Measures (Version 1.10)

This set of BASIC programs is designed for IBM microcomputers and compatibles. These programs are available for a nominal handling charge of $5.00 from

Dr. Paul M. Kotila
Environmental Science Department
Franklin Pierce College
Rindge, NH 03461
Telephone: (603) 899-4255 FAX: (603) 899-6448

This package contains a module for calculation of species diversity indices, including

Margalef diversity
Menhinick diversity
Simpson dominance
Simpson diversity and evenness
Inverse Simpson dominance and evenness
Simpson values for nonrandom samples
Shannon diversity and evenness
Brillouin diversity and evenness
Sheldon evenness
Heip evenness

A second module enables computation of community similarity indices, including the following:

Jaccard coefficient
Sorenson coefficient
Percent similarity
Morisita similarity index
Horn similarity index

SOFTWARE AND HARDWARE SPECIFIC TO INDIVIDUAL EXERCISES

Exercise 2. Life Science Network

Detailed information on the Life Science Network can be obtained from

BIOSIS
2100 Arch Street
Philadelphia, PA 19103-1399
Telephone: (800) 523-4806 in USA and Canada
(215) 587-4847 worldwide

The costs (1994) to use this database network are about $2.00 to scan the databases to find which contain items of a desired content, $6.00 for a search yielding up to 10 citations ($6.00 extra for each additional 10 citations), and $2.25 per abstract. Complete versions of some articles can be obtained on-line (about $3.50 each), and others as photocopies by mail (about $16.00 each by regular mail or $39.00 each by express mail).

Exercise 2. Absearch

Detailed information on this abstract service can be obtained from

ABSEARCH
2457 Twin Road
Moscow, ID 83843
Telephone: (800) 867-1877

The cost of each basic database ranges from $75 (student) to $89 (nonstudent) for members of the societies, or $99 for nonmembers of the societies for which databases are now available.

Exercise 3. EXPERTiMENTAL DESIGN

EXPERTiMENTAL DESIGN (Version 3.2) is designed for IBM PC/XT/AT microcomputers and compatibles running DOS 2.10 or higher. It is available from

Statistical Programs
9941 Rowlett, Suite 6
Houston, TX 77075
Telephone: (713) 947-1551

The cost (1993) of the basic program is $295.00. Additional statistical packages containing programs for analyzing data for specific designs are also available.

Exercise 10. FORTRAN Program JOLLY

This program for IBM microcomputers and compatibles may be obtained by sending a formatted IBM diskette to

Dr. James E. Hines
Patuxent Wildlife Research Center
Laurel, MD 20708
Telephone: (301) 498-0389 FAX: (301) 498-0438
BITNET: EJH@NIHCU

Using a modem, the program can be downloaded without charge from the SouthEastern Software and Message Exchange (SESAME) bulletin board (telephone: 919-737-3990; 1200 baud, 8 data bits, no parity).

Exercise 11. MicroFish (Version 3.0)

This program and the user's guide may be obtained from the American Fisheries Society Computer User Section for $10.00 by sending a request and payment to

AFS Computer User Section
Steven Atran, Software Librarian
Gulf of Mexico Fisheries Management Council
Lincoln Center, Suite 331
5401 West Kennedy Boulevard
Tampa, FL 33609-2486
Telephone: (813) 228-2815 FAX: (813) 225-7015

MicroFish will run on IBM microcomputers and compatibles using DOS 2.0 or higher.

Exercise 12. DISTANCE

Course instructors can obtain program DISTANCE, with the user's guide, without charge, from

Dr. Jeffrey L. Laake
Alaska Fisheries Science Center
National Marine Mammal Laboratory
7600 Sand Point Way N.E., Bin C15700
Seattle, WA 98115-0070

The program runs on IBM PCs and compatibles, and requires about 1MB of disk space (plus space for input and output files) and at least 506K free conventional memory.

This program may be duplicated for use by students, but may not be sold for any purpose. The program is subject to periodic revision.

Exercise 13. CALHOME

CALHOME (Version 1.0, July 1994) is developed for IBM PC and compatible microcomputers with a hard disk and 640K of RAM. A math coprocessor considerably speeds the analysis. The screen graphics are displayed best on a color VGA monitor, but are also displayed on VGA and Hercules monochrome monitors. CALHOME is available from

John G. Kie
Forestry Sciences Lab
2081 East Sierra Avenue
Fresno, CA 93710
Telephone: (209) 487-5589

Exercise 13. McPAAL

McPAAL (Version 1.22, March 1992) is designed for PC-DOS 2.0 or higher and requires at least 192K of computer memory. A CGA or VGA adaptor is required for CGA graphics, but a Hercules video board can be used with a special compiled version of the program. The software and manual are available at a distribution cost of $30.00 (payable to the "National Zoo") from

Michael Stuwe
Conservation and Research Center
Front Royal, VA 22630

Exercise 15. Cold Plate Equipment

Cold plate equipment is available from

Thermoelectrics Unlimited, Inc.
1202 Harrison Avenue
Holly Oak Terrace
Wilmington, DE 19809–1910
Telephone: (302) 798-5360 FAX: (302) 798-5369

Model CP-2, which has a cooling capability to $-20°$ C, has a base price of $245.00.

Exercise 16. Pressure Chambers

The PMS Instrument Company produces a series of pressure chambers, with or without pressure supply tanks and having different capabilities. These instruments range in price from about $1,450.00 to $3,200.00.

PMS Instrument Company
480 SW Airport Avenue
Corvallis, OR 97333
Telephone: (503) 752-7926

A plant water status console, consisting of a pressure chamber, tank, and pressure gauge, priced at about $2,098.00, is produced by

Soilmoisture Equipment Corporation
P.O. Box 30025
Santa Barbara, CA 93105
Telephone: (805) 964-3525 FAX: (805) 683-2189

Exercise 17. Photosynthesis System

The LI-6200 Portable Photosynthesis System is available from

LI-COR, Inc.
4421 Superior Street
P.O. Box 4425
Lincoln, NE 68504
Telephone: (402) 467-3576 FAX: (402) 467-2819

The total cost of this system is in the general range of $15,000.00.

Exercise 19. Growth Curves Software

A program for computation of parameters of the logistic and Brody-Bertalanffy growth curves, written in Microsoft BASIC, is available on diskette from

Dr. Thomas A. Ebert
Department of Biology
San Diego State University
San Diego, CA 92182-0057

Exercise 20. Rogers Index of Preference

An IBM PC program for computing the Rogers index of preference can be obtained from

Dr. Charles J. Krebs
Department of Zoology
University of British Columbia
Vancouver, B.C. V6T 2A9 CANADA

Exercise 29. RAMAS/space

RAMAS/space is distributed by

Exeter Software
100 North Country Road
Setauket, NY 11733
Telephone: (800) 842-5892 FAX: (516) 751-3435

This program is designed for IBM PCs or compatibles with at least a DOS 3.0 operating system and 320K RAM. The basic program cost is $265.00, with a site license also being offered at a cost of $530.00.

Exercise 33. Cornell Ecology Programs, CANOCO, and CANODRAW

These three programs are distributed by

Microcomputer Power
111 Clover Lane
Ithaca, NY 14850
Telephone: (607) 272-2188 FAX: (607) 272-0782

The Cornell Ecology Programs are available for IBM PCs and compatibles only. The basic software package is priced at $295.00. The CANOCO package, including CANOPLOT, is available in both IBM and Macintosh formats, and carries a basic price for educational use of $249.00. Additional copies of these programs for use at a particular site are available at discount, and site licenses are also available.

Exercise 37. Dissolved Oxygen Meter

YSI dissolved-oxygen meters and the YSI 5905 BOD Probe are manufactured by

YSI Inc.
1725 Brannum Lane
Yellow Springs, OH 45387
Telephone: (800) 765-4974 FAX: (513) 767-3953

YSI products are marketed by most major scientific supply houses. Information on specifications, availability, and cost of dissolved-oxygen meters and probes can be obtained by contacting YSI or checking current supply-house catalogs.

Exercise 38. STELLA II

The STELLA II 3.0 modeling program is now available for Macintosh microcomputers and Windows 3.1 for IBM-compatible microcomputers. The program is produced in a regular "modeling" form and in an "authoring" form in which locked models can be produced for instructional use. These programs are available from

High Performance Systems, Inc.
45 Lyme Road
Hanover, NH 03755
Telephone: (603) 643-9636 FAX: (603) 643-9502

Stella II Version 3.0 for university use is priced at $295.00, and the authoring version at $399.00. Reduced prices are also available for laboratory packages and student purchases.

Exercise 39. macGIS

The macGIS program is available from

macGIS
Attn: David Hulse
5234 Department of Landscape Architecture
School of Architecture and Allied Arts
University of Oregon
Eugene, OR 97403-5234
FAX (503) 346-3626
E-mail: dhulse@aaa.uoregon.edu

The academic price for version 2.0 (including user's guide) is $30.00 plus $2.50 postage. In ordering, quote ISBN #006500969X and make checks or purchase orders payable to "macGIS—Dept. of Landscape Architecture." A version 3.0 of macGIS is under development.

Glossary

A

Acclimation

Compensatory change in physiological mechanisms of response to the environment by an organism due to prolonged exposure to particular conditions.

Adaptation

A characteristic which promotes the ability of an organism to produce viable offspring in a given environment. *Or* the state of possession of such characteristics, or the process by which such characteristics are improved.

Aggregation

A pattern of spatial distribution in which individuals tend to occur in groups more often than expected by chance.

Allopatry

The occupation of nonoverlapping geographical ranges by two or more species.

Alpha diversity

A quantitative measure of species richness within a given habitat.

Alternate hypothesis

The hypothesis accepted when a statistical null hypothesis is rejected.

Apomixis

The production of seeds or other nonvegetative reproductive bodies without fertilization.

Aspection

Seasonal change in the characteristics of biotic communities.

B

Basal area

The total cross-sectional area of plant stems or trunks per unit area of habitat.

Before-after control-impact pairs method

An experimental design in which response variables are measured at experimental and control sites on a series of occasions before and after a manipulation on the experimental site.

Beta diversity

A quantitative measure of the relative change in species composition from one habitat to another.

Biomass

The total mass of organic matter per unit area or volume of habitat, or an index thereof.

C

Carnivore

An organism that feeds on animals or their tissues.

CD-ROM

A compact disk containing stored information that can be accessed, but not modified, by a microcomputer terminal or workstation.

Climax community

A biotic community that is self-replacing under constant conditions of the external environment; the final stage of biotic succession.

Coefficient of detectability

The decimal fraction of individuals that are noted by an observer at a particular distance from the point of observation.

Cohort

A group of individuals of the same age that are followed through time to obtain data on demographic processes.

Commensalism

An interaction between species in which one member benefits and the other is unaffected.

Community

A group of species that characteristically occur together in a certain habitat.

Competition

Active demand by organisms for environmental resources in excess of the quantity available.

Competition coefficient

The competitive influence of one species on another, measured as the inhibiting effect of an individual of the first species on the second, relative to that of an individual of the second species on itself.

Competitive exclusion principle

The idea that two or more resource-limited species cannot occupy the same realized niche in a stable environment.

Contingency test

A statistical test of whether or not the classification of a set of items by one variable is dependent on the classes of a second variable into which they fall.

Control

An experimental unit identical to those receiving manipulative treatment except in the essential treatment variable.

Correlation

Analysis of the degree of correspondence in trends of two sets of values, both of which are assumed to be subject to random variability.

Counterintuitive response

Behavior of a system that is contrary to that expected from intuition based on everyday experience.

Critical thermal minimum

The low temperature at which normal body movement of an ectotherm becomes impossible.

D

Decomposers

A biotic component of the ecosystem comprising bacteria, fungi, and other organisms which participate in the breakdown of dead organic matter.

Degrees of freedom

The number of values, in one or more sets of data with fixed means or other parameters, that can be varied independently without determining the values of those remaining.

Demography

The study of the vital statistics of populations; originally in reference to human populations, but now commonly applied to the study of plant and animal populations.

Density

The number of organisms per unit area or volume of habitat.

Density-dependent factor

A relationship that affects mortality or natality processes in a population with differing intensity at different densities; the relationship is direct if intensity increases as density increases, inverse if intensity declines as density increases.

Density-independent factor

A relationship that affects mortality or natality processes in a population in a fashion unrelated to population density.

Descriptive hypothesis

A tentative statement of the existence of a difference in structure or appearance in two or more ecological systems, or of a deviation of these characteristics from some theoretical expectation.

Dispersal

The dissemination of reproductive propagules in space and time.

Dispersion

The pattern of distribution of organisms or groups of organisms in habitat space.

Diversity

The richness of a habitat or region in species, based on the absolute number of species present and the degree of equitability in their abundances.

Dominance

In community description, the quantitative measure of total biomass or areal coverage of a species per unit of habitat space. In community dynamics, the tendency of a species to modify conditions of the external environment and thereby exert a controlling influence on biotic composition of the community. In animal behavior, a behavioral status that confers priority of access of certain individuals to food, mates, and other needs.

E

Ecological isolation

Differentiation of two or more species or components of the population of a single species in ways that minimize their direct competition for resources.

Ecological niche

The overall pattern of resource use by a species due to its adaptations for exploiting different types of resources and occupying different habitats.

Ecophene

A distinctive body form resulting from interaction of developmental processes and environmental influences during growth of the individual.

Ecosystem

An environmental unit consisting of living and nonliving components that interact with an interchange of nutrients and energy.

Ecotype

A local population of a species with a distinctive pattern of genetic adjustment to its habitat.

Emigration

Movement of individuals out of a population.

Equitability

In species diversity theory, the evenness of species in their abundances.

Experimental unit

The object or system on which an experimental or control condition is imposed in an experiment.

Exponential curve

In population dynamics, a pattern of population growth in which the rate of growth increases as the population density increases.

F

Facilitation

A mechanism of biotic succession in which one set of species change habitat conditions so as to favor the invasion and establishment of another group of species.

Field capacity

The mass of water that can be held by a volume of soil against gravitational drainage.

Flux rate

In ecosystem dynamics, the flow of materials or energy from one component to another per unit time.

Food chain

A linear scheme of feeding relationships among a set of species.

Food web

A depiction of the interlocking network of feeding relationships in a biotic community.

Frequency

In community description, the fraction of sampling units in which a particular species occurs.

Functional hypothesis

A tentative explanation of the cause or ecological significance of some ecological difference or relationship.

Fundamental niche

The overall pattern of resource use of a species in the absence of competitors.

G

Gamma diversity

The total richness of a geographical region in species due to the number of different habitats present, the richness of species within each, and the degree of species turnover from habitat to habitat.

Genotype

The genetic constitution of an individual or gamete.

Geographic Information System (GIS)

Computer hardware and software for input, storage, manipulation, and output of different types of spatial data referenced by a common set of geographical coordinates.

Goodness-of-fit test

A statistical comparison of the degree of correspondence of a set of observed data to some theoretical expectation.

Gradient analysis

A class of procedures for placing communities in graphical space so that their positions reflect differences in composition or compositional relationships with habitat conditions. *Indirect* gradient analysis positions communities strictly on the basis of their composition; *direct* gradient analysis explicitly relates composition to measured habitat conditions.

Gross primary production

Total production of photosynthate or chemosynthate by producer organisms.

H

Habitat

In reference to a species, the physical and biotic characteristics of locations occupied by individuals of the species. In reference to a community, the physical conditions of the community site.

Herbivores

Animals that feed exclusively on plants.

Heterogeneity

Nonuniformity of conditions in space or time.

Home range

The total area or volume of habitat used by an individual, a pair, or a group of individuals.

Hydrometer

A device for measuring the specific gravity of liquids.

Hypothesis

A tentative statement about the existence, cause, or significance of an ecological relationship.

I

Immigration

Movement of individuals into a population.

Importance value

A synthetic statistic describing the prominence of a species in a community, calculated as the sum of its relative density, relative dominance, and relative frequency (or of any two of these measures).

Interference

A direct detrimental effect of one organism or species on another. An aspect of competition involving such direct detrimental effects.

Interspecific

Relating to relationships between different species.

Intraspecific

Relating to relationships between individuals of one species.

K

K value

The defined upper limit of population growth in the logistic equation. A value taken to represent the carrying capacity of the environment occupied by an actual population.

L

Latin square design

An experimental arrangement in which two kinds of treatments are assigned to a grid of plots so that each specific treatment level occurs only once in each grid row and column.

Leslie matrix

An algebraic method for projecting the movement of individuals of different age classes to older classes, together with their reproduction, in a population in which probabilities of survival and reproduction differ with age.

Level of confidence

A chosen value representing the percentage of decisions about a statistical null hypothesis that are required to be made correctly.

Life table

A tabular summary of age-specific vital statistics of a population.

Lincoln index

A population estimate derived by marking a group of individuals and later determining the ratio of marked to total individuals in the population.

Logistic curve

A population growth curve in which potential exponential growth is modified by a degree-of-realization expression that decreases in straight-line fashion with increasing population size.

M

Macrophytes

Plants of large size; usually in reference to aquatic plants that are larger in size than algae.

Manipulative experiment

The systematic measurement of some condition in sets of experimental units, some of which have received a treatment altering their natural structure or dynamics.

Mathematical model

An equation or set of equations that describes the behavior of a system through time.

Mean life span

The average age at death of individuals that begin life in a population.

Mensurative experiment

The systematic measurement of some condition in experimental units in order to estimate one or more parameters of an ecological system.

Metapopulation

A regional population of a species that consists of semi-isolated subpopulations among which limited dispersal occurs.

Mortality rate

The number of deaths per unit time in a population.

N

Natality rate

The number of individuals entering a population by reproduction per unit time.

Natural selection

A change in the genetic composition of a population resulting from differential survival and reproduction of genotypes due to differences in their adaptive performance.

Net primary production

The production of photosynthate or chemosynthate by producers in excess of that used in their own cellular respiration; production evident in the form of new growth.

Niche

See *Ecological niche.*

Niche breadth

The range of habitat and/or resource categories utilized by a species.

Niche overlap

The degree of similarity of two or more species in utilization of habitat and/or resource categories.

Nonparametric test

A statistical test which does not incorporate mathematical assumptions about size-frequency relationships of data values in the test procedures.

Null hypothesis

A statement of no difference between sets of values, or of no deviation of values from some specified condition, formulated for the purpose of statistical testing.

Null model

A mathematical relationship that predicts conditions that should prevail according to chance.

Nutrient

A specific element or compound required for the life activities of an organism.

O

One-tailed test

A statistical test in which the probability zone for rejection of the null hypothesis is concentrated at one of the two extremes of the distribution of the test statistic.

Ordination

The positioning of communities in a graphical system of one to several axes so that distances between communities reflect differences in composition.

P

Parameter

In statistics, the true value of some mathematical characteristic for the entire statistical universe or sampled population.

Parametric test

A statistical test that incorporates specific mathematical assumptions about the distribution of data values in the test procedures.

Permanent wilting point

The soil water potential at which a test plant such as the domestic sunflower wilts without recovery overnight.

Petersen index

See *Lincoln index.*

Photosynthesis

The fixation of carbon, using light energy, by producer organisms.

Phytoplankton

Single-celled and colonial producer organisms that live free-floating in freshwater or marine ecosystems.

Plankton

Small organisms that live free-floating in freshwater or marine ecosystems; termed phytoplankton if they are plants, zooplankton if animals.

Poisson distribution

A mathematical series describing the frequency of co-occurrence of independent events of low probability.

Pressure bomb

A device that measures the water potential of a plant by exerting positive pressure on xylem fluid until the existing tension of water in the xylem vessels is balanced.

Primary production

The photosynthetic or chemosynthetic production of new organic matter by producer organisms.

Primary succession

Biotic succession which begins on a site that has not been occupied by a community of organisms.

Productivity

The total storage of energy or storage of energy in new tissues by a specified group of organisms per unit time.

Pseudoreplication

The location of experimental replicates so that they are not spread in an unbiased way throughout the experimental system. Pseudoreplication is *simple* when multiple measurements within one experimental unit are treated as if coming from different units, *temporal* if measurements at different times from the same unit are treated as being from different units, and *sacrificial* when opportunities to partition variability within and between experimental units exists but is ignored.

Q

Quadrat

A sampling plot of defined size and shape.

R

Randomized block design

An arrangement in which experimental units are divided into groups in which the units are randomly assigned the different control and experimental treatments.

Randomness

A condition in which the occurrence of an event is independent of that of other events of the same sort.

Raster format

A form of computer data storage in which an area is divided into grid units, or *pixels,* with each pixel being assigned a value reflecting the condition of some variable.

Realized niche

The utilization of resources and/or habitats by a species when it is in competition with other species.

Recruitment

The entry of new individuals into a population.

Regression

Analysis of the degree of correspondence in trends of two sets of values, one set of which is assumed to determine the value of the other and also to be free of measurement error.

Relative density

The density of one species as a decimal fraction or percentage of the combined density of all species.

Relative dominance

The dominance of one species as a decimal fraction or percentage of the combined dominance of all species.

Relative frequency

The frequency of one species as a decimal fraction or percentage of the combined frequencies of all species.

Relative humidity

The absolute quantity of moisture in a volume of air as a percentage of the saturation capacity at that air temperature.

Replicates

Experimental units chosen or structured to be as nearly identical in characteristics as possible.

Resource

Environmental factors that are utilized by organisms in such a way as to make the factors temporarily or permanently unavailable to other organisms.

Respiration

The oxidative breakdown of organic molecules with the release of energy for metabolic work.

Response variable

The characteristic chosen to reflect the influence of manipulations carried out in an experiment.

S

Sample

In statistics, a subset of the overall population or statistical universe. A sample is *random* when all parts of the universe have an equal chance of representation, *systematic* when the selected parts are uniformly spaced throughout the universe, and *stratified random* when the universe is divided into equal subsections from which parts are chosen randomly.

Scavengers

Organisms that opportunistically feed on a wide variety of living and dead food items.

Sessile

Permanently attached to a substrate.

Shannon-Wiener function

A mathematical expression stating the degree of uncertainty of occurrence of a particular symbol at a point in a message (information theory); used in ecology as a diversity index expressing the uncertainty of predicting the species of an individual drawn randomly from a community.

Simulation model

A set of equations intended to imitate the behavior of a real system through time.

Standard deviation

A parameter defining the dispersion of values about the mean in a normal distribution.

Standard error

A parameter defining the dispersion of sample means about the mean of a universe in which values are normally distributed.

State variable

Any quantitative feature of a system for which a simulation model predicts variation through time.

Statistic

An estimate of a parameter obtained from a sample taken from a statistical universe.

Stochasticity

A random influence that affects a system or its members. For a population, stochasticity can be *demographic,* involving chance accidents of survival and reproduction, or *environmental,* involving random influences of environmental conditions on survival or reproduction.

Stomatal conductance

The volume of a gas that passes through the stomates of a plant per unit time.

Stomatal resistance

The degree to which gas movement through stomates is impeded.

Succession

Change through time in the structure or composition of a community under constant external conditions.

Survivorship

The number of individuals of a cohort beginning life together that are still alive at a specified age.

Symbiosis

Organisms of different species that live together in a mutually beneficial arrangement.

System

A set of components tied together by regular interactions.

Systems ecology

The branch of ecology concerned with the investigation and mathematical modeling of the dynamics of complex ecological systems.

T

Territory

The area of habitat that is actively defended against intrusion of conspecifics or competitors by an individual, pair, or group.

Top carnivore

An animal that feeds exclusively on first-level carnivores. Also, the trophic level consisting of all animals with such feeding habits.

Transect

An elongate sampling unit along which observations, measurements, or samples are taken. A *line* transect is linear (without width), whereas a *strip* transect has a constant or variable width.

Transpiration

Loss of water by plants through evaporation from the stomata.

Treatment

The manipulation or lack of manipulation imposed on experimental units in a manipulative experiment.

Trophic level

All organisms that are the same number of energy conversions away from the energy input (e.g., solar radiation) to an ecosystem.

Tullgren funnel

An apparatus for extracting small organisms from soil or litter samples.

Two-tailed test

A statistical test in which the probability zone for rejection of the null hypothesis is split into equal areas at the two extremes of the distribution of the test statistic.

Type I and II errors

A type I error is the rejection of the null hypothesis when it is true; a type II error is the acceptance of the null hypothesis when it is false.

U

Universe

The total set of phenomena, characterized by certain quantitative parameters, for which a sample, with its quantitative statistics, is an estimate.

V

Variable

Any defined entity that may have different quantitative values. A variable is *continuous* if it can be measured with different degrees of precision, and *discrete* if it can only be enumerated. In regression analysis a *dependent* variable is a variable the value of which is partly or wholly determined by one or more *independent* variables.

W

Water potential

The capacity of water in an actual situation to do work, relative to the same amount of unconstrained pure water in the same location.

Y

Yates' correction

The practice of reducing the absolute magnitude of deviations of observed from expected by 0.5 in chi-square tests with 1 degree of freedom.

Metric Conversions

Metric Unit	Conversion × (Nonmetric)
Length	
kilometer (km)	1.609 (mile)
meter (m)	0.914 (yard)
	0.305 (foot)
centimeter (cm)	2.540 (inch)
millimeter (mm)	25.40 (inch)
Area	
square kilometer (km^2)	2.590 (square mile)
hectare (ha)	0.405 (acre)
square meter (m^2)	0.836 (square yard)
	0.093 (square foot)
square centimeter (cm^2)	6.452 (square inch)
square millimeter (mm^2)	645.2 (square inch)
Volume	
cubic meter (m^3)	1.233×10^3 (acre-foot)
	0.764 (cubic yard)
	0.035 (bushel)
	3.785×10^{-3} (gallon, U.S.)
liter (l)	3.785 (gallon, U.S.)
	28.32 (cubic foot)
	0.946 (quart, U.S.)
	0.473 (pint, U.S.)
Mass	
ton (t)	1.106 (ton, long)
	0.907 (ton, short)
kilogram (kg)	0.454 (pound)
gram (g)	28.35 (ounce)
Energy	
joule (J)	4184 (kilocalorie)
	4.184 (calorie)
	1054 (British thermal unit)
Power	
watt (W)	4184 (kilocalorie/second)
	69.73 (kilocalorie/minute)
	4.184 (calorie/second)
	0.070 (calorie/minute)
	735.5 (horsepower)
	1054 (British thermal unit/second)
	17.57 (British thermal unit/minute)

Heat
     joule per kilogram (J/kg)   4.184 (kilocalorie/gram)
                                               4184 (calorie/gram)
                                             2324 (British thermal unit/pound)

Speed
     kilometer per hour (km/h)   1.609 (mile/hour)
     meter per second (m/s)   0.447 (mile/hour)
                                             0.305 (foot/second)

Volume per Unit Time
     cubic meters per second (m³/s)   0.013 (cubic yards, minute)
                                             0.028 (cubic feet/second)

Temperature
     degrees Celsius (C)   1/1.8 (degree Fahrenheit − 32)
Pressure
     pascal (Pa)   133.3 (millimeter of mercury)
                                             100.0 (millibar)
                                           1.013×10^5 (atmosphere)

Metric Unit   **Conversion × (Nonmetric)**

Map Legend

The accompanying map is of the plant community of a bajada, or desert alluvial fan, in Borrego Valley, San Diego County, California. The map covers an area of 4800 m² (60 × 80 m) and gives the exact locations of all individuals of 17 species of desert shrubs and cacti. Sizes of these individuals are indicated by circles of the following diameters, in meters: 0.1, 0.2, 0.4, 0.6, 0.8, and 1.0, and 0.5 increments to 4.0. Canopy coverage (m²) for plants of these diameter classes is indicated by a key on the map. A meter scale is marked on the plot boundaries to facilitate the location of sampling sites (e.g., by random coordinates) and to permit the calibration of rulers or tape measures used in sampling activities.

This map can be used in connection with exercises on sampling design (4), distance sampling (12), vegetation analysis (14), intrapopulation dispersion (21), plant competition (27), interspecific association (31), species diversity (32), community similarity and ordination (33), and species-area curves (35). These exercises utilize data collected by quadrats, transects, or point-to-plant or plant-to-plant distance measurements. Locations of quadrats, transect beginning and endpoints, and sampling points can be obtained from the random numbers table (A.1) or in other ways, and found by reference to the meter scale on the plot boundaries. Quadrat outlines can be drawn on clear plastic material, or actual quadrats fashioned out of wire or other material. Transect lines can be laid out with string held in place by pins, and intercept distances and other measurements made with rulers or tape measures. Sampling points can be marked lightly in pencil and the appropriate distances measured in similar fashion. Comparisons can be made of the results of different sampling techniques, of wash and nonwash zones of the mapped area, or of species (dispersion) or species pairs (association). More specific comparisons are suggested in several of the above exercises.

Scientific names of the mapped species are desert senna, *Cassia armata;* chuparosa, *Beloperone californica;* creosote bush, *Larrea tridentata;* desert trumpet, *Eriogonum inflatum;* brittlebush, *Encelia farinosa;* barrel cactus, *Ferocactus acanthodes;* golden cholla, *Opuntia echinocarpa;* hedgehog cactus, *Echinocereus engelmannii;* indigo bush, *Dalea schottii;* jojoba, *Simmondsia chinensis;* krameria, *Krameria grayi;* pencil cholla, *Opuntia ramosissima;* desert straw, *Stephanomeria pauciflora;* bur sage, *Ambrosia dumosa;* teddy bear cholla, *Opuntia bigelovii;* beavertail cactus, *Opuntia basilaris;* and ocotillo, *Fouqueria splendens.*

Index

---

*Note: Page numbers in *italics* indicate illustrations; page numbers followed by t indicate tables.